AF601492

Modern Age Waste Water Problems

Mohammad Oves • Mohammad Omaish Ansari
Mohammad Zain Khan • Mohammad Shahadat
Iqbal M.I. Ismail
Editors

Modern Age Waste Water Problems

Solutions Using Applied Nanotechnology

Editors
Mohammad Oves
Centre of Excellence in Environmental Studies
King Abdul Aziz University
Jeddah, Saudi Arabia

Mohammad Zain Khan
Section of Industrial Chemistry, Department of Chemistry
Aligarh Muslim University
Aligarh, Uttar Pradesh, India

Iqbal M.I. Ismail
Department of Chemistry
King Abdulaziz University
Jeddah, Saudi Arabia

Mohammad Omaish Ansari
Center of Nanotechnology
King Abdulaziz University
Jeddah, Saudi Arabia

Mohammad Shahadat
Department of Biochemical Engineering and Biotechnology
Indian Institute of Technology Delhi
New Delhi, India

Department of Textile Technology
Indian Institute of Technology Delhi
New Delhi, India

ISBN 978-3-030-08282-6 ISBN 978-3-030-08283-3 (eBook)
https://doi.org/10.1007/978-3-030-08283-3

This Springer imprint is published by the registered company Springer Nature Switzerland AG.
The registered company address is: Gewerbestrasse 11, 6330 Cham, Switzerland

Preface

Water is the main constituent of Earth and is present in rivers, lakes, and seas. It is the vital source of all living and nonliving things besides any other organic nutrients or calorie sources. It circulates continuously in different forms by evaporation, transpiration, condensation, and precipitation and finally reaches to the sea. Thus, there exists a large amount of water in seas and oceans (~ 71%), while groundwater, glaciers, and ice caps account for only 3.4% water. A tiny quantity (0.0001%) of water exists in the form of vapor in air and clouds. The massive amount of water associated with mineral (as water of hydration) is considered to be essential for the flora and fauna. Dumping of industrial wastewater (80% of the world's wastewater) into the natural water resources has become the leading cause of water pollution. Consumption of pollutant-containing water causes a number of diseases in human beings and animals. Around one billion people are facing the problem of safe drinking water, while more than 2.8 billion still needs proper sanitation. In a recent scenario, it is anticipated that due to the rising world population, by 2060 (ten billion), huge water crisis-related problem (lack of fresh water) will be seen all over the world. Thus, it becomes an important task to treat industrial water before discharging it into natural water resources. Therefore, treatment of water has been carried out worldwide by using a number of techniques and materials such as filtration, biological treatment, photocatalysis, membrane technology, solar disinfection, oxidation precipitation, electrolysis, reverse osmosis, biodegradable nanocomposite, materials for adsorption, etc. However, adsorption is found to be the most versatile method owing to its ease of operation, low cost, and high efficiency. The selection of adsorbent is generally on the basis of high selectivity, cost-effectiveness, as well as substantial adsorption capacity.

Additionally, nontoxic behavior and regeneration capacity of adsorbents together with zero waste/sludge production have attracted adsorbents in water treatment. There has been a tremendous increase in the number of research papers published on adsorption during the period of 2000–2018; however, still more work is needed to improve adsorption capacity and regenerative potential of exhausted adsorbents. Nowadays, the most crucial accountability of the research community is to develop innovative, cost-effective, reliable technique for the analysis as well as removal of environmental contaminants. The present book deals with the basic principles and methodologies, including advanced techniques of water treatment. In addition, water-related bacterial infection, development of water treatment plants, and sensing behavior of nanocomposite for the detection of gases have been discussed. Various nanotechnological approaches have been presented for the effective treatment of contaminated water.

Nanomaterial-implemented devices, including carbon nanotube-alumina fibers, nanoscopic pores in the zeolite filtration membrane, polymer nanocomposites involving conducting and nonconducting polymers, nanocatalysts, magnetic nanoparticles, etc., generally used to clean wastewater have also been discussed. Alumina-based nanofiber has been used for the treatment of negatively charged impurities (bacteria, viruses, organic/inorganic colloids). Besides water treatment, nanotechnology can also be employed for the removal and recovery of chemical effluents, pathogens, bacteria, sediments, viscous liquid impurities, etc. In the future, it is assumed that nanomaterial-supported water treatment devices have the potential to capitalize novel nanohybrid materials. Treatment of industrial wastewaters using nanohybrid membranes is another advanced initiative. Different types of membranes have been used to improve water quality in the enhancement of oil recovery (EOR) operation. Membrane technology has been effectively employed to analyze sulfate-based impurities. A lot of work is carried out to develop nanocomposite membrane. However, some issues regarding energy efficiency, e.g., fouling characteristics of membrane, have to be resolved to obtain renewable energy associated with membrane technology. To meet the industrial demands in terms of oil and gas recovery, new ways for the fabrication of potential membrane used on a large scale are discussed. Till date, membrane technology has been developed as a steadfast cost-effective technology for the water treatment. It is anticipated that more than 50 million m3/day of electrodialysis reversal (EDR) capacity and reverse osmosis (RO) units have been installed all over the world. In addition, another 20 million m3/day of membrane filtration installation capacity units have been set up in desalination plants. To utilize naturally available resources, treatment of petroleum pollutants using bioremediation technique is another interesting topic which is discussed in detail.

The present book deals with the treatment of wastewater using biological treatment technology to control water pollution and prevent human beings from waterborne diseases. It also focuses on the removal and recovery of organic and inorganic contaminants from wastewater composite and nanocomposite. Disinfection of wastewater using solar energy and photocatalytic degradation is another interesting topic of the book. Chitosan-supported nanocomposites are discussed for the analysis of environmental pollutants. To recycle industrial effluents, cost-effective, eco-friendly fuel cell technology is also discussed in detail. Last but not the least, the present book covers advanced technologies which have been effectively employed in present day's wastewater treatment.

We are really thankful to all the contributing authors from different countries who shared their scientific knowledge in the form of book chapters. Book chapters provided by all authors are well written together with the most relevant references of literature. We are highly thankful to the world-renowned institutes: Department of Biochemical Engineering and Biotechnology, Textile Technology, Indian Institute of Technology (IIT Delhi), Hauz Khas, New Delhi; Aligarh Muslim University, Aligarh, UP, India; and Center of Excellence and Environmental Studies & Center of Nanotechnology, King Abdulaziz University, Saudi Arabia, whose research scholar and research scientists contributed their ideas and knowledge to give an appropriate shape of this book which is extremely acknowledged. A special thanks to our family members for proving constant support during the book preparation period. We are pleased with the hard work of the Springer book publishing team for their prompt reply and constructive suggestions during the whole project. We would also like to mention that in case this book contains some mistakes, printing errors, and inaccuracies, we feel regret in anticipation. However, if it is pointed out at any stage, we will certainly try to improve them in the subsequent print/edition. Readers are most welcome to provide critical analysis and suggestions related to the content presented in the present book.

Jeddah, Saudi Arabia — Mohammad Oves
Jeddah, Saudi Arabia — Mohammad Omaish Ansari
New Delhi, India — Mohammad Shahadat
Aligarh, Uttar Pradesh, India — Mohammad Zain Khan
Jeddah, Saudi Arabia — Iqbal M.I. Ismail

Contents

Contributors

Shaikh Ziauddin Ahammad Department of Biochemical Engineering and Biotechnology, Indian Institute of Technology Delhi, New Delhi, India

Akil Ahmad School of Industrial Technology, Universiti Sains Malaysia, Gelugor, Penang, Malaysia

Hilal Ahmad Centre for Nanoscience and Nanotechnology, Jamia Millia Islamia (A Central University), New Delhi, India

M. S. Ahmad Department of Chemistry, Aligarh Muslim University, Aligarh, Uttar Pradesh, India

Nafees Ahmad Environmental Research Laboratory, Department of Chemistry, Aligarh Muslim University, Aligarh, Uttar Pradesh, India

S. Wazed Ali Department of Textile Technology, Indian Institute of Technology Delhi, New Delhi, India

Sajad Ali CORD, University of Kashmir, Srinagar, Jammu and Kashmir, India

Mohammad Omaish Ansari Center of Nanotechnology, King Abdulaziz University, Jeddah, Saudi Arabia

O. Ansari Department of Chemistry, Aligarh Muslim University, Aligarh, Uttar Pradesh, India

Abdul Hakeem Anwer Department of Chemistry, Aligarh Muslim University, Aligarh, Uttar Pradesh, India

M. T. Baig Department of Chemistry, Aligarh Muslim University, Aligarh, Uttar Pradesh, India

Sayfa Bano Environmental Research Laboratory, Department of Chemistry, Aligarh Muslim University, Aligarh, Uttar Pradesh, India

Rani Bushra Department of Chemistry and Center of Excellence for Innovation in Chemistry, Faculty of Science, Mahidol University, Bangkok, Thailand

Juhi Chaudhary College of Agricultural and Life Sciences, University of Florida, Gainesville, Florida

Neha Dhingra Department of Zoology, University of Delhi, New Delhi, India

Sajad Ahmad Ganai Department of Chemistry, NIT Sri Nagar, Kashmir, India

Ashitha Gopinath DRDO – BU – CLS, Bharathiar University Campus, Coimbatore, Tamil Nadu, India

Uzma Haseen Department of Chemistry, Aligarh Muslim University, Aligarh, India

Syed Zajif Hussain Department of Chemistry and Chemical Engineering, SBA School of Science & Engineering (SBASSE), Lahore University of Management Sciences (LUMS), Lahore, Pakistan

Suzylawati Ismail School of Chemical Engineering, Universiti Sains Malaysia, Gelugor, Pulau Pinang, Malaysia

Asim Jilani Advanced Membrane Technology Research Centre, Universiti Teknologi Malaysia, Johor Bahru, Johor, Malaysia

School of Chemical and Energy Engineering, Faculty of Engineering, Universiti Teknologi Malaysia, Johor Bahru, Johor, Malaysia

Center of Nanotechnology, King Abdul-Aziz University, Jeddah, Saudi Arabia

Rajkumar Joshi Department of Civil Engineering, Jamia Millia Islamia, New Delhi, India

Krishna Gounder Kadirvelu DRDO-BU Center for Life Sciences, Bharathiar University, Coimbatore, India

Chinnannan Karthik DRDO – BU – CLS, Bharathiar University Campus, Coimbatore, Tamil Nadu, India

Amjad Mumtaz Khan Department of Chemistry, Aligarh Muslim University, Aligarh, India

Mohammad Danish Khan Department of Chemistry, Aligarh Muslim University, Aligarh, Uttar Pradesh, India

Mohammad Zain Khan Environmental Research Laboratory, Department of Chemistry, Aligarh Muslim University, Aligarh, Uttar Pradesh, India

Kadirvelu Krishna DRDO – BU – CLS, Bharathiar University Campus, Coimbatore, Tamil Nadu, India

Krishnasamy Lakshmi DRDO-BU Center for Life Sciences, Bharathiar University, Coimbatore, India

Zahoor A. Mir NRCPB, IARI, New Delhi, India

A. Mohammad Discipline of Chemistry, Indian Institute of Technology Indore, Indore, India

Mohamad Nasir Mohamad Ibrahim School of Chemical Sciences, Universiti Sains Malaysia, Pulau Pinang, Malaysia

Mohd Hafiz Dzarfan Othman Advanced Membrane Technology Research Centre, Universiti Teknologi Malaysia, Johor Bahru, Johor, Malaysia

School of Chemical and Energy Engineering, Faculty of Engineering, Universiti Teknologi Malaysia, Johor Bahru, Johor, Malaysia

Mohammad Oves Centre of Excellence in Environmental Studies, King Abdul Aziz University, Jeddah, Saudi Arabia

Tabassum Parveen Department of Civil Engineering, Indian Institute of Technology, Roorkee, India

Talat Parween Department of Bioscience, Jamia Millia Islamia, New Delhi, India

Mohd. Rafatullah School of Industrial Technology, Universiti Sains Malaysia, Gelugor, Penang, Malaysia

Mukul Raizada Department of Chemistry, Aligarh Muslim University, Aligarh, India

Suhail Sabir Environmental Research Laboratory, Department of Chemistry, Aligarh Muslim University, Aligarh, Uttar Pradesh, India

Farasha Sama Department of Chemistry, Aligarh Muslim University, Aligarh, India

Mohammad Shahadat Department of Biochemical Engineering and Biotechnology, Indian Institute of Technology Delhi, New Delhi, India

Department of Textile Technology, Indian Institute of Technology Delhi, New Delhi, India

M. Shahid Department of Chemistry, Aligarh Muslim University, Aligarh, India

Abid Hussain Shalla Department of Chemistry, Islamic University of Science & Technology (IUST), Pulwama, Jammu and Kashmir, India

Ranju Sharma Indian Institute of Technology, New Delhi, India

Ngangbam Sarat Singh University of Delhi to Department of Zoology, Dr. SRK Government Arts College, Yanam, Puducherry (UT), India

M. Siraj Department of Chemistry, Aligarh Muslim University, Aligarh, Uttar Pradesh, India

T. R. Sreekrishnan Department of Biochemical Engineering and Biotechnology, Indian Institute of Technology Delhi, New Delhi, India

Saima Sultana Environmental Research Laboratory, Department of Chemistry, Aligarh Muslim University, Aligarh, Uttar Pradesh, India

M. Tauqeer Department of Chemistry, Aligarh Muslim University, Aligarh, Uttar Pradesh, India

Anjali Tyagi CSIR-National Physical Laboratory, New Delhi, India

Anshika Tyagi NRCPB, IARI, New Delhi, India

Khalid Umar School of Chemical Sciences, Universiti Sains Malaysia, Pulau Pinang, Malaysia

Venkatramanan Varadharajan DRDO-BU Center for Life Sciences, Bharathiar University, Coimbatore, India

Chapter 1
Recent Advancement in Wastewater Decontamination Technology

Mohammad Shahadat, Akil Ahmad, Rani Bushra, Suzylawati Ismail, Shaikh Ziauddin Ahammad, S. Wazed Ali, and Mohd. Rafatullah

Abstract Water contamination has become a worldwide severe environmental problem owing to the existence of heavy metal ions. Extension of industrializations is releasing heavy metals ions containing effluents into water bodies and causes damage to the aquatic environment. Treatment of industrial wastewaters using ion-exchange adsorbents has achieved attractiveness in comparison to other treatment methods. The present chapter deals with the preparation and characterization of Polyaniline (PANI)-Titanium-supported nanocomposites ion-exchanger materials. It generally focuses on the ion exchange behavior of nanocomposites for the detection of heavy metal ions in wastewater, industrial effluents, and synthetic mixtures. These nanomaterials have been characterized using advanced techniques of characterizations. Physico-chemical properties; ion uptake efficiency, pH titration, the effect of temperature as well as concentration and kinetic studies have been examined to establish the significant performance of these materials to

M. Shahadat (✉)
Department of Biochemical Engineering and Biotechnology, Indian Institute of Technology Delhi, New Delhi, India

Department of Textile Technology, Indian Institute of Technology Delhi, New Delhi, India

A. Ahmad
School of Industrial Technology, Universiti Sains Malaysia, Gelugor, Penang, Malaysia

R. Bushra
Department of Chemistry and Center of Excellence for Innovation in Chemistry, Faculty of Science, Mahidol University, Bangkok, Thailand

S. Ismail
School of Chemical Engineering, Universiti Sains Malaysia, Gelugor, Pulau Pinang, Malaysia
e-mail: chsuzy@usm.my

S. Z. Ahammad (✉)
Department of Biochemical Engineering and Biotechnology, Indian Institute of Technology Delhi, New Delhi, India

S. Wazed Ali
Department of Textile Technology, Indian Institute of Technology Delhi, New Delhi, India

M. Rafatullah
School of Industrial Technology, Universiti Sains Malaysia, Gelugor, Penang, Malaysia

M. Oves et al. (eds.), *Modern Age Waste Water Problems*,
https://doi.org/10.1007/978-3-030-08283-3_1

achieve maximum adsorption towards heavy metal ions. These nanomaterials demonstrated significant ion uptake efficiency, high thermal and chemical stability as compared to pure organic or inorganic ion-exchanger adsorbents. Based on high ion-exchange capacity, the nanomaterials can be successfully used in the wastewater treatment. In spite of the detection of metal pollutants in contaminated waters, these nanocomposites ion-exchange adsorbents can also be effectively utilized in other fields (e.g., Photochemical degradation of organic contaminants, antimicrobial agents, and conducting material). On the basis of excellent performance of titanium-supported nanocomposite in terms of metal removal efficiency, it is anticipated that these nanomaterials could be open, innovative ways to show their excellent uses in diverse fields.

Keywords Polyaniline · Heavy metal ions · Conducting material · Ion-exchanger · Nanocomposite

1.1 Introduction

Water pollution due to the hazardous pollutants has become a serious worldwide environmental problem. The most common hazardous contaminant can be divided into two categories; naturally hazardous contaminants and man made toxicants (anthropogenic). Anthropogenic contaminants are obtained by the chemical reaction or combustion of natural and synthetic materials. In addition, natural gas, oil, and coal release a huge quantity of risky and global warming gases into the environment which has directly and indirectly affected the life of living beings. Among all fossil fuels, natural gas (methane) is considered as the cleanest source of all the fossil fuels according to EPA (Environmental Protection Agency). In contrast to natural gas, other energy sources (coal and oil) are made up of complex composition molecules along with the high percentage of carbon, nitrogen and sulphur contents. Thus, the combustion of these complex molecules emits a higher level of toxic emissions, associated with higher percentage of carbon emissions, oxides of sulphur and nitrogen. These harmful gases, as well as transport-related pollutants including greenhouse gases, have harshly affected the ozone level, which resulted in global warming (Engstrom et al. 2007) and environmental pollution (Engstrom et al. 2009). Ammonia is one of the most extensive used reagents in the world which also generates nitrogen-based pollutants to the environment (Engstrom et al. 2010; Institute of Medicine, Food and Nutrition Board 2001). The contamination through heavy metal ions is found risky compared to non-metal pollutants, owing to not biodegradable nature of heavy metals ions. The heavy metal ions becomes very harmful when it gets accumulated in the soft organ of the body (kidney, brain, lungs) without metabolized in the body; and thus interrupts normal body functions resulting in a number of infections. Due to the fast extension of industrializations, these heavy metals are frequently released into the water bodies and damage the aquatic environment (Long et al. 2002; Pereira et al. 2010; Chen et al. 2006; Barbieri et al. 2000; Rawat et al. 2009). Directly or indirectly when these heavy metal ions are inhaled or are consumed through food or water, they cause severe damage to the body parts. Among heavy metal series, inorganic arsenic is found to be the most contaminating element because of its harmful effect even at low concentration (Amin 2008; Mui et al. 2010; Ip et al. 2010). In potable and other natural water, the toxicity of arsenic has been detected around the world (Dizge et al. 2008) which leads to a number of diseases (Sun et al. 2010; Sud et al. 2008; Irmak and Ozturk 2010; Khalil et al. 2010). Zinc is an essential element and plays a key role in a number of cellular metabolic activity (Mahmoud et al. 2012; Hasfalina et al. 2012; Azelee et al. 2014; Sharif et al. 2013). However, higher intake of zinc disturbs the metabolic activity and results in outbursts of various diseases (Misha et al. 2015). The utilization of lead at industrial scale has severely affected the atmosphere (Gan et al. 2015). The poisoning of lead affects different organs of the body, the symptoms in kids causes development in growth and mental retardation. In China, Approximately 2000 children living near the metal industry zone (zinc and manganese) have been severely exaggerated by the poisoning of lead (Meng Ng et al. 2015). A limited concentration of copper is essential for metabolic activity in the body of human beings (Othman and Akil 2008). Nevertheless, the higher amount above a specific concentration can't lead to serious health problems (Brzonova et al. 2014; Jeun et al. 2015). Nickel is also considered a nutritionally essential trace element. Consequently, its minute or excess concentration can be poisonous to human beings (Andreu et al. 2013). Directly or indirectly human beings gets exposed to nickel through the polluted water and foodstuff and leads to gastrointestinal distress together with neurological consequences (Cuerda-Correa et al. 2008). The pollution of mercury in the atmosphere has been amplified owing to the releasing of wastes from a number of industries namely; hydroelectric, tanning, pulp-paper industries accompanied by the flaming of municipal agricultural and medical desecrates. In the air, the oxide of these wastes reacts with water vapor and gets transformed into methylmercury which is highly toxic. The aquatic animal (e.g. Fishes, prawn, crab, etc.), which are cultivated in mercury-polluted waters are found the most widespread syndrome when eaten by human and infects the human body (Andreu and Vidal 2013). The hazardous effect of chromium is seen in the Hollywood blockbuster movie "Erin Brockovitch". In India, many industries have been fined to discharge chromium to the environment. The use of cadmium in plant, animal pests and commercial fertilizers has become the main source of cadmium toxicity which produces several diseases and other health disorder (John et al. 2010; Nacos et al. 2006). The utilization of aluminium is frequently not hazardous; however, its excess amount is injurious to health (Hao et al. 2013; Inagaki et al. 2004; Chena et al. 2013; Bushra et al. 2014a). Thus, exposure of a trace concentration of heavy metal ions may cause health problem in human beings. Thus, flora and fauna have been badly affected by the detrimental outcome of these metal ions pollutants. So, it becomes a global concern to separate metal ions from wastewater before any constructive use (Nabi et al. 2011a; Shahadat et al. 2015; Ahmad et al. 2012). In this regards, a number of methods including ion exchange, chemical precipitation, flotation, coagulation-flocculation, and filtration through the membrane have been used for the management of these metal ions contaminants (Kurniawan et al. 2006). Among these separating methods, the ion-exchange adsorbents have been found to be the best for treating materials because of their more selective and less expensive behaviour. For the separation of metal ions, a number of organic and inorganic adsorbents ion-exchange have been prepared, however, these materials are associated with some limitations (Shahadat and Bushra 2015). Hence, to remove all the drawbacks of organic and inorganic ion exchange adsorbents, "organic-inorganic" composite ion-exchange adsorbents have been synthesized. The composite materials show better physico-chemical properties in terms of thermal and chemical stability and can be effectively utilized in diverse fields such as a gas sensor (Khan and Baig 2013), photocatalyst-conducting material (Shahadat et al. 2012), ion selective electrode (Khan and Innamuddin 2006), and in the form of chelating agents for detection of radioactive isotopes (Raman et al. 1996). The present chapter reports synthesis, characterization of titanium-based novel ion exchange materials coupled with improved chemical and thermal stability as well as their practical utility for the analysis of metal pollutants in wastewaters.

1.2 Titanium-Supported Ion-Exchange Nanocomposites

To improve the utilization of titanium associated with organic polymer, attention has been paid to synthesize PANI-Titanium supported nanocomposite materials. Among other metals, titanium shows excellent corrosion prevention property, low density together with the highest strength to weight ratio. Apart from these, it is also lighter (45%) than other metals. Owing to these properties, it can also be used as in the form of strong steel (in unalloyed form). Titanium can be effectively used in water bodies (e.g., sea, river, halides-rich atmosphere, etc.) because of its high corrosion resistance behavior. Therefore, it can also be utilized in aerospace and automobiles industry, information technology instruments and in mobile phones parts. Organic polymers namely; polyaniline, polythiophene (POT), polyacrylonitrile (PACN), etc. have been widely employed in the preparation of ion-exchange nanocomposite adsorbents. The main significant characteristics of organic polymers are their ease of protonation, cost-effectiveness, significant electrical properties together with significant chemical and thermal stability. Thus, various organic-inorganic hybrid ion-exchange adsorbent have been prepared by incorporating transition metals in the matrix of PANI, POT, and PACN, etc. However, titanium-supported ion-exchange composites in comparison to other nanohybrid materials possesses improved thermal and chemical stability. Apart from these, besides the treatment of wastewater, these nanocomposites could be used in diverse fields.

1.3 PANI-Titanium Supported Nano-Composite Ion-Exchange Adsorbents

In the field of hybrid ion-exchange materials, PANI-Titanium based nano-composite materials play key in the treatment of industrial wastewaters. In this regard, different organic and inorganic hybrid ion-exchange adsorbents have been prepared. Initially, the preparation of inorganic ion exchangers was carried out. These materials misplaced their effectiveness after the discovery of organic resins. Generally, native organic and inorganic ion-exchanger materials have good ion-uptake efficiency. However, they are associated with some drawbacks. The main severe limitations associated with organic exchangers are their lower thermal and chemical stability. Although, inorganic exchanger achieved in powder form suffers from the drawback of being stuck in the column during treatment process (Arrad and Sasson 1989). To remove drawbacks of pure organic and inorganic exchanger, researchers have synthesized PANI-Titanium supported nanocomposite owing to their outstanding ion-exchange efficiency, selectivity, and other physico-chemical properties. A proposed outline for the preparation, characterization, and application of PANI-titanium supported nanomaterials is shown in Scheme 1.1.

1.4 Preparation of PANI-Titanium Supported Nano-Composite

In the preparation of composite materials, polyaniline based nano-composite ion-exchange materials show improve ion-uptake efficiency for the analysis of metal ions. These nano-composite ion-exchange materials can be prepared using various methods including; electrochemical deposition, ultrasound-assisted (Cabrera et al. 2010), sol-gel (Vatutsina et al. 2007), thermal and chemical decomposition, co-precipitation (Xie et al. 2014) and the electrospinning method. Among these methods, the sol-gel method is the most commonly used (Mascolo et al. 2013). In this method, the gel of organic polymers (e.g., PANI, PPy, PAA, etc.) was prepared using an oxidizing agent. The inorganic precipitate of metal ions can be developed under controlled stoichiometry of different reactants solutions of compounds with continuous mixing (using magnetic stirrer) at room temperature. The PANI-titanium-supported nanocomposite is prepared by the mixing of PANI gel in the matrix of inorganic precipitate of titanium with constant stirring. The sol-gel chemical route yields the samples of nano-composite material with good size control which shows excellent behavior regarding ion-uptake efficiency, thermal and chemical stability. Thus, an easy sol-gel synthetic route can be employed for the synthesis of PANI-Ti

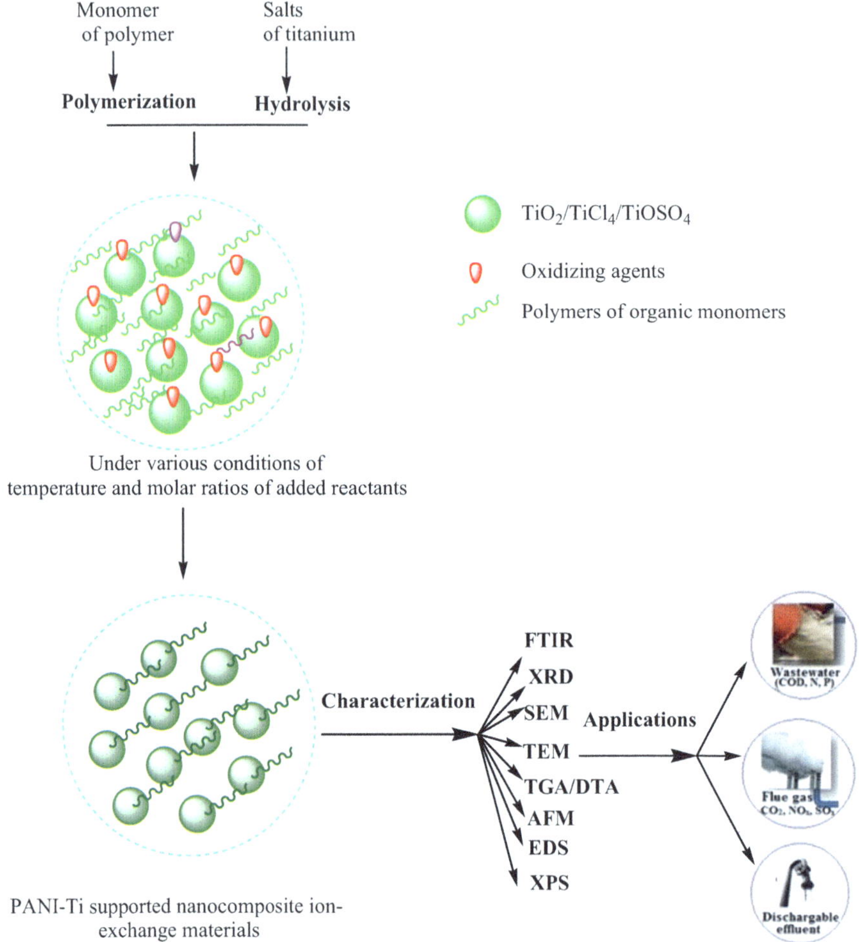

Scheme 1.1 Preparation of PANI-Ti-supported nanocomposite ion-exchange adsorbents for the treatment of pollutants

supported nanocomposite adsorbents. Based on better ion uptake efficiency together with good yield (%), chemical and thermal stability, the selected sample can be extensively utilised to scrutinize the physicochemical properties for the maximum adsorption of metal ions.

1.4.1 Physico-chemical Properties

The complete description of PANI-Titanium-supported nanocomposite ion-exchange materials can be performed by studying the following physical and chemical properties:

1.4.1.1 Ion Exchange Efficiency

The ion uptake efficiency of an ion exchange adsorbent describes the quantity of uptake of exchangeable ions under the specific conditions. The theoretical capacity of the ion exchanger is found higher than the apparent capacity which strongly depends on solutions concentration (reactants) and its pH. The ion uptake potential is calculated by a non-equilibrium process, i.e., using column operation. In this process, a specific amount of ion-exchange materials (in H^+ form) is taken into a column (glass) of particular diameter (e.g., Internal diameter 0.5 cm) plugged with glass wool at the bottom. The alkaline earth metal nitrate solutions of different concentration are utilized to elute H^+ ions from ion exchange material by keeping a constant flow rate (e.g., 0.5 mL min^{-1}). The hydrogen ions (H^+) concentration in the collected effluent is calculated using by standard titration method of sodium hydroxide.

1.4.1.2 The pH Titration

To establish the behavior of the ionogenic group present in nanocomposite ion exchange materials, the pH titrations studies are carried out using the Topp and Pepper method (Topp and Pepper 1949). It helps to calculate the number of replaceable H^+ ions per molecule of the material. In this way, it can be established whether the exchanger is mono, bi or polyfunctional or has a combination of weak and strong functional groups. In this method, a fixed amount (0.5 g) of ion-exchange materials (in H^+ form) is taken in conical flasks (50 mL) by the mixing of an equimolar concentration of alkali and alkaline metal chlorides and their corresponding hydroxides (e.g. NaCl–NaOH, KCl–KOH and $CaCl_2$–$Ca(OH)_2$, $BaCl_2$–$Ba(OH)_2$ systems). The volume of solution is maintained as 50 mL to adjust the ionic strength of the solution.

1.4.1.3 The Effect of Eluent Concentration

To determine the consequence of eluent strength, a constant volume (250 mL) of different sodium nitrate solutions are employed for complete removal of hydrogen ions from the glass column containing a fixed amount of composite material (0.5 g in H^+ form). The effluent is titrated against the standard method of sodium hydroxide.

1.4.1.4 The Elution Behavior

The elution characteristic of ion-exchange material is carried out to eliminate hydrogen ions completely from the column containing a fixed amount of ion-exchange material. The removal of hydrogen ions from the material generally depends on the concentration (strength) of eluent. The optimal concentration of $NaNO_3$ solution for the maximum elution of H^+ ions from the ion-exchange material depends on the nature of the exchangeable group which exists in the exchanger which in turn depends upon the pK_a values of the acids used in its preparation. The effluent of H^+ ions was collected in 10.0 mL fraction at a fixed flow rate and each fraction was titrated against a standard solution of sodium hydroxide.

1.4.1.5 Chemical and Thermal Stability

Chemical and thermal stability tests was carried out to establish whether the material can sustain in a particular solvent or not during operation. However, thermal stability explains the degradation of material at a particular temperature.

1.4.1.6 Distribution Coefficients (K_d Values)

The distribution coefficient (K_d) of metal ions such as Mg^{2+}, Ca^{2+}, Sr^{2+}, Ba^{2+}, Pb^{2+}, Hg^{2+}, Cd^{2+}, Zn^{2+}, Mn^{2+}, Cu^{2+}, Ni^{2+}, Fe^{3+}, Al^{3+}, Bi^{3+}, and La^{3+} can be calculated using the batch method in various non-ionic, anionic and cationic surfactants solvent systems. The K_d value is determined to examine the overall potential of the exchanger to remove the ions of interest under different conditions. In this method, different fractions (300 mg each) of composite exchanger is taken in Erlenmeyer flasks to which a fixed volume (30 mL) of different metal nitrate solution is added in the required medium and consequently stirred in a temperature controlled shaker for 6 h at room temperature (25 ± 2 °C) to attain the equilibrium. The concentration of different metal ions before and after the equilibrium is determined using standard EDTA titration method. The K_d value is the measure of the functional uptake of metal ions competing for H^+ ions from a solution containing a fixed amount of ion-exchange adsorbent.

The distribution coefficients of different metal ions on ion-exchange adsorbent is calculated using the equation:

$$K_d = \frac{\text{milli equivalent of metal ions/g of ion-exchanger}}{\text{milli equivalent of metal ions/mL of solution}} \text{mL g}^{-1}$$

$$K_d = \frac{\text{I} - \text{F}}{\text{F}} \times \frac{\text{V}}{\text{M}} \text{mL g}^{-1}$$

where K_d is the distribution coefficient metal ions, I is the initial concentration of the metal ion in the solution, F is the final concentration of metal ion in the solution after treatment with the composite material, V is the volume of the solution (mL), and M is the taken amount of composite ion exchanger (g). The sorption of metal ions involves the exchange of hydrogen ions in exchanger phase with that of metal ions in the solution phase as given below:

$$\underset{\text{Exchanger phase}}{2\text{R-H}^+} + \underset{\text{Solution phase}}{\text{M}^{2+}} \rightleftharpoons \underset{\text{Exchanger phase}}{\text{R}_2\text{-M}} + \underset{\text{Solution phase}}{2\,\text{H}^+}$$

1.5 Characterization of PANI-Titanium Nano-Composite

With the development of modern analytical instruments, it becomes easy to understand the chemistry of ion-exchange materials. All these nano-composite materials can be characterized using advanced analytical techniques, namely; Scanning electron microscopy (SEM), Transmission electron microscopy (TEM), Fourier transform infrared (FTIR), X-ray diffraction (XRD) analysis, thermogravimetric analysis (TGA), Atomic force microscopy (AFM) and elemental analysis. The potential applications of the composite materials have been examined by obtaining separations of some critical metal ions quantitatively indicating its utility to control environmental pollution.

1.6 Analytical Application of PANI-Titanium Supported Nano-Composite Ion-Exchange Adsorbents for the Treatment of Wastewaters

An easy sol-gel chemical method was employed for the development of PANI/TiW adsorbent (Bushra et al. 2014b). The composite ion-exchanger was fabricated at pH 1.0, and it was characterized using XRD, FTIR and SEM analyses. FTIR analysis of PANI/TiW revealed the existence of groups (e.g., C–C, –CH, O=W=O, Ti–O–W, –OH, H–O–H, and W–O, etc.) in the form of different frequencies. XRD analysis did not showed any sharp peak, which confirmed the amorphous nature of the material. SEM analysis at different magnifications also showed the amorphous morphology. The proposed scheme for the preparation of PANI/TiW and its mechanism is shown in Fig. 1.1. The ion-uptake efficiency of the composite was calculated for alkali and alkaline earth metal ions. The ion-exchange composite established improved ion uptake efficiency along with thermo-chemical stability. In terms of thermal stability, it has the potential to retain 98% initial ion-uptake capacity up to 300 °C. The K_d values of metal ions were studied in different concentrations of surfactant systems (non-ionic, cationic and anionic). Based on higher K_d values, the composite showed selective behavior towards essential metal ions including heavy metal pollutants;

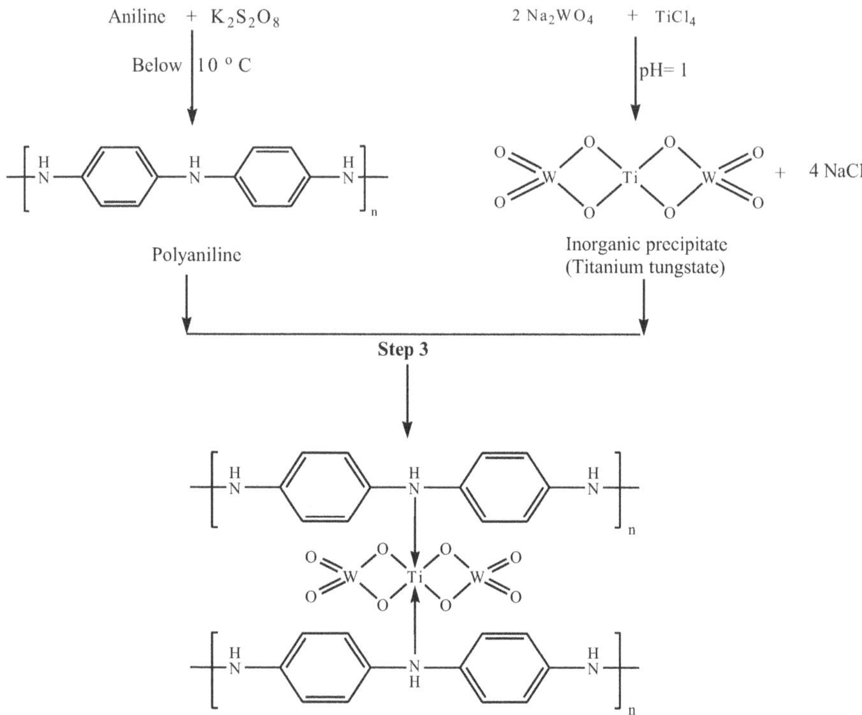

Fig. 1.1 A proposed scheme for the preparation of PANI/TiW supported nano-composite adsorbent (Duval 1963)

Bi^{3+}, Pb^{2+}, Hg^{2+}, and Zr^{4+}. For Pb^{2+} ion, the value of limit of detection (LOD) and limit of quantification (LOQ) was achieved to be 0.85 and 2.85 μg L^{-1}, respectively. Some important binary separation in industrial effluents and natural wastewaters were successfully obtained by using this material. Thus, PANI/TiW can be effectively utilized for the separation of metal ions in industrial and wastewater bodies.

Semi-crystalline PANI/Ti/Arsenate nanocomposite ion-exchange material was synthesized, characterized for the treatment of industrial effluents (Bushra et al. 2016). The material was categorized using FTIR, XRD, TGA/DTA, SEM and TEM analyses. The FTIR analysis of PANI/Ti/Arsenate showed an extensive peak in the region of 3550–3000 cm^{-1} owing to the existence of –OH group (Rao 1963a). A broadband at 1616 cm^{-1} was pointed due to H–O–H bending while the peak at 1442 cm^{-1} corresponded to the existence of –NH– of PANI in nanocomposite adsorbent. Other peaks at 1385, 1136 and 614 cm^{-1} were pointed towards the in-plane bending of –CH vibration (Silverstein et al. 1981). The TGA curve of PANI-Ti(IV)As provides an idea regarding weight loss of the composite material concerning temperature. Up to 100 °C, a nonstop loss in weight (about 4%) was found owing to the removal of external water molecules from composite (Duval 1963). Decline in weight (100–200 °C) was because of the elimination of interstitial water from the material. Onwards 816 °C, weight loss occurred as a result of the complete breakdown of the organic component and metal oxides formation of a PANI/Ti/Arsenate. The XRD pattern of the material exhibited tiny peaks of weak intensity which confirmed the semi-crystalline nature. The SEM images for PANI, Ti/Arsenate and PANI/Ti/Arsenate were obtained at different magnifications. The SEM photographs of PANI, Ti/Arsenate, exposed the zigzag particle irregular morphology. It was found that after binding Ti(IV)As with PANI, the morphology of the PANI/Ti/Arsenate material has been altered which confirmed semi-crystalline morphology. TEM studies established that Ti/Arsenate and PANI/Ti/Arsenate composite showed particles size in the range of 20 nm. The ion-exchange material demonstrated an ion uptake efficiency of 1.37 meq g^{-1} Na^+.

The nanocomposite material was also examined for its electrical resistivity and photochemical breakdown of an organic pollutant (AB-29). The electrical conductivity of PANI/Ti/Arsenate improved with increasing % of Ti/Arsenate. The findings of photochemical degradation confirmed the efficiency of PANI/Ti/Arsenate for the degradation of organic dye. Besides the ion exchange adsorbent, PANI/Ti/Arsenate can also be employed as a conducting material, a photocatalyst as well as an ion exchange material for the management of industrial wastewater.

A novel ion exchange material; PANI/Ti/W was developed and used to exclude Cs^+ ions in industrial wastewaters (Naggar et al. 2012). A very simple sol-gel method was employed; by the addition of PANI into the precipitate of titanotungstate. The physico-chemical properties such as the effect of metal ions concentration, size of the nanocomposite particle and temperature were studied. The PANI/Ti/W demonstrated better selectivity for Cs^+ in comparisons to other composites. To examine the thermodynamic and kinetic factors including Gibbs free energy (ΔG), enthalpy (ΔH), and entropy (ΔS), batch sorption studies were carried out. At higher temperatures, the value of Gibbs free energy decreased, which demonstrated the feasible sorption (spontaneous) behavior of the composite at prominent temperatures. The positive values of enthalpy change confirmed endothermic reaction and suggested that a chemisorption reaction mechanism is involved. Both models; pseudo first and pseudo-second-order kinetics models were used, however, the pseudo-second-order models (homogeneous particle diffusion) were found to fit best. In addition, by using a linearized form of the Arrhenius equation, the values of activation energy, self-diffusion coefficient, and entropy of activation were also determined which recommended an associative mechanism for diffusion-controlled reaction as well as a sorption mechanism together with each adsorbent. The adsorption mechanism of PANI/Ti/W composite was observed for the separation of Cs^+ in milk using a scintillation detector head technique (El-Naggar et al. 2012). The K_d values for various metal ions were calculated in the solution of different solvents system. The ion-exchange material exhibited maximum ion uptake efficiency for the

Cs^+ ion (27 mg g^{-1}). Based on high K_d value, PANI/Ti/W was found to be highly selective for cesium ions and followed the trend; Cs^+>>> Zr^{4+} > Mo^{6+} > V^{5+} > As^{5+} > Cr^{3+} > Co^{2+} > Cu^{2+} > Zn^{2+} > Cd^{2+}. Chemical kinetics study results confirmed the best fitting of the Freundlich isotherm model.

Different solutions were employed to measure the breakthrough capacity by changing the bed depths of ion-exchange material. The obtained data of the breakthrough efficiency revealed that with increasing bed depth the time of half breakthrough increased proportionally. In terms of good ion-exchange efficiency, PANI/Ti/W can be effectively employed for the separation of Cs^+ in the sample of milk as well as in other wastewater using a scintillation detector head.

Treatment of wastewaters has been effectively carried out for the removal of toxic metal ions in aqueous medium using immobilized TiO_2 (Tennakone and Wijayantha 1998). Generally, in most cases, it is very difficult to utilize TiO_2 powder owing to some drawbacks (drop in pressure and filtration). Therefore, this drawback was removed by the immobilization of TiO_2 powder with an organic polymer. Recently, these inspirations have accelerated research to synthesize organic-inorganic nanohybrid adsorbents for industrial and environmental applications. In this regards, the spherical beads of PACN were synthesized for the detection of Pb^{2+} in aqueous medium (Kim et al. 2003). A technique (dual nozzle technique) was applied for the preparation of PACN composite. By modeling the experimental results, the physicochemical properties namely the effect of pH on the adsorption of Pb^{2+} ions and kinetic parameters were investigated. It was established that at pH 5.6, composite materials containing 70% TiO_2 demonstrated maximum viscosity (950 cP) and ion uptake capacity. Porosity of composite adsorbent was also compared to inorganic beads, and it was revealed that composite material exhibited more than 80% porosity than zeolite. The homogeneous diffusion efficiency of PACN was found to be approximately 10^{-8} cm^2 s^{-1}. Thus, PACN composite can be profitably applied to treat Pb^{2+} present in wastewater and industrial effluents.

Polystyrene/Titanium based composite membranes were prepared using sol-gel chemical route together with the phase inversion process (Yang and Wang 2006). These composite membranes were used for the treatment of industrial wastewaters. In addition, the characterization was conducted using advanced techniques. The mages of SEM analysis established the formation of porous morphology with homogeneous distribution of TiO_2 particles in composite membrane. The hydrophilicity and permeability characteristics of composite membrane augmented significantly by increasing amount of TiO_2 content up to 0–9.3% (by wt) without altering any retention behaviour. The aggregation of the nanoparticles increases with enhancing the TiO_2 content which causes decline in hydrophilicity and permeability. Owing to the excellent thermal properties together with their extraordinary hydrophilicity, high porosity and superior permeability, hybrid membranes of PS/TiO_2 can be applied for the treatment of wastewaters.

Nano-composite of PANI-Zr/Ti/P was synthesized using sol-gel chemical route by the addition of PANI and zirconium-titanium phosphate (Khan and Paquiza 2011a). The XRD, SEM and TEM analyses were used to characterize the material which established semicrystalline nature of the material. The nano-composite material showed an unexpectedly high ion uptake efficiency (4.52 meq g^{-1}) along with better thermo-chemical characteristics than that of reported nanocomposite adsorbents.

The K_d values of metal ions were investigated in various organic and acidic solvents solution (by varying concentrations). Based on high K_d value, the composite adsorbent was found to be highly selective towards the separation of metal pollutants. The composite adsorbent was examined by accomplishing removal and recovery of heavy metal ions in synthetic mixtures (binary and ternary separations). In addition, a comparative study in term of the ion-exchange properties of other reported PANI supported composites, and PANI-Zr/Ti/P was executed. The outstanding ion uptake efficiency of PANI-Zr/Ti/P proves its functional use in the treatment of heavy metal ions; Hg^{2+} and Pb^{2+} from industrial wastewaters. To use this composite material as a membrane, a PANI-Zr/Ti/P supported ion-selective membrane electrode was constructed to determine potentiometric sensing behavior towards Hg^{2+} ions (Khan and Paquiza 2011b). The prepared composite membrane electrode covered a fast response (time) for the detection of Hg^{2+} ions and covered a wide dynamic range; 10^{-10} to 10^{-1} mol m^{-3}. This membrane can be activated for 4 months without any considerable deviation in its potential response. Moreover ion exchanger, PANI supported PANI-Zr/Ti/P composite membrane electrodes can be used as an indicator electrode in complexation titrations for the analysis of Hg^{2+} in wastewaters.

Another titanium supported PANI/Ti/P nanocomposite adsorbent was prepared using solution casting route (Khan and Baig 2012). The material was prepared by the mixing of PVC and PANI/TiP in a fixed ratio. On the basis of mixing volume ratio, a composite membrane of 1:1 ratio of PVC and PANI-TiP was selected to show the best water uptake capacity, electrical resistivity, porosity, thickness as well as swelling efficiency. The K_d values of metal ions were also calculated in DMW and different concentrations of HNO_3, H_2SO_4, CH_3COOH using PANI-TiP nanocomposite. Based on higher K_d values, the PANI/TiP was found to be selective for the environmental pollutant (Pb^{2+} ions). The composite was also used to develop an ion selective electrode for the analysis of Pb^{2+} in wastewater. The fabricated ion selective membrane exhibited a fast time response, which covered a wide range together with mechanical stability and it can be used for at least 5 months. PANI-TiP based cation exchange nanocomposite membrane demonstrated a better potentiometric response in compression to existing electrodes. The PANI-TiP can be used as an ion-exchanger as well as a potential ion selective membrane for the analysis of Pb^{2+} in wastewaters without any interference of charge species.

A novel Polystyrene-supported PS/titanium-vanadium composite material was synthesized to develop composite membrane (Arfin et al. 2011). The physical-chemical property of composite membranes (ionic potentials) was measured and compared with the equation derived by Terrell-Meyer-Sievers (TMS). The validation of utilized theory was confirmed by the investigation of membrane parameters; ion exchange efficiency, porosity, thickness, and swelling. The values of membrane potential decreased with enhancing the concentration which follows the order for alkali metal as $Li^+ > Na^+ > K^+$. The distribution coefficient values also demonstrated an identical trend of KCl > NaCl > LiCl. By increasing the hydration radii of the electrolytes, the values of fixed charge density and charge effectiveness was found to be increased as compared to the reported data. Thus, the Polystyrene-based PS/ titanium-vanadium hybrid adsorbent can be employed for the separation of alkali metals from the environmental wastewaters.

Polystyrene-titanium molybdate composite ion-exchanger was utilized for the exclusion of metal pollutants (Ishrat and Rafiuddin 2012). The membrane was examined for the separation of alkali metal ions (K^+), and the physico-chemical properties (fixed charge density and electrochemical) of the composite membrane were examined using the TMS method which exhibited functionality towards K^+ ions in solutions with the highest electrolytes concentration. The fixed charge density showed an instinctively stable behavior of the membrane which can be worked over a wide pH range; 2.0–7.0. The structural characteristics of composite membrane demonstrated useful properties regarding pore radius, charge density and a better membrane thickness to membrane porosity. Thus it can be concluded that polyaniline based POM composite ion-exchange materials membranes can be successfully utilized for the separation of K^+ in industrial effluents containing different metal.

Polyamidoamine (PAMAM) dendrimers with ethylenediamine cores (G4-OH) were fabricated using sol-gel synthetic route (Barakat et al. 2013). The material was examined for the removal of metal pollutants in the synthetic mixture containing Cu^{2+}, $Cr^{3+}$and Ni^{2+}. To get an idea regarding the maximum ion adsorption capacity of metal ion on the composite material, the parameters including the effect of pH, the retention time, the effect of concentration were studied. Due to the chelates formation of metal (Cu^{2+}, $Cr^{3+}$and Ni^{2+} ions) with PAMAM/G4, the batch remediation experiment was performed. It was found that the optimal metal ion efficiency was observed for Cu^{2+} and Cr^{3+} ions at pH = 7 upon equilibration after 1 h for a while and at pH = 9.0 it was optimum for the Ni(II) ion. The effective treatment of Cu^{2+}, Ni^{2+}, and Cr^{3+} from the synthetic mixtures confirmed that the PAMAM/G4-OH could be effectively used for the exclusion of Cu^{2+}, Ni^{2+} and Cr^{3+} ions from wastewaters.

PANI-supported, PANI-Ti(IV) arsenophosphate (PANI-Ti(IV)As(IV)P) nanocomposite ion-exchange material was synthesized for the analysis of metal ions (Bushra et al. 2014c). The nanocomposite material was characterized on the basis of advanced techniques. The spectra were studied at different temperatures (25–500 °C). A broad peak in the region of 3550–3000 cm^{-1} as well as a medium intensity band at 1600 cm^{-1} attributed to the occurrence of interstitial water and a hydroxyl group, respectively (Rao 1963a). A medium intensity peak at 1479 cm^{-1} was recognized due to the existence of secondary aromatic amine while other bands at 1382 cm^{-1}, together with merged broadband from 1050 to 1028 cm^{-1} established the in-plane –CH bending (Silverstein et al. 1981). The peaks at ~ 874, 814 and ~511 cm^{-1} were ascribed to υ (P = O) (Rao 1963b), υ (As–O–Ti) (Zhang et al. 2009) and υ (Ti–O) vibration (Nakamoto 1986), respectively. The TGA studies of PANI-Ti(IV)As(IV)P confirmed 12% weight loss up to 150 °C owing to the elimination of water moieties. The region from 200 to 300 °C showed a 10% weight loss due to the elimination of interstitial water molecules as a result of condensation. Furthermore, weight loss up to 580 °C was found owing to the complete degradation of the organic back bone of nano-composite. Onwards, 600 °C the weight of material remained constant due to the formation of oxides of nano-composite material (Duval 1963). The XRD spectra of PANI-Ti(IV)As(IV)P did not showed exhibit any sharp peak which confirmed the amorphous nature. In addition, SEM micrographs also confirm the amorphous morphology. TEM images established evidence for aggregation of PANI-Ti(IV)As(IV)P nanoparticles in the range of 20–50 mm.

The K_d values for different metal ions were measured in DMW and the different concentrations of $HClO_4$ solutions by using this material. Based on high K_d some important separations of metal contaminants were obtained from synthetic solution and tap water samples. The composite material was also examined for the other applications namely; conducting property, antimicrobial activity and photocatalytic degradation of organic pollutants. Thermodynamics and kinetics parameters were examined to measure the optimum adsorption of organic dye. Significant results of electrical conductivity, antimicrobial activity and a high rate of photochemical degradation established that PANI/Ti/As/P can be considered an exceptional conducting material for the removal and recovery of organic dye and metal ions pollutants. One of the most critical features of titanium-based nanocomposites is that besides the treatment of heavy metal ions, these materials can be employed in other fields. Thus, PANI-titanium supported nanomaterials can be fruitfully used for the treatment of wastewaters. Some important methods utilized for the preparation and characterization of PANI-titanium supported nanocomposite materials and their significant applications in the treatment of industrial wastewaters are shown in Table 1.1. A comparative study regarding ion-uptake efficiency capacity of PANI-titanium based composite with other polyaniline supported materials is shown in Table 1.2. Among all nanocomposite materials, the PANI-Titanium supported materials demonstrated higher ion-exchange capacity.

1.7 Summary

Different chemical methods were used for the preparation of titanium supported nanocomposite ion-exchange materials. Among various methods of synthesis, the sol-gel technique is considered as one of the best routes as a result of cosmetic preparation of the desired composite with high purity at a moderately low temperature. It was also observed that the incorporation of TiO_2 into the matrix of organic polymeric back bone not only improve the ion exchange efficiency but also enhanced the thermal-chemical properties of the nano-hybrid material composite. The improvement in ion-exchange capacity was found because of the development of ion exchange active sites (functional groups) on the surfaces of the nanocomposite materials. These nanomaterials, not only exclude the inorganic pollutants (metal ions) but have also established significant applications in various fields. Thus, fabrication of titanium supported nano-hybrid materials has constructed innovative use regarding their commercial application primarily in the treatment of wastewater.

Table 1.1 A summary of the preparation of titanium-supported nano-hybrid adsorbents and their significant uses in the treatment of wastewater

Nano-hybrid material	Preparation route	Characterization	Significant applications	Reference
Polyaniline/ TiW	Sol and gel chemical method	SEM, FTIR, XRD, CHNO, TGA, and FAAS	Analysis of heavy metal ions	Bushra et al. (2014b)
Polyaniline/ Ti(IV)As	Sol and gel chemical method	SEM, XRD, and TEM	Removal of metal ions	Bushra et al. (2016)
Polyaniline/ TiW	Sol and gel chemical method	FTIR, XRD and, TGA	Detection of cesium	Naggar et al. (2012) and El-Naggar et al. (2012)
Polyaniline/ TiO_2	Dual nozzle method	–	Detection of Pb^{2+} ions	Tennakone and Wijayantha (1998) and Kim et al. (2003)
Polystyrene/ TiO_2	Sol and gel chemical method	TGA and SEM	Removal of metal ions	Yang and Wang (2006)
PANI/Zirco-nium tita-nium phosphate	Sol and gel chemical method	SEM, TGA/DTA, AAS, XRD, FTIR potentiometer and TEM	Analysis of Pb^{2+} and Hg^{2+}	Khan and Paquiza (2011a, b)
PANI/Ti/ phosphate	Sol and gel chemical method	Potentiometer, TGA/ DTA, FTIR and SEM	Electrode membranfor the detection of Pb^{2+}	Khan and Baig (2012)
Polystyrene/ Ti/Vanadium	Sol and gel chemical method		Electrode mem-brane for Li^+, Na^+ and K^+	Arfin et al. (2011)
Polystyrene/ Titanium/ molybdate	Sol and gel chemical method	SEM, XRD and FTIR,	Analysis of K^+	Ishrat and Rafiuddin (2012)
PAMAM-G4	Sol and gel chemical method	SEM, TEM, FTIR and XRD,	Removal of Cr^{3+}, Cu^{2+}, Ni^{2+}	Barakat et al. (2013)
PANI/Tita-nium/ Arseno-Phosphate	Sol and gel chemical method	XRD, SEM, FTIR, and TEM	Removal and recovery of pollutants	Bushra et al. (2014c)

On the basis of outstanding results for titanium supported nano-hybrid, continuous attention to this field remains enforced.

The present book chapter states that titanium-supported nano-hybrid materials are indubitably considered as imminent innovative nano-hybrid materials for the management of wastewater. There will also probably be a growing demand for the synthesis of these nanocomposites having customized properties for a particular environmental applications (separation of inorganic pollutants). Moreover, so far the work on the application of nanocomposites in wastewater and

Table 1.2 A comparative study of ion-uptake efficiency of titanium supported nanocomposite ion-exchange adsorbents

PANI based composite ion-exchange materials	Colour	Ion-uptake efficiency for Na^+ (meq g^{-1})	References
PANI-Zr-Titanium phosphate	Black shiny	4.52	Khan and Paquiza (2011b)
PANI-Ti(IV)phosphate	Greenish black	2.59	Khan et al. (2011)
PANI-Ti(IV) molybdophosphate	Greenish black	2.46	Khan et al. (2013)
PANI-Sn(IV)phosphate	Greenish	1.96	Khan and Innamuddin (2006)
PANI-Zr(IV) molybdophosphate	Black	1.70	Khan and Shaheen (2015)
PANI-Sn(IV)tungstoarsenate	Greenish	1.67	Khan and Alam (2003)
PANI-Sn(IV) tungstomolybdate	Green	1.77	Bushra et al. (2015)
PANI-Sn(IV)tungstophosphate	Greenish	1.10	Nabi et al. (2014)
PANI-Ce(IV)molybdate	Greenish	1.45	Al-Othman et al. (2011a)
PANI-Zr(IV)selenoiodate	White	2.1	Khan et al. (2015)
PANI-Th(IV)phosphate	Black	1.56	Khan and Inamuddin (2005)
POT-Zr(IV) phosphate	White	1.46	Khan and Akhtar (2008)
POAN and POT-Sn(IV) tungstate	Greenish	2.50	Khan and Shaheen (2012)
PANI-Zr(IV) molybdophosphate	Black	2.10	Khan and Shaheen (2014)
PANI-Zr(IV)sulphosalicylate	Green granular	1.80	Nabi et al. (2011b)
POT-Zr(IV)tungstate	Green	1.47	Nabi et al. (2012a)
PANI-Zr(IV)Arsenate	Dark green	1.33	Nabi et al. (2012b)
Poly-o-toluidineZr(IV)iodate composite	Green	0.85	Nabi et al. (2012c)
PANI-Sn(IV)iodophosphate	Green	1.20	Khan et al. (2014)
PANI-Th(IV) tungstomolybdophosphate	Green	1.07	Sharma et al. (2014)
PPy-Th(IV) phosphate	Blackish	1.56	AL-Othman et al. (2011b)

wastewater sludge treatment is still in the research phase. These ion-exchange nanocomposites demonstrated significant potential compared to inorganic ion-exchanger and organic resins. Subsequently, these materials might soon become one of the most influential nano-hybrid composites towards making a green aquatic environment.

Acknowledgments The authors express their appreciations to, [1]Department of Biochemical Engineering and Biotechnology, [2]Department of Textile Technology, Indian Institute of Technology, Delhi, New Delhi-110016, India. One of the authors (Dr. Md. Shahadat) is very much thankful to SERB-DST (SB/FT/CS-122/2014), Govt. of India for awarding a Postdoctoral research grant to carry out research at IIT Delhi.

References

Ahmad R, Kumar R, Haseeb S (2012) Adsorption of Cu^{2+} from aqueous solution onto iron oxide coated eggshell powder: evaluation of equilibrium, isotherms, kinetics, and regeneration capacity. Arab J Chem 5:353–359

AL-Othman ZA, Inamuddin, Naushad M (2011a) Inamuddin, forward (M^{2+}–H^{+}) and reverse (H^{+}–M^{2+}) ion exchange kinetics of the heavy metals on polyaniline Ce(IV) molybdate: a simple practical approach for the determination of regeneration and separation capability of ion exchanger. Chem Eng J 171:456–463

AL-Othman ZA, Inamuddin, Naushad M (2011b) Adsorption thermodynamics of trichloroacetic acid herbicide on polypyrrole Th(IV) phosphate composite cation-exchanger. Chem Eng J 169:38–42

Amin NK (2008) Removal of reactive dye from aqueous solutions by adsorption onto activated carbons prepared from sugarcane bagasse pith. Desalination 223:152–161

Andreu G, Vidal T (2013) Laccase from pycnoporus cinnabarinus and phenolic compounds: can the efficiency of an enzyme mediator for delignifying kenaf pulp be predicted? Bioresour Technol 131:536–540

Andreu G, Barneto AG, Vidal T (2013) A new biobleaching sequence for kenaf pulp: influence of the chemical nature of the mediator and thermogravimetric analysis of the pulp. Bioresour Technol 130:431–438

Arfin T, Jabeen F, Kriek RJ (2011) An electrochemical and theoretical comparison of ionic transport through a polystyrene based titanium-vanadium (1:2) phosphate membrane. Desalination 274(1–3):206–211

Arrad O, Sasson Y (1989) Commercial ion exchange resins as catalysts in solid-solid-liquid reactions. J Org Chem 54:4993–4998

Azelee NIW, Jahim JM, Rabu A, Murad AMA, Bakar FDA, Illias RM (2014) Efficient removal of lignin with the maintenance of hemicellulose from kenaf by two-stage pretreatment process. Carbohydr Polym 99:447–453

Barakat MA, Ramadan MH, Alghamdhi MA, Algarny SS, Woodcock HL, Kuhn JN (2013) Remediation of Cu(II), Ni(II) and Cr(III) ions from simulated wastewater by dendrimer titania composites. J Environ Manag 117:50–57

Barbieri L, Bonamartini AC, Lancellotti I (2000) Alkaline and alkaline-earth silicate glasses and glass-ceramics from municipal and industrial wastes. J Eur Ceram Soc 20:2477–2483

Brzonova I, Kozliak E, Kubatova A, Chebeir M, Qin W, Christopher L, Ji Y (2014) Kenaf biomass biodecomposition by basidiomycetes and actinobacteria in submerged fermentation for production of carbohydrates and phenolic compounds. Bioresour Technol 173:352–360

Bushra R, Shahadat M, Nabi SA, Raeissi AS, Oves M, Umar K, Muneer M, Ahmad A (2014a) Synthesis, characterization, antimicrobial activity and applications of composite adsorbent for the analysis of organic and inorganic pollutants. J Hazard Mater 264:481–489

Bushra R, Mohammad, Shahadat M (2014b) Synthesis, characterization and applications of nanocomposite materials in diverse fields. Adv Environ Res 35:105–130

Bushra R, Shahadat M, Ahmad A, Nabi SA, Raeissi AS, Umar K, Oves M, Muneer M (2014c) Synthesis, characterization, antimicrobial activity and applications of composite adsorbent for the analysis of organic and inorganic pollutants. J Hazard Mater 264:481–489

Bushra R, Naushad M, Adnan R, AL-Othman ZA, Rafatullah M (2015) Polyaniline supported nanocomposite cation exchanger: synthesis, characterization and applications for the efficient removal of Pb^{2+} ion from aqueous medium. J Ind Eng Chem 21:1112–1118

Bushra R, Ahmed A, Shahadat Md (2016) Mechanism of adsorption on nanomaterials. Advanced environmental analysis, pp 90–111

Cabrera L, Gutierrez S, Herrasti P, Reyman D (2010) Sonoelectrochemical syn- thesis of magnetite. Phys Procedia 3:89–94

Chen GC, He ZL, Stoffella PJ, Yang XE (2006) Leaching potential of heavy metals (Cd, Ni, Pb, Cu and Zn) from acidic sandy soil amended with dolomite phosphate rock (DPR) fertilizers. J Trace Elem Med Biol 20:127–133

Chena YD, Chena WQ, Huang B, Huang MJ (2013) Process optimization of $K_2C_2O_4$-activated carbon from kenaf core using box–Behnken design. Chem Eng Res Des 91:1783–1789

Cuerda-Correa EM, Antonio Macías-Garcia A, Diez MAD, Ortiz AL (2008) Textural and morphological study of activated carbon fibers prepared from kenaf. Microporous Mesoporous Mater 111:523–529

Dizge N, Aydiner C, Demirbas E, Kobya M, Kara S (2008) Adsorption of reactive dyes from aqueous solutions by fly ash: kinetic and equilibrium studies. J Hazard Mater 150:737–746

Duval C (1963) Inorganic thermogravimetric analysis. Elsevier, Amsterdam, p 315

El-Naggar IM, Zakaria ES, Ali IM, Khalil M, El-Shahat MF (2012) Chemical studies on polyaniline titanotungstate and its uses to reduction cesium from solutions and polluted milk. J Environ Radioact 112:108–117

Engstrom K, Broberg K, Concha G, Nermell B, Warholm M, Vahter M (2007) Genetic polymorphisms influencing arsenic metabolism: evidence from Argentina. Environ Health Perspect 115(4):599–605

Engstrom K, Nermell B, Concha G, Stromberg U, Vahter M, Broberg K (2009) Arsenic metabolism is influenced by polymorphisms in genes involved in one-carbon metabolism and reduction reactions. Mutat Res 667(1–2):4–14

Engstrom KS, Vahter M, Lindh C, Teichert F, Singh R, Concha G (2010) Low 8-oxo-7, 8-dihydro-2′-deoxyguanosine levels and influence of genetic background in an Andean population exposed to high levels of arsenic. Mutat Res 683:98–105

Gan S, Zakaria S, Chia CH, Padzil FNM, Ng P (2015) Effect of hydrothermal pretreatment on solubility and formation of kenaf cellulose membrane and hydrogel. Carbohydr Polym 115:62–68

Hao A, Zhao H, Chen JY (2013) Kenaf/polypropylene nonwoven composites: the influence of manufacturing conditions on mechanical, thermal, and acoustical performance. Compos Part B 54:44–51

Hasfalina CM, Maryam RZ, Luqman CA, Rashid M (2012) Adsorption of copper (II) from aqueous medium in fixed-bed column by kenaf fibres. APCBEE Procedia 3:255–263

Inagaki M, Nishikawa T, Sakuratani K, Katakura T, Konno H, Morozumi E (2004) Carbonization of kenaf to prepare highly-microporous carbons. Lett Ed/Carbon 42:885–901

Institute of Medicine, Food and Nutrition Board (2001) Dietary reference intakes for vitamin A, vitamin K, arsenic, boron, chromium, copper, iodine, iron, manganese, molybdenum, nickel, silicon, vanadium, and zinc. National Academy Press, Washington, DC

Ip AWM, Barford JP, McKay G (2010) Acomparative study onthe kinetics andmechanisms of removal of reactive black 5 by adsorption onto activated carbons and bone char. Chem Eng J 157:434–442

Irmak S, Ozturk I (2010) Hydrogen rich gas production by thermocatalytic decomposition of kenaf biomass. Int J Hydrog Energy 35:5312–5317

Ishrat U, Rafiuddin (2012) Synthesis, characterization and electrical properties of titanium molybdate composite membrane. Desalination 286:8–15

Jeun J, Min BL, Young JL, Hyun KP, Park JK (2015) An irradiation-alkaline pretreatment of kenaf core for improving the sugar yield. Renew Energy 79:51–55

John MJ, Bellmann C, Anandjiwala RD (2010) Kenaf–polypropylene composites: effect of amphiphilic coupling agent on surface properties of fibres and composites. Carbohydr Polym 82:549–554

Khalil HPSA, Yusra AFI, Bhat AH, Jawaid M (2010) Cell wall ultrastructure, anatomy, lignin distribution, and chemical comp osition of Malaysian cultivated kenaf fiber. Ind Crop Prod 31:113–121

Khan AA, Akhtar T (2008) Preparation, physico-chemical characterization and electrical conductivity measurement studies of an organic–inorganic nanocomposite cation-exchanger: poly-o-toluidine Zr(IV) phosphate. Electrochim Acta 53:5540–5548

Khan AA, Alam MM (2003) Synthesis, characterization and analytical applications of a new and novel 'organic–inorganic' compositematerial as a cation exchanger and Cd(II) ion-selective membrane electrode: polyaniline Sn(IV) tungstoarsenate. Rect Funct Polym 55:277–290

Khan AA, Baig U (2012) Electrically conductive membrane of polyaniline–titanium(IV)phosphate cation exchange nanocomposite: applicable for detection of Pb(II) using its ion-selective electrode. J Ind Eng Chem 18(6):1937–1944

Khan AA, Baig U (2013) Electrical conductivity and ammonia sensing studies on in situ polymerized poly(3-methythiophene)-titanium(IV)molybdophosphate cation exchange nanocomposite. Saens Actuators: B 177:1089–1097

Khan AA, Inamuddin AMM (2005) Determination and separation of Pb^{2+} from aqueous solutions using a fibrous type organic–inorganic hybrid cation-exchange material: Polypyrrole thorium (IV) phosphate. Rect Funct Polym 63:119–133

Khan AA, Innamuddin (2006) Application of Hg(II) sensitive polyaniline Sn(IV) phosphate composite cation exchange material in determination of Hg2+ from aqueous solutions and in making ion selective membrane electrode. Saens Actuators B 120:10–18

Khan AA, Paquiza L (2011a) Characterization and ion-exchange behavior of thermally stable nanocomposite polyaniline zirconium titanium phosphate: its analytical application in separation of toxic metals. Desalination 265(1–3):242–254

Khan AA, Paquiza L (2011b) Analysis of mercury ions in effluents using potentiometric sensor based on nanocomposite cation exchanger polyaniline–zirconium titanium phosphate. Desalination 272(1–3):278–285

Khan AA, Shaheen S (2012) Thermal stability and electrical properties of conducting polymer based 'polymeric–inorganic' composites: poly-o-anisidine and poly-o-toluidine Sn(IV) tungstate. Mater Res Bull 47:4414–4441

Khan AA, Shaheen S (2014) Chronopotentiometric and electroanalytical studies of Ni(II) selective polyaniline Zr(IV) molybdophosphate ion exchange membrane electrode. J Rect Funct Polym Chem 714–715:38–44

Khan AA, Shaheen S (2015) Preparation, characterization and kinetics of ion exchange studies of Ni^{2+} selective polyaniline–Zr(IV)molybdophosphate nanocomposite cation exchanger. J Ind Eng Chem 26:157–166

Khan S, Cao Q, Zheng YM, Huang YZ, Zlm YG (2008) Health risks of heavy metals in contaminated soils and food crops irrigated with wastewater in Beijing, China. Environ Pollut 152:686–692

Khan AA, Baig U, Khalid M (2011) Ammonia vapor sensing properties of polyaniline–titanium (IV)phosphate cation exchange nanocomposite. J Hazard Mater 186(2–3):2037–2042

Khan AA, Baig U, Khalid M (2013) Electrically conductive polyaniline-titanium(IV) molybdophosphate cation exchange nanocomposite: synthesis, characterization and alcohol vapour sensing properties. J Ind Eng Chem 19:1226–1233

Khan MDA, Akhtar A, Nabi SA, Khan MA (2014) Synthesis, characterization, and photocatalytic activity of polyaniline-Sn(IV)iodophosphate nanocomposite: its application in wastewater detoxification. Ind Eng Chem Res 53:15253–15260

Khan AA, Rao RAK, Alam N, Shaheen S (2015) Formaldehyde sensing properties and electrical conductivity of newly synthesized Polypyrrole-zirconium(IV)selenoiodate cation exchange nanocomposite. Saens Actuators B 211:419–427

Kim HT, Lee CH, Shul YG, Moon JK, Lee EH (2003) Evolution of PAN-TiO_2 compoiste adsorbent for the removal of Pb(II) ions in aqueous solution. Sep Sci Technol 38(3):695–713

Kurniawan TA, Chan GYS, Hung LW, Babel S (2006) Physico–chemical treatment techniques for wastewater laden with heavy metals. Chem Eng J 118:83–98

Long XX, Yang XE, Ni WZ (2002) Current status and prospective on phytoremediation of heavy metal polluted soils. Chin J Appl Ecol 13:757–762

Mahmoud DK, Salleh MAM, Karim WA, Idris A, Abidin ZZ (2012) Batch adsorption of basic dye using acid treated kenaf fibre char: equilibrium, kinetic and thermodynamic studies. Chem Eng J 181–182:449–457

Mascolo MC, Pei Y, Ring TA (2013) Room temperature co-precipitation synthesis of magnetite nanoparticles in a large pH window with different bases. Materials 6:5549–5567

Meng Ng H, Sin LT, Ting TT, Bee ST, Hui D, Yu LC, Rahmat AR (2015) Extraction of cellulose nanocrystals from plant sources for application as reinforcing agent in polymers. Compos Part B 75:176–200

Misha S, Mat S, Ruslan MH, Salleh E, Sopian K (2015) Performance of a solar assisted solid desiccant dryer for kenaf core fiber drying under low solar radiation. Sol Energy 112:194–204

Mui ELK, Cheung WH, McKay G (2010) Tyre char preparation from waste Tyre rubber for dye removal from effluents. J Hazard Mater 175:151–158

Nabi SA, Shahadat M, Bushra R, Shalla AH, Azam A (2011a) Synthesis and characterization of nano-composite ion-exchanger; its adsorption behaviour. Colloids Surf B: Biointerfaces 87:122–128

Nabi SA, Shahadat M, Bushra R, Oves M, Ahmed F (2011b) Synthesis and characterization of polyanilineZr(IV)sulphosalicylate composite and its applications (1) electrical conductivity, (2) and antimicrobial activity studies. Chem Eng J 173:706–714

Nabi SA, Bushra R, Shahadat M, Raeissi AS (2012a) Development of nano-composite adsorbent for removal of metals from industrial effluent and synthetic mixtures; its conducting behaviour. Desalination 289:1–11

Nabi SA, Bushra R, Shahadat M (2012b) Removal of toxic metal ions by using composite cation-exchange material. J Appl Polym Sci 125:3438–3446

Nabi SA, Raeissi AS, Shahadat M, Bushra R, Khan AMT (2012c) Synthesis and characterization of novel cation exchange adsorbent for the treatment of real samples for metal ions. Chem Eng J 200–202:426–432

Nabi SA, Akhtar A, Khan MDA, Khan MA (2014) Synthesis, characterization and electrical conductivity of polyaniline-Sn(IV)tungstophosphate hybrid cation exchanger: analytical application for removal of heavy metal ions from wastewater. Desalination 340:73–83

Nacos MK, Katapodis P, Pappas C, Daferera D, Tarantilis PA, Christakopoulos P, Polissiou M (2006) Kenaf xylan – a source of biologically active acidic oligosaccharides. Carbohydr Polym 66:126–134

Naggar IME, Zakaria ES, Ali IM, Khalil M, El-Shahat MF (2012) Kinetic modeling analysis for the removal of cesium ions from aqueous solutions using polyaniline titanotungstate. Arab J Chem 5(1):109–119

Nakamoto K (1986) Infrared and Raman spectra of inorganic and coordination compounds. Part B: Applications in coordination, organometallic, and bioinorganic chemistry, 6th edn. Wiley, New York, p 154

Othman MR, Akil MH (2008) The CO_2 adsorptive and regenerative behaviors of Rhizopus oligosporus and carbonaceous Hibiscus cannabinus exposed to thermal swings. Microporous Mesoporous Mater 110:363–369

Pereira FV, Gurgel LVA, Gil LF (2010) Removal of Zn2þ from aqueous single metal solutions and electroplating wastewater with wood sawdust and sugarcane bagasse modified with EDTA di-Anhydride (EDTAD). J Hazard Mater 176:856–863

Raman NK, Anderson MT, Brinker CJ (1996) Template-based approaches to the preparation of amorphous, nanoporous silicas. Chem Mater 8:1682–1701

Rao CNR (1963a) Application of infrared spectroscopy. Academic, New York, p 355

Rao CNR (1963b) Chemical applications of infrared spectroscopy. Academic, New York, p 338
Rawat M, Ramanathan AL, Subramanian V (2009) Quantification and distribution of heavy metals from small-scale industrial areas of Kanpur City, India. J Hazard Mater 172:1145–1149
Shahadat M, Bushra R (2015) Synthesis, characterization and significant applications of PANI-Zr (IV)sulphosalicylate nanocomposite. Adv Nanotechnol 13:81–206
Shahadat M, Nabi SA, Bushra R, Raeissi AS, Umar K, Ansari MO (2012) Synthesis, characterization, photolytic degradation, electrical conductivity and applications of nanocomposite adsorbent for the treatment of pollutants. RSC Adv 2:7207–7220
Shahadat M, Teng TT, Rafatullah M, Arshad M (2015) Titanium-based nanocomposite materials: a review of recent advances and perspectives. Colloids Surf B: Biointerfaces 126:121–137
Sharif J, Mohamad SF, Othman NAF, Bakaruddin NA, Osman HN, Güven O (2013) Graft copolymerization of glycidyl methacrylate onto delignified kenaf fibers through pre-irradiation technique. Radiat Phys Chem 91:125–131
Sharma G, Pathania D, Naushad M, Kothiyal NC (2014) Fabrication, characterization and antimicrobial activity of polyaniline Th(IV) tungstomolybdophosphate nanocomposite material: efficient removal of toxic metal ions from water. Chem Eng J 251:413–421
Silverstein RM, Bassler GC, Morrill TC (1981) Chapter 3: Spectrometric identification of organic compounds, 4th edn. Wiley, New York, p 111
Sud D, Mahajan G, Kaur MP (2008) Agricultural waste material as potential adsorbent for sequestering heavy metal ions from aqueous solutions – a review. Bioresour Technol 99:6017–6027
Sun D, Zhang WY, Liu X (2010) Adsorption of anionic dyes fromaqueous solution on fly ash. J Hazard Mater 181:335–342
Tennakone K, Wijayantha KGU (1998) Heavy-metal extraction from aqueous medium with immobilized TiO_2 photocatalyst and a solid sacrificial agent. J Photochem Photobiol A Chem 113(1):89–92
Topp NE, Pepper KW (1949) Properties of ion-exchange resins in relation to their structure. Part I. Titration curves. J Chem Soc 1949:3299–3303
Vatutsina OM, Soldatov VS, Sokolova VI, Johann J, Bissen M, Weissenbacher A (2007) A new hybrid (polymer/inorganic) fibrous sorbent for arsenic removal from drinking water. Rect Funct Polym 67:184–201
Xie L, Jiang R, Zhu F, Liu H, Ouyang G (2014) Application of functionalized magnetic nanoparticles in sample preparation. Anal Bioanal Chem 406:377–399
Yang Y, Wang P (2006) Preparation and characterizations of a new PS/TiO_2 hybrid membrane by sol-gel process. Polymer 47(8):2683–2688
Zhang M, Guangzhi H, Pan G (2009) Combined DFT and IR evidence on metastable-equilibrium adsorption of arsenate on TiO_2_surfaces. J Colloid Interface Sci 338:284–286

Chapter 2
Nanocomposite Materials for Wastewater Decontamination

M. Tauqeer, M. S. Ahmad, M. Siraj, A. Mohammad, O. Ansari, and M. T. Baig

Abstract Modernization and industrialization of human life make the fresh water continuously contaminated due to the use of a variety of organic and inorganic pollutants. Contamination of clean water may be because of chemicals, pesticides, soil erosion, heavy metals, dyes, etc. Heavy metal ions and dyes are the most critical pollutants for contamination of water/wastewater. Decontamination of different pollutants is one of the significant challenges for the water industry to produce high quality drinking water. There is a continuous need to develop more appropriate material to be used as water purification materials with high separation capacity, cheap, porosity, and recyclability. Different methods such as adsorption, coagulation, oxidative-reductive degradation and membrane separation, etc. have been developed for the decontamination of water/waste water. Adsorption among them is the most essential method for the purification of water. Variety of different adsorbents like activated carbon and other carbon-based materials (graphene, carbon nanotubes etc.) are used for the decontamination of water. A recent investigation on remediation of water involves the use of, nanomaterials. However, these nanoparticles can agglomerate which may limit its use. Nanocomposites can be used instead of nanoparticles with high purifying capability. Contamination of water by the adsorption of different pollutants e.g. heavy metals ions, dyes, etc. is nowadays effectively removed by these nanocomposites because of their characteristic properties like metal ion affinity, small size, and high surface area-to-volume ratio. The present chapter is a compilation of different nanocomposites and their uses for decontamination of water/waste available till date.

Keywords Nanocomposites · Hazardous dyes · Toxic heavy metals · Waste-water decontamination

---

M. Tauqeer (✉) · M. S. Ahmad · M. Siraj · O. Ansari · M. T. Baig
Department of Chemistry, Aligarh Muslim University, Aligarh, Uttar Pradesh, India
e-mail: mohdtauqeer.ch@amu.ac.in

A. Mohammad
Discipline of Chemistry, Indian Institute of Technology Indore, Indore, India

M. Oves et al. (eds.), *Modern Age Waste Water Problems*,
https://doi.org/10.1007/978-3-030-08283-3_2

2.1 Introduction

Water, one of the most abundant natural resources on earth, is essential for all aspects of human life including health, food, energy, and economy. It is estimated that only about 1% of water is used in human consumption (Grey et al. 2013; Adeleye et al. 2016). However, fresh water resources are continuously contaminated by pollutants due to rapid industrialization (Schwarzenbach et al. 2006; Mohammad et al. 2018; Ferroudj et al. 2013). Because of waterborne and nonfatal infection, the estimation of around 10–20 million people die every year with almost about 5000–6000 children die every day (Leonard et al. 2003; Ashbolt 2004; Hutton et al. 2007). More than 1.1 billion people all over the world do not have sources to get clean water, (Wigginton et al. 2012) which results in major health issues. The increment in the access of fresh water is limited because of healthy demand for fresh water with an increase in population throughout the world and with an increase in water- related problem (Adeleye et al. 2016). Pollution of fresh water may be due to soil erosion, inadequate sanitation and by algal blooms, detergents, fertilizers, pesticides, chemicals, heavy metals, etc. for contaminating ground/surface water (Ritter et al. 2002; Fawell and Nieuwenhuijsen 2003; Rodriguez-Mozaz et al. 2004; Falconer and Humpage 2005).

The most critical pollutants includes hazardous metal and dyes that are mainly various industrial pollutant wastes like textile, food processing, plastics, dyes, steel and metal processing plants, agricultural industry, manufacturing industrial wastes. These pollutants significantly affect the pollution in water/waste water which may be due to their high stability towards the light, temperature and other. Contamination due to dye causes an undesirable colour to the water that reduces the sunlight penetration which resists the photochemical and biological effects to aquatic life (Wang et al. 2009; Liu et al. 2012; Tian et al. 2013; Jia et al. 2013; Pereira and Alves 2012; Gajda 1996; Ivanov 1996; Kabdaşli et al. 1996; Bensalah et al. 2009; Wróbel et al. 2001; Dawood et al. 2014; Wong 2004; Hoffmann et al. 1995; Robles et al. 2013). The exact amount of contaminated particles of dyes and metals in water is not known, but the contamination causes serious challenge to the environmental scientists. Acute and chronic effects of dyes may be seen when the organism is exposed to these pollutants (Hoffmann et al. 1995; Robles et al. 2013; Clarke and Anliker 1980; Mishra and Tripathy 1993; Banat 1996; Gupta et al. 1990).

The water industry is in the process to develop different cost-effective and stable materials and method to produce high -quality drinking water. There is a need to modernize, update or modified the already existing methods to be more stable, economical and efficient. It is essential to save the water by recycling of water, in addition, to tackle the day by day contamination of drinking water. The primary need for any appropriate material to be used as water purification materials is their high separation capacity, cheap, porosity, and recyclability (Malaviya and Singh 2011; Bhatnagar and Sillanpää 2009). Treatment due to biological wastewater is widely used but these are slow and limited because of non-biodegradable contaminants which sometime cause toxicity (Zelmanov and Semiat 2008; Anjum et al. 2016).

There are different methods which have been developed for the remediation of such pollutants. These methods include adsorption, coagulation, oxidative-reductive degradation and membrane separation and so on (Daraei et al. 2013; Dutta et al. 2001; Moghaddam et al. 2010; Gao et al. 2012). Variety of different fascinating materials has been used for the decontamination of water/wastewater. Remediation of water can be done by the use of different adsorbents like activated carbon, carbon-based materials (graphene, carbon nanotubes etc.) and their hybrids because of their high adsorption capacity and good chemical activity (Wang et al. 2006a, b; Gandhi et al. 2016; Upadhyay et al. 2014). A tremendous effort has been made for the decontamination of water by enhancing the properties of the catalyst, electro-catalyst, photo electro-catalyst, disinfectant and desalination agent (Perreault et al. 2015; Chatterjee and Dasgupta 2005; Moo et al. 2014; Wang et al. 2013; Mahata et al. 2007; Chang and Wu 2013).

Recently, Nanotechnology emerged as a most advanced area for providing useful materials to be used for water decontamination by optimizing their properties like their hydrophobic and hydrophilic character, high toughness, high strength and porosity (Daer et al. 2015). Nanoparticles may be useful in the contribution of wastewater decontaminations because of their high surface area, but its agglomeration may limit its use (Ray and Shipley 2015). This agglomeration can be minimized when a nanomaterial is converted to nanocomposites (Twardowski 2007).

In nanocomposites, incorporation of nanoparticles occurred within different functionalized materials like activated carbon, reduced graphene oxide, multiwalled carbon nanotubes and different polymeric matrices. Contamination of water by the adsorption of various pollutants, e.g., heavy metals ions, dyes, etc. is nowadays effectively removed by these nanocomposites because of their characteristic properties like high surface area, metal ion affinity, small size, low or no internal diffusion resistance, and high surface area-to-volume ratio. Nanocomposites have emerged as a highly efficient water decontaminating agent than nanoparticles due to their high adsorption selectivity, capacity and stability.

This book chapter provides a basic understanding of the research of nanocomposites for water/wastewater treatment and their recycling by focusing on the challenges of future research.

2.2 Nanocomposite Materials for Waste Water Decontamination of Heavy Metal Ions

Contamination of wastewater in human society by different pollutant is increasing rapidly with increasing industrialization. Among different pollutants in wastewater, heavy metal ions such as Hg^{2+}, Ni^{2+}, Cd^{2+} etc. is one of the severe pollutants which have high toxicity and non-biodegradability (Ozay et al. 2009). The existence of heavy metal ions in wastewater is by many industrial wastes like chemical industry, battery industry etc., and by contamination of drinking water, soil, and the food chain that may be due to precipitation and adsorption of these pollutants in waste water. These toxic metal ions in water are very harmful to human health and can cause various health related issues like cadmium can cause chronic disorder and acute pain, copper (II) can cause iron disorder and cell membrane damage, etc. Thus, the decontamination of these heavy metal ions is very necessary and primary issue. A plethora of research has been done for the remediation of waste water. Variety of methods has been applied for the decontamination of heavy metal ions such as adsorption, Ion-exchange, membrane filtration, electrochemical treatment, solvent extraction, reverse osmosis and many others (Gupta and Suhas 2009; Gupta et al. 2006; Xu et al. 2011). However, adsorption among them is considered as an effective and economical method) (Ali and Gupta 2007; Ge et al. 2014; Feng et al. 2010; Mittal et al. 2008; Gupta and Nayak 2012). This method is very efficient and can be applied to the production of high-quality water on a larger scale without producing or less waste sludge contaminants, etc. Recently the adsorption process is keeping attention because of its simplicity and economical technique with cheaper and commercial adsorbents for of heavy metal ions decontamination from waste water (Gupta et al. 2007, 2011).

Decontamination of toxic heavy metal ions by using nanocomposites has become a fascinating area of research. There are different nanocomposites which are used for water decontamination of heavy metal ions. Different adsorbents have been used which enhances metal ions decontamination from wastewater (Table 2.1) (Min and Hering 1998; Shubha et al. 2001; Manju et al. 2002; Greenleaf et al. 2006; Abbasizadeh et al. 2013; Kaygun and Akyil 2007; Manju and Anirudhan 2000; Riccardo and Muzzarelli 2011; Keshtkar et al. 2013; Simex et al. 2003; Borai et al. 2015; Puttamraju and SenGupta 2006; Zayed et al. 2013; Esmat et al. 2017; Lin et al. 2012; Niu et al. 2009; Sahoo Sahoo 2017).

Some other adsorbents have also been used for decontamination purpose as presented in Table 2.2 (Yusuf et al. 2015; Sui et al. 2012; Murugesan et al. 2006; Gupta et al. 2009; Li et al. 2012; Dong et al. 2015; Shahzad et al. 2017; Sharma and Kumar Sharma and Kumar 2018). The important adsorbent used such as EDTA-MCS/GO nanocomposite (EDTA-functionalized magnetic chitosan graphene oxide nanocomposites) for decontamination of water/wastewater represents high removal efficiency for metal ions than the previously reported adsorbents (Shahzad et al. 2017). High q_{max} values of 206.52, 207.26, and 42.75 mg g^{-1} for Pb^{2+}, Cu^{2+}, and As^{3+}, respectively, is achieved using the minimum amount of sorbent (Shahzad et al. 2017). This may be because of EDTA, GO and CS conglomeration in a single composite.

Many reports are available for decontamination of heavy metal ions from waste water based on graphene oxide and EDTA. A recent investigation of EDTA-functionalized magnetic chitosan graphene oxide (EDTA-MCS/GO) nanocomposites shows the effective decontamination of heavy metals from waste water (Shahzad et al. 2017). Magnetic properties are introduced into the nanocomposites which causes the easy removal of metal ion from the adsorbent.

pH is a significant parameter in water/waste water treatment because it can affect both adsorbent structure and distribution of heavy metal pollutants. pH effect is

Table 2.1 Adsorbents used for the decontamination of waste-water

Heavy Metal ion	Nanocomposites as adsorbent	References
As(V)	Fe(III)-Doped Alginate Gels	Min and Hering (1998)
Pb(II), Cd (II), Hg(II)	Carboxylate functionalized polyacrylamide grafted hydrous tin (IV) oxide gel	Shubha et al. (2001; Manju et al. (2002)
	Carboxylate functionalized polyacrylamide-grafted hydrous iron(III) oxide	
As(III) and As(V)	Hydrated ferric oxide-ion exchange fibre	Greenleaf et al. (2006)
Thorium	Polyvinyl alcohol/titanium oxide, PAN/ Zeolite	Abbasizadeh et al. (2013) and Kaygun and Akyil (2007)
Cd(II) and Pb (II)	Polyacrylamide coated hydrous iron(III) oxide having carboxylate functional group	Manju and Anirudhan (2000)
Uranium	Polyvinyl alcohol/titanium oxide, chitin/ chitosan complex, PVA/TEOS/APTES hybrid nanofibre	Abbasizadeh et al. (2013), Riccardo and Muzzarelli (2011), Keshtkar et al. (2013)
Pb^{2+}, UO_2^{+}, Ra^{2+}, Tl^{+} and Ac^{3+}	Polyacrylamide-bentonite composite	Simex et al. (2003)
Cs, Co and Eu ions	TiO_2/Poly(acrylamide-styrene sodium sulfonate)	Borai et al. (2015)
Cu(II)	Metal oxide-Ion exchanger support	Puttamraju and SenGupta (2006)
Cd(II), Pb (II), Ni(II)	Fe_3O_4/Cyclodextrin polymer nanocomposite	Zayed et al. (2013)
Cu^{2+}, Fe^{3+}, As^{3+}	Alginate-based nanocomposites	Esmat et al. (2017)
As(III), As (V)	Magnetic γ-Fe_2O_3 nanoparticles	Lin et al. (2012)
As(III), As (V)	Protonated titanate nanotubes	Niu et al. (2009)

taken into consideration for the removal of heavy metal ions from wastewater using the EDTA-MCS/GO nanocomposites adsorbent (Shahzad et al. 2017). The uptake efficiency shows that the adsorption of metal ion onto the EDTA-MCS/GO nanocomposite heavily depends upon the initial pH. Increase in Pb^{2+} removal is observed with increase in pH upto 5.0, while at higher pH a sudden decrease is observed. Cu^{2+} removal shows a gradual increase with an increase in pH with maximum adsorption at pH 5.5, but the optimal pH of 8.0 is observed for As^{3+} removal. (Shahzad et al. 2017). Results indicate that for all the metal ions (cationic) removal, the adsorption capacity is highest for higher pH and vice versa. The removal of metal ions may be because of hydrophilic functional groups (–OH, –COOH, and –NH_2) in EDTA-MCS/GO surface which is in contrast to decontamination of different metal cations by GO-based adsorbents that occur at pH value >5.0.26.

Table 2.2 Comparison of different adsorbents with synthesized EDTA-MCS/GO nanocomposites

Adsorbents	Metal ions	References
Graphene Oxide–Calcium–Zinc nanocomposite	Cr(III) and Cu (II)	Yusuf et al. (2015)
Graphene– CNT hybrid aerogels	Pb^{2+}	Sui et al. (2012)
Fungal biomass	As(III)	Murugesan et al. (2006)
Iron-chitosan composites	As(III), As(V)	Gupta et al. (2009)
GO/Fe_3O_4	Cu^{2+}	Li et al. (2012)
Polyethylenimine grafted polydopamine-mediated graphene oxide composites (PEI-PD/GO)	Cu^{2+}, Cd^{2+}, Pb^{2+}, Hg^{2+}	Dong et al. (2015)
Ethylenediaminetetraacetic acid (EDTA)-functionalized magnetic chitosan (CS) graphene oxide (GO) nanocomposites (EDTA-MCS/GO)	Pb^{2+}, Cu^{2+}, As^{3+}	Shahzad et al. (2017)

The effect of pH on the removal efficiency of Cu^{2+}, Fe^{3+} and As^{3+} is examined using the different adsorbents (Esmat et al. 2017). Increase in the removal efficiency of $Cu^{2+}$and Fe^{3+}ions is observed with an increase in solution pH from 2 to 6.5, which may be due to the carboxylic acid neutralization of alginate microparticles at lower pH. Subsequently, there is an increase in the electrostatic interaction between metal ions and composites which in turn increase the removal efficiency of the adsorbents. However, a decrease in removal efficiency occurs when pH increases above 6.5 (Esmat et al. 2017).

Use of polymer-based nanocomposite is one of the most promising areas of the latest research and development because of their advantageous properties like the ability to form a film, dimensional variability, and actin of different functionalities (Jordan et al. 2005; Lofrano et al. 2016; Pandey et al. 2017). A review has been developed with economic adsorbents for heavy metal decontamination from water/wastewater (Lim and Aris 2014). The polymer-based nanocomposite decontaminate the metal ions more efficiently in various forms like candle, membrane, beads, etc. (Chowdhury and Balasubramanian 2014). Preparation of the nanocomposite-based adsorbents involve the infusion of inorganic nanoparticles onto the polymers like cellulose, (Guo and Chen 2005) alginate, (Zouboulis and Katsoyiannis 2002) porous resins, (Katsoyiannis and Zouboulis 2002) and ion-exchangers (Blaney et al. 2007). Because of their high mechanical strength and adjustable surface chemistry, porous PNCs adsorbents or ion exchangers is emerged as an ideal hybrid adsorbents to avoid issues related to ultra-fine particle size (Pan et al. 2007).

Nano iron oxide loaded polystyrene is used for decontamination of heavy metal ions from water. Adsorption is affected by the presence of counter ions which causes the formation of the tertiary complex around the nanocomposite (Qiu et al. 2012). Intercalation of Fe_3O_4nanoparticles is coated with polyethyleneimine (PEI) polymer into sodium rich montmorillonite (MMT) layers under an acidic condition of pH 2.

This composite behave as magnetic adsorbents and is used for the decontamination of Cr(VI) (Vijayaraghavan et al. 2006).

Radioactive waste is the other metallic pollutant which is mainly discharged during nuclear plant operations (Valsala et al. 2009). Caesium, americium, neptunium, uranium and curium are some of the vital constituents of radioactive waste. Due to its higher rate of fission yield, long half-life, and severe environmental effects, caesium radioisotopes is one of the critical fission products and is the primary pollutant in contamination of water (Nilchi et al. 2007). This may cause severe social and environmental impacts (Chang et al. 2008). Therefore, decontamination of nuclear waste is essential for the ecological problem. An important method for remediation of these pollutants from water is based mainly through solid phase extraction with the use of polymer nanocomposite ion exchangers because of its high selectivity, fast and easy to separate and large thermal and radiation stabilities (Sheha and El-Shazly 2010). Report on the different adsorbents used in the decontamination of RW metals (Cs, Sr, Co) are magnetic nanocomposites, (Zach-Maor et al. 2011) polymer nanocomposites, (Park et al. 2010), etc. Cellulose nano fibre is developed through the use of (2,2,6,6-tetramethylpiperidin-1-yl)oxyl (TEMPO)/NaBr/NaClO process for oxidation of wood pulp following mechanical treatment which has the negatively charged carboxylate groups on the surface and is useful in adsorption of UO_2^{2+} in water with adsorption capacity of 167 mg/g that may be at least two times better adsorbent than an adsorbent without carbonyl group (Ma et al. 2012).

2.3 Nanocomposite Materials for Waste Water Decontamination of Dyes and Other Organic Pollutants

Dyes, one of the severe organic pollutants, are discharged in water because of its use in many industries such as paper, paints, plastics, etc. (Hunger 2003; Christie 2007; Hai et al. 2007; Husain 2006). More than 700,000 tons dyes are produced annually, with 10%–15% discharge in wastewater (Horng and Huang 1993; Kuo 1992; Sujoy et al. 2006; Legrini et al. 1993). One of the major environmental issues is the contamination of wastewater from dye which may be increased due to the growing use of a variety of dyes (Fig. 2.1). These dyes are very toxic and can cause problems

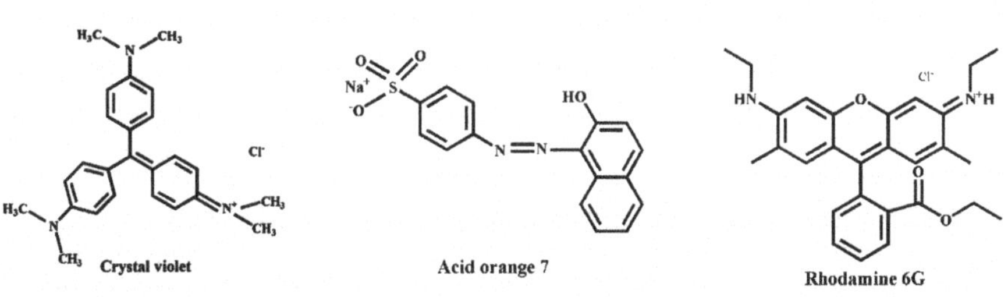

Crystal violet

Acid orange 7

Rhodamine 6G

Fig. 2.1 Different types of dyes present in the environment as the possible pollutant

even in trace level in the environment (Bekiari and Lianos 2006; Horng and Huang 1993; Kuo 1992; Sujoy et al. 2006). These dye contaminations deteriorate the water quality by giving undesirable colour to the water body which will influence sunlight penetration that ultimately affects the aquatic life (Legrini et al. 1993; Mohanty et al. 2006). In addition, dyes as water contaminated pollutants can cause food chain contamination which results in effecting the human and animal health. Therefore, decontamination of these dyes from wastewater is the essential requirement for the purification of water. Different methods of dye decontamination from industrial wastewaters such as oxidation, coagulation, adsorption, electrochemical techniques and flotation are known (Chen et al. 2007; Pandit and Basu 2004; Yunus et al. 2009; Zhao et al. 2009; Blackburn 2004). Adsorption is the preferred method in dyes decontamination due to its high efficiency, the economic occurrence of different adsorbents and their easy use of handling (Crini 2005). Because of their stability towards light and oxidation reaction, the difficulty in degrading dye materials has formed the basis for the development of new and high performing adsorbents for the efficient decontamination of dyes from wastewater (Hoffmann et al. 1995; Mohammad et al. 2016).

Recently, there are few reports on the use of Metal-organic frameworks (MOFs) for the removal of dye pollutants (Chen et al. 2012; Rajak et al. 2017). Different nanocomposite or nano adsorbents have been used for the efficient decontamination of dyes and other organic pollutants and have listed in Table 2.3 (Zhou et al. 2014; Bhattacharyya and Kumar-Ray 2015; El-Zahhar et al. 2014; Sun et al. 2011; Mahdavinia et al. 2013; Zhang et al. 2012a; Shirsath et al. 2011; Li et al. 2015; Marinovic et al. 2014; Ganigar et al. 2010).

2.3.1 Co/Co(OH)2 Nanocomposites and Magnetite/Reduced Graphene Oxide (MRGO) for Decontamination of Dyes from Waste Water

$Co/Co(OH)_2$ nanocomposite is used for the decontamination of Congo Red (CR) dyes from wastewater. The room temperature solution-phase reduction method was used to synthesize lamellar structured $Co/Co(OH)_2$ nanocomposite. This composite is high efficiency for the removal of dye contaminant from waste-water and is known as green-synthetic chemical method because of its pollution-free, low cost and simple process of preparation. Removal of dye contaminant, CR from wastewater is used to evaluate a $Co/Co(OH)_2$ nanocomposite. It shows high removal efficiency for CR dye and is removed 150 ppm of CR dye from an aqueous solution within 10 min with the absorption maximum of 2058 mg g^{-1} (Wu et al. 2014).

MRGO nanocomposites have been produced by a simple single step, non-toxic with low cost solvo-thermal method (Mahdavinia et al. 2013). Transmission electron microscopy (TEM), powder XRD and FTIR were used to investigate the morphology and compositional analysis of GO and MRGO and also matched with previous

Table 2.3 Different nanocomposite adsorbents used for decontamination of different dyes

Dyes and other organics as pollutant species	Nanocomposites adsorbents	References
Methylene blue and methyl orange	Montmorillonite (MMT) and layered double hydroxides (LDHs) based adsorbents	Zhou et al. (2014)
Congo red and methyl violet	Poly acrylic acid and polyethylene glycol based nano clay filled composite hydrogels	Bhattacharyya and Kumar-Ray (2015)
Bromophenol blue (BPB)	Poly(acrylamide-co-acrylic acid)/Kaolinite P(AAm-AA)-Kao) composite adsorbent	El-Zahhar et al. (2014)
Rhodamine B and malachite green	Magnetite/reduced graphene oxide nanocomposite	Sun et al. (2011)
Crystal violet	Nanocomposite hydrogels composed of *kappa*-carrageenan (Car), sodium alginate (Alg) and sodium montmorillonite (CarAlg/MMt) Polyacrylamide/Laponite nanocomposite hydrogels	Mahdavinia et al. (2013) and Zhang et al. (2012a)
Basic fuchsin		
Crystal violet	Poly(acrylic acid)-bentonite-FeCo (PAA-B-FeCo) based hydrogel nanocomposite	Shirsath et al. (2011)
Orange II	Al-pillared Fe-smectite	Li et al. (2015)
4-nitrophenol, 2-nitrophenol and 2-chloro-4 dinitrophenol.	Functionalized nanocomposite of poly (glycidyl methacrylate-co-ethylene glycol dimethacrylate) and acid modified bentonite	Marinovic et al. (2014)
Trichlorophenol and trinitrophenol	Polycation–clay mineral nanocomposites	Ganigar et al. (2010)

literature (Sun et al. 2011; Sheshmani and Mashhadi 2018; Cong et al. 2009). The GO sheets show wrinkle like features while MRGO shows the existence of embedded nanoparticles onto graphene's sheet surface (Sun et al. 2011; Cong et al. 2009). The XRD analysis show the presence of crystalline nature with well-defined several peaks and also the successful formation of graphene was confirmed by the disappearing the peak of GO in the composite (Ma et al. 2016; Cong et al. 2009; Zhang et al. 2010; Zhu et al. 2010a). Further, the mean diameter of about 8.8 nm of the Fe_3O_4 nanoparticles is determined from most intense peak which coincides well with the found measurements from TEM images (Sun et al. 2011; Lu et al. 2005, 2008; Warren 1969). Moreover, further confirmation for the formation of GO and MRGO was done using Fourier transform infrared spectroscopy (FTIR) spectra (Sun et al. 2011). The spectrum of GO shows the characteristic peaks of GO and shown the significant differences with the FTIR spectrum of MRGO (Li et al. Li et al. 2013; Sun et al. 2011; Guo et al. 2009; Ge et al. 2007). These MRGO nanocomposites show a dynamic behaviour as an adsorbent for decontamination of dye pollutant from waste water by the advantageous properties of graphene and magnetic nanoparticles. Importance of this adsorbent is; (i) a convenient single step and non-toxic synthetic route, (ii) Controlled loading of magnetic nanoparticles is possible on graphene surface, (iii) Complete removal of adsorbent is secure from water by an applied magnetic field, (iv) Simple washing with ethylene glycol (EG) is used to regenerate the adsorbent and about 80% of dyes can be removed after reusing five times (Sun et al. 2011).

2.3.2 Polymer Nanocomposites

Polymer nanocomposites emerge as one of the essential adsorbents for decontamination of dyes. The process of adsorption is the widely used technique for decontamination of dyes (Bansal and Goyal 2005). Adsorption based on Ion exchange-process has been found suitable for remediation of dyes (Gertsen and Sønderby 2009; Raval et al. 2016; Dulman et al. 2016). Some of the ion exchange resins like polystyrene phosphonate, sulfonated polystyrene, sulfonated phenolic resin, polystyrene-based trimethyl benzyl ammonium, epoxy-polyamine etc. are highly efficient for the decontamination of dyes through the ion-exchange mechanism (Yu et al. 2001; Zhang et al. 2006; Vakili et al. 2014; Qing et al. 2017). Some other polymer nanocomposites used as adsorbents for the removal of dyes are listed in Table 2.4 (Debrassi et al. 2012; Yaoji et al. 2014; Wang et al. 2010; Zhu et al. 2010b; Annadurai et al. 2002; Ghorai et al. 2014; McKay et al. 1997). Results indicate that decontamination involves not only the adsorption but also electrostatic attraction between nanocomposite and molecules of dyes.

Mechanistic steps involved in the adsorption of dyes on a polymer (Mahmoodian et al. 2012). The adsorption process completes in three levels: (i) film diffusion or boundary layer diffusion or outer diffusion which involve the transfer of impurities across the boundary of liquid to the exterior surface of the adsorbent surface (b) intra-particle diffusion or inner diffusion (c) finally, on the inner and outer

Table 2.4 Polymer nanocomposites used for the decontamination of different dye pollutants

Polymer nanocomposites (PNC) adsorbents	Dyes as pollutant	References
L-Cht/γ-Fe_2O_3	Remazol red 198	Debrassi et al. (2012)
Poly(acrylic acid-coacrylamide)/Kaolin hydrogel	Methyl violet	Yaoji et al. (2014)
CNTs/activated carbon fibre	Basic violet 10	Wang et al. (2010)
Chitosan/Fe_2O_3/MW-CNTs	Methyl orange	Zhu et al. (2010b)
Banana and orange peels	Methyl orange, methylene blue, Rhodamine B (RB), Congo red etc.	Annadurai et al. (2002)
Hydrolyzed NanoSilica incorporated	Methylene blue	Ghorai et al. (2014)
Polyacrylamide grafted xanthum gum	Methyl violet	
Bagasse pith	Acid and basic blue	McKay et al. (1997)

surfaces of the adsorbent, the adsorption of adsorbate is considered to be achieved (Li and Wang 2009). This step is not treated as the rate determining step because it is supposed to be very fast.

2.3.3 Biochar-Based Nanocomposites for Removal of some Organic Pollutants from Wastewater

Biochar is produced by pyrolysis of biomass in the absence of or little oxygen (Lehmann and Joseph 2012). Some research efforts are known on biochar-based nanocomposites as adsorbents for decontamination of water because of its high potential to adsorb water contaminants, cheap and favorable chemical and physicalsurface characteristics (Ahmad et al. 2014; Tan et al. 2015, 2016a).

Several mechanisms (p–p interactions, hydrogen bond, ionic interaction and hydrophobic interaction) may involve in adsorption by interactions between contaminants and biochar's functional groups (Tan et al. 2016b). Additionally, adsorption capacity can be increased by the nanoparticle coating on the carbon structure for the decontamination of water/waste water (Sun et al. Sun et al. 2015; Zhang et al. 2013a). Different biochar based adsorbents and others which are used as a decontaminating agent of wastewater are presented in Table 2.5 (Inyang et al. 2014; Zhang et al. 2012b, 2013b; Sun et al. 2015; Tang et al. 2015; Inyang et al. 2015; Karakoyun et al. 2011).

Table 2.5 Biochar- based nanocomposites exposed to various organic contaminants

Feedstock	Nanomaterials	Organic contaminants	References
Hickory chips and bagasse biochars	Hybridized CNT–biochar nanocomposite	Methylene blue	Inyang et al. (2014)
Cotton wood	Biochar/γ-Fe_2O_3 composite	Methylene blue	Zhang et al. (2013b)
Corn stalks	Fe_3O_4-coated biochar	Crystal violet	Sun et al. (2015)
Milled cotton wood	Graphene-coated biochar	Methylene blue	Zhang et al. (2012b)
Wheat straw	Graphene/biochar composite	Phenanthrene	Tang et al. (2015)
Hickory chips	Carbon nanotube–biochar nanocomposite	Sulfapyridine	(Inyang et al. (2015)
Chicken Wood Tire	P(acrylamide)-chicken biochar (p(AAm)-CB), p(acrylamide)-wood biochar (p(AAm)-WB), and p(acrylamide)-tire biochar (p(AAm)-TB) hydrogel composites	Phenol	Karakoyun et al. (2011)

2.3.4 Polymer-Clay Nanocomposites for Decontamination of Phenolic Compounds from Water/Wastewater

Other organic pollutants like phenol are found to exist through various industrial applications like a synthesis of aspirin, dyes, pesticides, paints, etc. Traces of phenol can cause toxicity and foul odour to the water that maybe because of the formation of chlorophenol and nitrophenol when it reacts with chlorine and nitrate, respectively, in the soil. The maximum allowed concentration of phenol in water/waste water is less than 1 ppm in many countries (Mahamuni and Pandit 2005; Maleki et al. 2005). Phenol and its by-products cause toxicity to living beings and the environment. Therefore, the decontamination of phenol from water/waste water is necessary. In other works, zeolites and MOF were also used as the adsorbents for the adsorptive removal of phenols (Khalid et al. 2004; Xie et al. 2015).

2.4 Nanocomposite Materials for Waste Water Decontamination of Other Pollutants

Other water pollutant species present in water are microorganism, pesticides, pathogens, and other organic materials. The problem related to these pollutants involves the disinfected by-product formation and drug resistance bacterial formation which promptly explore the advanced disinfection method (Verbyla and Mihelcic 2015). Pathogens are killed by the nano structured composite which liberates toxic chemicals. Advancements can achieve improvement in bactericidal effect of metal in nanotechnology and its use as a viable disinfectant with improved efficiency. However, a severe health problem is also seen in the case of a metal particle's use as a disinfectant (Zhang et al. 2016). The detail disinfection mechanism is not fully understood, but it may involve the metal atom interaction with base pairs of DNA which disrupts hydrogen bonding that leads to denaturation of DNA molecule of the cell (Klueh et al. 2000). Advancement in nanotechnology made nanocomposite an efficient water decontaminator which replaces the current chemical disinfectants (Jain and Pradeep 2005). The metal bound copolymer beads and cellulose acetate fibres embedded with Ag nanoparticles are used for removing the bacterial strains (Rieger et al. 2016). Some of the adsorbents used for the decontamination of waste water is presented in Table 2.6 (Paul et al. 2007; Chalageri et al. 2016; Wang et al. 2016; Yu et al. 2015; Zadaka et al. 2009).

Table 2.6 Nanocomposite adsorbents and other pollutants

Other pollutants	Nanocomposites	References
Bacteria and virus	Layered double hydroxide (LDH) nanocomposite	Paul et al. (2007)
Nitrate	Graphene, graphene oxide and nanocomposite	Chalageri et al. (2016)
Phosphate	Magnetite core/zirconia shell nanocomposite (Fe_3O_4@ZrO_2)	Wang et al. (2016)
Oil	Fe_3O_4/polystyrene magnetic nanocomposite	Yu et al. (2015)
Atrazine (pesticide)	Polycation–clay composites	Zadaka et al. (2009)

2.4.1 Layered Doubled Hydroxide (LDH) Nanocomposites for Decontamination of Viruses and Bacteria from Waste Water

Layered double hydroxides (LDH) nanocomposites, a combination of anionic clay-like materials, have the general formula of $[M^{2+}{}_{1-x}\ (OH)_2]^{x+}\ X^{m-}\ H_2O$, where M^{2+} and M^{3+} are bivalent and trivalent cations, respectively, in octahedral positions and Xm is an anion positioned between the interlayers (Miyata 1975, 1980, 1983; Rhee et al. 1997; You et al. 2003).

Viruses and bacteria are found abundantly in different water systems and are composed of a polypeptide chain containing weakly acidic and basic groups under natural pH (Gerba and Rose 1990; Redman et al. 1997). Reports on LDH coated sand to absorb viruses make LDH as a promising candidate for the decontamination of biological molecules like bacteria and viruses from waste water (Jin et al. 2004).

2.4.2 Fe_3O_4/Polystyrene Magnetic Nanocomposites as a Sorbent for Clean-up of Oil from Contaminated Water

Decontamination of water due to sewage containing oil causes the oil spill accidents and damage to our environment. This not only adversely affects the people's health but also damage the normal business. The behaviour of self-purification of the oceans and lakes is limited. The toxic oil constituent like benzene and toluene, destroy the food chain which ultimately destroys lower algae to higher mammals. Weight measurement study shows the oil adsorption capacity of Fe_3O_4/Polystyrene nanocomposites. Fe_3O_4/Polystyrene nanocomposites then reuse the nanocomposites to decontaminate oil from water surface many times (Yu et al. 2015).

2.4.3 Decontamination of Nitrate from Wastewater

An excessive amount of nitrate in the environment can cause health-related issues such as diabetes, methemogloglobinema, stomach cancer and other serious infectious diseases (Sharma and Sobti 2012). EPA and WHO guidelines the maximum uptake limit of nitrate concentration is 50 mg/L by nitrate and 10 mg/L by nitrogen (Fataei et al. 2013). Decontamination of around 82% of nitrate removal and 66% of N_2 selectivity is achieved using a catalyst such as Pd-Cu held by graphene with Fe^0 reductants (Yupan et al. 2016). Decontamination of nitrate by Nanocomposites material shows efficient behaviour than nanoscale zero-valent iron (Guo et al. 2016). Graphene-based nanocomposites along with iron, cobalt, and nickel show an excellent removal of nitrate from waste (Motamedi et al. 2014).

2.4.4 Decontamination of Emerging Pollutants from Water/ Waste Water

Emerging pollutants are generally non-regulated pollutants which mainly constitute compounds of pesticides, herbicides, pharmaceutical products and personal hygiene products (Gil et al. 2012). Decontamination of Atrazine (pesticide) and 3,6-dichloropyridine-2-carboxylic acid(Clopyralid) from water is achieved by Poly(4-vinylpyridine-costyrene) and polydiallyl dimethylammonium chloride with Montmorillonite Clay nanocomposites (Zadaka et al. 2009) and montmorillonite (SWy-2)–chitosan bio-nanocomposites (SW–CH) adsorbents, respectively.

2.5 Conclusions and Future Perspective

Presently, decontamination of water/waste water is still needed development of more advanced materials to acquire a high quality of pure water. Adsorbents process is found the most suitable process for the effective decontamination of water. In this regards, variously advanced adsorbents have been developed for the effective decontamination of pollutants from water. The method of remediation depends on materials properties and its types. So, to enhance the removal efficiency, the specific modification is needed in adsorbents. Adsorbent based on nanotechnology is emerges as the most advanced candidates for the treatment of water. There are different nanomaterials which have been used in applications related to waste water treatment. However, these nanoparticles can agglomerate which may limit its use. Nanocomposites can be used instead of nanoparticles with high purifying capability. Various nanocomposites based adsorbents have been known and discussed for the efficient decontamination of different pollutants from water/ waste water. Nanocomposites based adsorbents are found as one of the most advanced and useful material for the purification of water. However, these nanocomposites based adsorbents may contaminate the water/waste water during sample treatment processes which can seriously affect the environment and human being. Therefore, quality material is needed to develop for future research with minimum toxicity, cost-effective and maximum removal capability of contaminated pollutants. So that, a fascinating approach for the purification of water would be possible.

Acknowledgments We would like to kindly acknowledge the discussed work in the present chapter all references and from other sources. M. Tauqeer gratefully acknowledges the Department of Chemistry, A.M.U., Aligarh. M. Sarfaraz Ahmad, M. Siraj, O. Ansari and M. T. Baig are grateful to the Department of Chemistry and Department of Industrial Chemistry, A.M.U., Aligarh. A. Mohammad is grateful to IIT Indore for Institute Post-doctoral fellowship. We gratefully acknowledge Dr. Shaikh M Mobin, IIT Indore for the fruitful discussions and suggestions.

References

Abbasizadeh S, Keshtkar AR, Mousavian MA (2013) Preparation of a novel electrospun polyvinyl alcohol/titanium oxide nanofiber adsorbent modified with mercapto groups for uranium(VI) and thorium(IV) removal from aqueous solution. Chem Eng J 220:161–171

Adeleye AS, Conway JR, Garner K, Huang Y, Su Y, Keller AA (2016) Engineered nanomaterials for water treatment and remediation: costs, benefits, and applicability. Chem Eng J 286:640–662

Ahmad M, Rajapaksha AU, Lim JE, Zhang M, Bolan N, Mohan D, Vithanage M, Lee SS, Ok YS (2014) Biochar as a sorbent for contaminant management in soil and water: a review. Chemosphere 99:19–33

Ali I, Gupta VK (2007) Advances in water treatment by adsorption technology. Nat Protoc, 1:2661–2667

Anjum M, Miandad R, Waqas M, Gehnay F, Barakat MA (2016) Remediation of wastewater using various nanomaterials. Arab J Chem. https://doi.org/10.1016/j.arabjc.2016.10.004

Annadurai G, Juang RS, Lee DJ (2002) Use of cellulose-based wastes for adsorption of dyes from aqueous solutions. J Hazard Mater 92:263–274

Ashbolt NJ (2004) Microbial contamination of drinking water and disease outcomes in developing regions. Toxicology 198:229–238

Banat IM (1996) Microbial decolorization of textile-dye containing effluents: a review. Bioresour Technol 58:217–227

Bansal RC, Goyal M (2005) Activated carbon adsorption. CRC Press/Taylor & Francis Group. ISBN 9780824753443

Bekiari V, Lianos P (2006) Ureasil gels as a highly efficient adsorbent for water purification. Chem Mater 18:4142–4146

Bensalah N, Alfaro M, Martínez-Huitle C (2009) Electrochemical treatment of synthetic wastewaters containing alphazurine A dye. Chem Eng J 149:348–352

Bhatnagar A, Sillanpää M (2009) Applications of chitin- and chitosanderivatives for the detoxification of water and wastewater—a short review. Adv Colloid Interf Sci 152:26–38

Bhattacharyya R, Kumar-Ray S (2015) Removal of Congo red and methyl violet from water using nano clay filled composite hydrogels of poly acrylic acid and polyethylene glycol. Chem Eng J 260:269–283

Blackburn RS (2004) Natural polysaccharides and their interactions with dye molecules: applications in effluent treatment. Environ Sci Technol 38:4905–4909

Blaney LM, Cinar S, Sengupta AK (2007) Hybrid anion exchanger for trace phosphate removal from water and wastewater. Water Res 41:1603–1613

Borai EH, Breky MME, Sayed MS, Abo-Aly MM (2015) "Synthesis, characterization and application of titanium oxide nanocomposites for removal of radioactive cesium, cobalt and europium ions", J. Colloid Interface Sci 450:17–25

Chalageri BD, Rajeswari A, Kulkarni M (2016) Denitrification of groundwater using graphene, graphene oxide and nanocomposite. IJRET Int J Res Eng Technol 05:10–14

Chang H, Wu H (2013) Graphene-based nanocomposites: preparation, functionalization, and energy and environmental application. Energy Environ Sci 6:3483–3507

Chang C-Y, Chau L-K, Hu W-PW-P, Wang W-P, Liao J-H (2008) Nickel hexacyanoferrate multilayers on functionalized mesoporous silica supports for selective sorption and sensing of cesium. Micro Meso Mater 109:505–512

Chatterjee D, Dasgupta SJ (2005) Photochem. Visible light induced photocatalytic degradation of organic pollutants. Photobiol C Photochem Rev 6:186–205

Chen W, Lu W, Yao Y, Xu M (2007) Highly efficient decom-position of organic dyes by aqueous-fiber phase transfer and in situ catalytic oxidation using fiber-supported cobalt phthalocyanine. Environ Sci Technol 41:6240–6245

Chen C, Zhang M, Guan Q, Li W (2012) Kinetic and thermodynamic studies on the adsorption of xylenol Orange onto MIL-101 (Cr). Chem Eng J 183:60–67

Chowdhury S, Balasubramanian R (2014) Recent advances in the use of graphene-family nanoadsorbents for removal of toxic pollutants from wastewater. Adv Colloid Interf Sci 204:35–56

Christie RM (2007) Environmental aspects of textile dyeing. Woodhead Publishing, Cambridge

Clarke E, Anliker R (1980) Organic dyes and pigments. Handb Environ Chem 3:181–215

Cong HP, He JJ, Lu Y, Yu SH (2009) Water-soluble magnetic-functionalized reduced graphene oxide sheets: in situ synthesis and magnetic resonance imaging applications. Small 6:169–173

Crini G (2005) Recent developments in polysaccharide-based materials used as adsorbents in wastewater treatment. Pol Sci 30:38–70

Daer S, Kharraz D, Giwa A, Hasan SA (2015) Recent applications of nanomaterials in water desalination: A critical review and future opportunities. Desalination 367:37–48

Daraei P, Madaeni SS, Salehi E, Ghaemi N, Ghari HS, Khadivi MA, Rostami E (2013) Novel thin film composite membrane fabricated by mixed matrix nanoclay/chitosan on PVDF microfiltration support: preparation, characterization and performance in dye removal. J Membr Sci 436:97–108

Dawood S, Sen TK, Phan C (2014) Synthesis and characterization of novel-activated carbon from waste biomass pine cone and its application in the removal of Congo red dye from aqueous solution by adsorption. Water Air Soil Pollut 225:1–16

Debrassi V, Baccarin T, Demarchi C, Nedelko N, Ślawska-Waniewska A, Dłużewski P, Bilska M, Rodrigues CA (2012) Adsorption of Remazol Red 198 onto magnetic N-lauryl chitosan particles: equilibrium, kinetics, reuse and factorial design. Environ Sci Pollut Res 19:1594–1604

Dong Z, Zhang F, Wang D, Liu X, Jin J (2015) Polydopamine-mediated surface-functionalization of graphene oxide for heavy metal ions removal. J Solid State Chem 224:88–93

Dulman V, Cucu-Man S-M, Bunia I, Dumitras M (2016) Batch fixed bed column studies on removal of Orange G acid dye by a weak base functionalized polymer. Desalin Water Treat 57:14708–14727

Dutta K, Mukhopadhyay S, Bhattacharjee S, Chaudhuri B (2001) Chemical oxidation of methylene blue using a Fenton-like reaction. J Hazard Mater 84:57–71

El-Zahhar AA, Awwad NS, El-Katori EE (2014) Removal of bromophenol blue dye from industrial waste water by synthesizing polymer-clay composite. J Mol Liq 199:454–461

Esmat M, Farghali AA, Khedr MH, El-Sherbiny IM (2017) Alginate-based nanocomposites for efficient removal of heavy metal ions. Int J Biol Macromol 102:272–283

Falconer IR, Humpage AR (2005) Health risk assessment of cyanobacterial (blue-green algal) toxins in drinking water. Int J Environ Res Public Health 2:43–50

Fataei E, Shariff AS, Kourandeh HHP, Sharifi AS, Safavyan STS (2013) Nitrate removal from drinking water in laboratory-scale using iron and sand nanoparticles. Int J Biosci 3(10):256–261
Fawell J, Nieuwenhuijsen MJ (2003) Contaminants in drinking water. Br Med Bull 68:199–208
Feng Y, Gong JL, Zeng GM, Niu QY, Zhang HY, Niu CG et al (2010) Adsorption of Cd (II) and Zn (II) from aqueous solutions using magnetic hydroxyapatite nanoparticles as adsorbents. Chem Eng J 162:487–494
Ferroudj N, Nzimoto J, Davidson A, Talbot D, Briot E, Dupuis V, Abramson S (2013) Maghemite nanoparticles and maghemite/silica nanocomposite microspheres as magnetic Fenton catalysts for the removal of water pollutants. App Catal B Environ 136:9–18
Gajda S (1996) Synthetic dyes based on environmental considerations. Dyes Pigments 30:1–20
Gandhi MR, Vasudevan S, Shibayama A, Yamada M (2016) Graphene and graphene-based composites: A rising star in water purification – a comprehensive overview. Chem Select 1:4358–4385
Ganigar R, Rytwo G, Gonen Y, Radian A, Mishael YG (2010) Polymer-clay nanocomposites for the removal of trichlorophenol and trinitrophenol from water. Appl Clay Sci 49:311–316
Gao Y, Pu X, Zhang D, Ding G, Shao X, Ma J (2012) Combustion synthesis of graphene oxide–TiO2 hybrid materials for photodegradation of methyl orange. Carbon 50:4093–4101
Ge J, Hu Y, Biasini M, Beyermann WP, Yin Y (2007) Superparamagnetic magnetite colloidal nanocrystal clusters. Angew Chem Int Ed 46:4342–4345
Ge Y, Li Z, Kong Y, Song Q, Wang K (2014) Heavy metal ions retention by bi-functionalized lignin: synthesis, applications, and adsorption mechanisms. J Ind Eng Chem 20(6):4429–4436
Gerba CP, Rose JB (1990) Viruses in source and drinking water. In: McFeters GA (ed) Drinking water microbiology. Springer, New York, pp 380–396
Gertsen N, Sønderby L (2009) Water purification. Nova Science Publishers, New York
Ghorai S, Sarkar A, Raoufi M, Panda AB, Schönherr H, Pal S (2014) Enhanced removal of methylene blue and methyl violet dyes from aqueous solution using a nanocomposite of hydrolyzed polyacrylamide grafted xanthan gum and incorporated nanosilica. ACS Appl Mater Interfaces 6:4766–4777
Gil MJ, Soto AM, Usma JI, Gutiérrez OD (2012) Contaminantesemergentes en aguas, efectos y posiblestratamientos. Producción + Limpia 7:52–73
Greenleaf JE, Lin JC, SenGupta AK (2006) Two novel applications of ion exchange fibers: arsenic removal and chemical-free softening of hard water. Environ Prog 25(4):300–311
Grey D, Garrick D, Blackmore D, Kelman J, Muller M, Sadoff C (2013) Water security in one blue planet: twenty-first century policy challenges for science. Philos Trans Roy Soc London A Math Phys Eng Sci 371(2012):0406
Guo X, Chen F (2005) Removal of arsenic by bead cellulose loaded with iron oxyhydroxide from groundwater. Environ Sci Technol 39:6808–6818
Guo HL, Wang XF, Qian QY, Wang FB, Xia XH (2009) A green approach to the synthesis of graphene nanosheets. ACS Nano 3:2653–2659
Guo L, Yaqi Z, Zhaoyang L, Junjie Z, Binbin T, Shaogui Y, Cheng S (2016) Efficient nitrate removal using micro-electrolysis with zero valent iron/activated carbon nanocomposite. J Chem Technol Biotechnol 91:2942–2949
Gupta VK, Nayak A (2012) Cadmium removal and recovery from aqueous solutions by novel adsorbents prepared from orange peel and Fe_2O_3 nanoparticles. Chem Eng J 180:81–90
Gupta VK, Suhas (2009) Application of low-cost adsorbents for dye removal—a review. J Environ Manag 90:2313–2342
Gupta G, Prasad G, Singh V (1990) Removal of chrome dye from aqueous solutions by mixed adsorbents: fly ash and coal. Water Res 24:45–50
Gupta VK, Rastogi A, Saini V, Jain N (2006) Biosorption of copper(II) from aqueous solutions by Spirogyra species. J Colloid Interface Sci 296(1):59–63
Gupta VK, Ali I, Saini VK (2007) Defluoridation of wastewaters using waste carbon slurry. Water Res 41(15):3307–3316

Gupta A, Chauhan VS, Sankararamakrishnan N (2009) Preparation and evaluation of iron–chitosan composites for removal of As(III) and As(V) from arsenic contaminated real life groundwater. Water Res 43:3862–3870

Gupta VK, Jain R, Saleh TA, Nayak A, Malathi S, Agarwal S et al (2011) Equilibrium and thermodynamic studies on the removal and recovery of safranine-T dye from industrial effluents. Sep Sci Technol 46(5):839–846

Hai FI, Yamamoto K, Fukushi K (2007) Hybrid treatment systems for dye wastewater. Crit Rev Environ Sci Technol 37:315–377

Hoffmann MR, Martin ST, Choi W, Bahnemann DW (1995) Environmental applications of semiconductor photocatalysis. Chem Rev 95:69–96

Horng JY, Huang SD (1993) Phosphates, phosphites, and phosphides in environmental samples. Environ Sci Technol 27:1169–1174

Hunger K (2003) Industrial dyes: chemistry, properties, applications. Wiley-VCH, Weinheim

Husain Q (2006) Potential applications of the oxidoreductive enzymes in the decolorization and detoxification of textile and other synthetic dyes from polluted water: a review. Crit Rev Biotechnol 26:201–221

Hutton G, Haller L, Bartram J (2007) Economic and health effects of increasing coverage of low cost household drinking water supply and sanitation interventions. World Health Organization

Inyang M, Gao B, Zimmerman A, Zhang M, Chen H (2014) Synthesis, characterization, and dye sorption ability of carbon nanotube–biochar nanocomposites. Chem Eng J 236:39–46

Inyang M, Gao B, Zimmerman A, Zhou Y, Cao X (2015) Sorption and cosorption of lead and sulfapyridine on carbon nanotube-modified biochars. Environ Sci Pollut Res 22:1868–1876

Ivanov K (1996) Possibilities of using zeolite as filler and carrier for dyestuffs in paper. Papier-Zeitschrift fur die Erzeugung von HolzstoffZellstoffPapier und Pappe 50:456–460

Jain P, Pradeep T (2005) Potential of silver nanoparticle-coated polyurethane foam as an antibacterial water filter. Biotechnol Bioeng 90:59–63

Jia WN, Wu X, Jia BX, Qu FY, Fan HJ (2013) Self-assembled porous ZnS nanospheres with high photocatalytic performance. Sci Adv Mater 5:1329–1336

Jin S, Brown TH, Vance GF, You YW (2004) Charge-based water filtration systems, US Patent Pending

Jordan J, Jacob KI, Tannenbaum R, Sharaf MA, Jasiuk I (2005) Experimental trends in polymer nanocomposites—a review. Mater Sci Eng A 393:1–11

Kabdaşli I, Tünay O, Orhon D (1996) Wastewater control and management in a leather tanning district. Water Sci Technol 40:261–267

Karakoyun N, Kubilay S, Aktas N, Turhan O, Kasimoglu M, Yilmaz S, Sahiner N (2011) Hydrogel-biochar composites for effective organic contaminant removal from aqueous media. Desalination 280:319–325

Katsoyiannis IA, Zouboulis AI (2002) Removal of arsenic from contaminated water sources by sorption onto iron-oxide-coated polymeric materials. Water Res 36:5141–5155

Kaygun AK, Akyil S (2007) Study of the behaviour of thorium adsorption on PAN/zeolite composite adsorbent. J Hazard Mater 147:357–362

Keshtkar AR, Irani M, Moosavian MA (2013) Removal of uranium(VI) from aqueous solutions by adsorption using a novel electrospun PVA/TEOS/APTES hybrid nanofiber membrane: comparison with casting PVA/TEOS/APTES hybrid membrane. J Radioanal Nucl Chem 295:563–571

Khalid M, Joly G, Renaud A, Magnoux P (2004) Removal of phenol from water by adsorption using zeolites. Ind Eng Chem Res 43:5275–5280

Klueh U, Wagner V, Kelly S, Johnson A (2000) Efficacy of silver-coated fabric to prevent bacterial colonization and subsequent device-based biofilm formation. J Biomed Mater Res 53:621–631

Kuo WG (1992) Decolorizing dye wastewater with Fenton's reagent. Water Res 26:881–886

Legrini O, Oliveros E, Braun AM (1993) Photochemical processes for water treatment. Chem Rev 93:671–698

Lehmann J, Joseph S (2012) Biochar for environmental management: science and technology. Routledge, London

Leonard P, Hearty S, Brennan J (2003) Advances in biosensors for detection of pathogens in food and water. Enzym Microb Technol 32:3–13

Li KL, Wang XH (2009) Adsorptive removal of Pb(II) by activated carbon prepared from Spartinaalterniflora: equilibrium, kinetics and thermodynamics. Bioresour Technol 100:2810–2815

Li J, Zhang S, Chen C, Zhao G, Yang X, Li J, Wang X (2012) Removal of Cu(II) and fulvic acid by graphene oxide nanosheets decorated with Fe_3O_4 nanoparticles. ACS Appl Mater Interfaces 4:4991–5000

Li M, Sun S, Yin P, Ruan C, Ai K (2013) Controlling the formation of rodlike V O nanocrystals on reduced graphene oxide for high-performance supercapacitors. ACS Appl Mater Interfaces 5 (21):11462–11470

Li H, Li Y, Xiang L, Huang Q, Qiu J et al (2015) Heterogeneous photo-Fenton decolorization of Orange II overAl-pillared Fe-smectite: response surface approach, degradation pathway, and toxicity evaluation. J Hazard Mater 287:32–41

Lim AP, Aris AZ (2014) A review on economically adsorbents on heavy metals removal in water and wastewater. Rev Environ Sci Biotechnol 13:163–181

Lin S, Lu D, Liu Z (2012) Removal of arsenic contaminants with magnetic-Fe_2O_3 nanoparticles. Chem Eng J 211–212:46–52

Liu B, Xu J, Ran SH, Wang ZR, Chen D, Shen GZ (2012) High-performance photodetectors, photocatalysts, and gas sensors based on polyol reflux synthesized porous ZnO nanosheets. Cryst Eng Comm 14:4582–4588

Lofrano G, Carotenuto M, Libralato G, Domingos RF, Markus A, Dini L, Gautam RK, Baldantoni D, Rossi M, Sharma SK, Chattopadhyaya MC, Giugni M, Meric S (2016) Polymer functionalized nanocomposites for metals removal from water and wastewater: an overview. Water Res 92:22–37

Lu LH, Randjelovic I, Capek R, Gaponik N, Yang JH, Zhang HJ, Eychmuller A (2005) Controlled fabrication of gold-coated 3D ordered colloidal crystal films and their application in surface-enhanced Raman spectroscopy. Chem Mater 17:5731–5736

Lu LH, Ai KL, Ozaki Y (2008) Environmentally friendly synthesis of highly monodisperse biocompatible gold nanoparticles with urchin-like shape. Langmuir 24:1058–1063

Ma H, Hsiao BS, Chu B (2012) Ultrafine cellulose nanofibers as efficient adsorbents for removal of UO2 2+ in water. ACS Macro Lett 1(1):213–216

Ma S, Wang Y, Wang X, Li Q, Tong S, Han X (2016) BifunctionalDemulsifier of ODTS modified magnetite/reduced graphene oxide nanocomposites for oil-water separation. Chem Select 1:4742–4746

Mahamuni NN, Pandit AB (2005) Effect of additives on ultrasonic degradation of phenol. Ultrason Sonochem 13:165–174

Mahata P, Aarthi T, Madras G, Natarajan S (2007) Photocatalytic degradation of dyes and organics with nanosized GdCoO. J Phys Chem C 111:1665–1674

Mahdavinia R, Aghaie G, Sheykhloie H, Vardini H, Etemadi MH (2013) Synthesis of CarAlg/MMt nanocomposite hydrogels and adsorption of cationic crystal violet. Carbohydr Polym 98:358–365

Mahmoodian H, Moradi O, Shariatzadeha B, Salehf TA, Tyagi I, Maity A, Asif M, Gupta VK (2012) Enhanced removal of methyl orange from aqueous solutions by poly HEMA–chitosan-MWCNT nanocomposite. J Mol Liq 202:189–198

Malaviya P, Singh A (2011) Physicochemical technologies for remediation of chromium-containing waters and wastewaters. Crit Rev Environ Sci Technol41:1111–1172

Maleki A, Malvi AH, Vaezi F, Nabizadeh R (2005) Ultrasonic degradation of phenol and determination of the oxidation byproduct toxicity. Iran J Environ Health Sci Eng 2:201–216

Manju GN, Anirudhan TS (2000) Batch lead and cadmium ions binding and exchange properties of polymer-coated hydrous iron(III) oxide. J Sci Ind Res 59(2):144–152

Manju GN, Krishnan KA, Vinod VP, Anirudhan TS (2002) An investigation into the sorption of heavy metals from wastewaters by polyacrylamide-grafted iron(III) oxide. J Hazard Mater 91(1–3):221–238

Marinovic SR, Milutinovic-Nikolic AD, Nastasovic AB, Žunic MJ, Vukovic ZM et al (2014) Sorption of different phenol derivatives on a functionalized macroporous nanocomposite of poly (glycidyl methacrylate-co-ethylene glycol dimethacrylate) and acid modified bentonite. J Serb Chem Soc 79:1249–1261

McKay G, El-Geundi M, Nassar MM (1997) Equilibrium studies for the adsorption of dyes on Bagasse Pith. Adsorpt Sci Technol 15:251–270

Min JH, Hering JG (1998) Arsenate sorption by Fe(III)-doped alginate gels. Water Res 32(5):1544–1552

Mishra G, Tripathy MA (1993) Critical review of the treatments for decolourization of textile effluent. Colourage 40:35–35

Mittal A, Gupta V, Malviya A, Mittal J (2008) Process development for the batch and bulk removal and recovery of a hazardous, water-soluble azo dye (metanil yellow) by adsorption over waste materials (bottom ash and de-oiled soya). J Hazard Mater 151(2–3):821–832

Miyata S (1975) Syntheses of hydrotalcite-like compounds and their structures and physico-chemical properties—I: the system Mg^{2+}-Al^{3+}-NO_3, Mg^{2+}-Al^{3+}-Cl, Mg^{2+}-Al^{3+}-ClO_4, Ni^{2+} -Al^{3+}-Cl, and Zn^{2+}-Al^{3+}-Cl. Clay Clay Miner 23:369–375

Miyata S (1980) Physico-chemical properties of synthetic hydrotalcites in relation to composition. Clay Clay Miner 28:50–55

Miyata S (1983) Anion-exchange properties of hydrotalcite-like compounds. Clay Clay Miner 31:305–311

Moghaddam SS, Moghaddam MA, Arami M (2010) Coagulation/flocculation process for dye removal using sludge from water treatment plant: optimization through response surface methodology. J Hazard Mater, 175:651–657

Mohammad A, Kapoor K, Shaikh MM (2016) Improved photocatalytic degradation of organic dyes by ZnO-nanoflowers. Chem Select 1:3483–3490

Mohammad A, Ahmad K, Rajak R, Mobin SM (2018) Remediation of water contaminants. Springer, Chapter 147-1

Mohanty K, Naidu JT, Meikap BC, Biswas MN (2006) Removal of crystal violet from wastewater by activated carbons prepared from rice husk. Ind Eng Chem Res 45:5165–5171

Moo JGS, Khezri B, Webster RD, Pumera M (2014) Graphene oxides prepared by Hummers', Hofmann's, and Staudenmaier's methods: dramatic influences on heavy-metal-ion adsorption. Chem Phys Chem 15:2922–2929

Motamedi E, Talebi AM, Kassaee MZ (2014) Comparision of nitrate removal from water via graphene oxide coated Fe, Ni and Co nanocomposites. Mater Res Bull 54:34–40

Murugesan GS, Sathishkumar M, Swaminathan K (2006) Arsenic removal from groundwater by pretreated waste tea fungal biomass. Bioresour Technol 97:483–487

Nilchi A, Atashi H, Javid AH, Saberi R (2007) Preparations of PAN based adsorbers for separation of cesium and cobalt from radioactive wastes. Appl Radiat Isot 65:482–487

Niu HY, Wang JM, Shi YL, Cai YQ, Wei FS (2009) Adsorption behavior of arsenic onto protonated titanate nanotubes prepared via hydrothermal method. Microporous Mesoporous Mater 122(1–3):28–35

Ozay O, Ekici S, Baran Y, Aktas N, Sahiner N (2009) Removal of toxic metal ions with magnetic hydrogels. Water Res 43:4403–4411

Pan BC, Zhang QR, Zhang WM, Pan BJ, Du W, Lv L, Zhang QJ, Xu ZW, Zhang QX (2007) Highly effective removal of heavy metals by polymer-based zirconium phosphate: a case study of lead ion. J Colloid Interface Sci 310:99–105

Pandey N, Shukla SK, Singh NB (2017) Water purification by polymer nanocomposites: an overview. Nano 3(2):47–66

Pandit P, Basu S (2004) Removal of ionic dyes from water by solvent extraction using reverse micelles. Environ Sci Technol 38:2435–2442

Park Y, Lee Y-C, Shin WS, Choi S-J (2010) Removal of cobalt, strontium and cesium from radioactive laundry wastewater by ammonium molybdophosphate–polyacrylonitrile (AMP–PAN). Chem Eng J 162:685–695
Paul S, Jeffrey HF, Morris M, Chen Q (2007) Removal of bacteria and viruses from water using layered double hydroxide nanocomposites. Sci Technol Adv Mater 8(1–2):67–70
Pereira L, Alves M (2012) In: Malik A, Grohmann E (eds) Environmental protection strategies for sustainable development, vol 4. Springer, New York, pp 111–162
Perreault F, Faria FDA, Elimelech M (2015) Environmental applications of graphene-based nanomaterials. Chem Soc Rev 44:5861–5896
Puttamraju P, SenGupta AK (2006) Evidence of tunable on-off sorption behaviors of metal oxide nanoparticles: role of ion exchanger support. Ind Eng Chem Res 45(22):7737–7742
Qing W, Meiying L, Yili X, Jianwen T, Qiang H, Fengjie D, Liucheng M, Qingsong Z, Xiaoyong Z, Yen W (2017) Facile and highly efficient fabrication of graphene oxide-based polymer nanocomposites through mussel-inspired chemistry and their environmental pollutant removal application. J Mater Sci 52(1):504–518
Qiu H, Zhang S, Pan B, Zhang W, Lv L (2012) Effect of sulfate on Cu(II) sorption to polymer-supported nano-iron oxides: behavior and XPS study. J Colloid Interfaces Sci 366:37–43
Rajak R, Saraf M, Mohammad A, Shaikh MM (2017) Design and construction of a ferrocene based inclined polycatenated Co-MOF for supercapacitor and dye adsorption applications. J Mater Chem A 5:17998–18011
Raval NP, Shah PU, Shah NK (2016) Adsorptive amputation of hazardous azo dye Congo red from wastewater: a critical review. Environ Sci Pollut Res 23(15):14810–14853
Ray PZ, Shipley HJ (2015) Inorganic nano-adsorbents for the removal of heavy metals and arsenic: a review. RSC Adv 5:29885–29907
Redman JA, Grant SB, Olson TM, Hardy ME, Estes MK (1997) Filtration of recombinant Norwalk virus particles and bacteriophage MS2 in quartz sand, importance of electrostatic interactions. Environ Sci Technol 31:3378–3383
Rhee SW, Kang MJ, Kim H, Moon CH (1997) Removal of aquatic chromate ion involving rehydration reaction of calcinated layered double hydroxide (Mg–Al–CO_3). Environ Technol 18:231–236
Riccardo A, Muzzarelli A (2011) Potential of chitin/chitosan-bearing materials for uranium recovery: an interdisciplinary review. Carbohydr Polym 84:54–63
Rieger KA, Cho HJ, Yeung HF, Fan W, Schiffman JD (2016) Antimicrobial activity of silver ions released from zeolites immobilized on cellulose nanofiber mats. ACS Appl Mater Interfaces 8 (5):3032–3040
Ritter L, Solomon K, Sibley P (2002) Sources, pathways, and relative risks of contaminants in surface water and ground water: a perspective prepared for the Walkerton inquiry. J Toxicol Environ Health 65:1–142
Robles AC, Martinez E, Alcantar IR, Frontana C, Gutierrez LG (2013) Development of an activated carbon-packed microbial bioelectrochemical system for azo dye degradation. Bioresour Technol 127:37–43
Rodriguez-Mozaz S, L'opez De Alda MJ, Barcel'o D (2004) Monitoring of estrogens, pesticides and bisphenol A in natural waters and drinking water treatment plants by solid-phase extraction-liquid chromatography-mass spectrometry. J Chromatogr A 1045:85–92
Sahoo TR (2017) Polymer nanocomposites for environmental applications. In: Tripathy D, Sahoo B (eds) Properties and applications of polymer nanocomposites. Springer, Berlin/Heidelberg
Schwarzenbach RP, Escher BI, Fenner K, Hofstetter TB, Johnson CA, Von Gunten U, Wehrli B (2006) The challenge of micropollutants in aquatic systems. Science 313:1072–1077
Shahzad A, Miran W, Rasool K, Nawaz M, Jang J, Lim SR, Lee DS (2017) Heavy metals removal by EDTA-functionalized chitosan graphene oxide nanocomposites. RSC Adv 7:9764
Sharma R, Kumar D (2018) Adsorption of Cr(III) and Cu(II) on hydrothermally synthesized graphene oxide–calcium–zinc nanocomposite. J Chem Eng Data 63(12):4560–4572

Sharma A, Sobti RC (2012) Nitrate removal from ground water: a review. E J Chem 9(4):1667–1675

Sheha RR, El-Shazly EA (2010) Kinetics and equilibrium modeling of Se(IV) removal from aqueous solutions using metal oxides. Chem Eng J 160:63–71

Sheshmani S, Mashhadi S (2018) Potential of magnetite reduced grapheme oxide/chitosan nanocomposite as biosorbent for the removal of dyes from aqueous solutions. Polym Compos 39:457–462

Shirsath SR, Hage AP, Zhou M, Sonawane SH, Ashokkumar M (2011) Ultrasound assisted preparation of nanoclayBentonite-FeCo nanocomposite hybrid hydrogel: a potential responsive sorbent for removal of organic pollutant from water. Desalination 281:429–437

Shubha KP, Raji C, Anirudhan TS (2001) Immobilization of heavy metals from aqueous solutions using polyacrylamide grafted hydrous tin (IV) oxide gel having carboxylate functional groups. Water Res 35(1):300–310

Simex S, Ulusoy U, Ceyhan Ö (2003) Adsorption of UO22+, Tl+, Pb2+, Ra2+ and Ac3+ onto polyacrylamide-bentonite composite. J Radio Anal Nucl Chem 256(2):315–321

Sui Z, Meng Q, Zhang X, Ma R, Cao B (2012) Green synthesis of carbon nanotube–graphene hybrid aerogels and their use as versatile agents for water purification. J Mater Chem 22:8767–8771

Sujoy KD, Jayati B, Akhil RD, Arun KG (2006) Adsorption behavior of rhodamine B on rhizopusoryzae biomass. Langmuir 22:7265–7272

Sun H, Cao L, Lu L (2011) Magnetite/reduced graphene oxide nanocomposites: one step solvothermal synthesis and use as a novel platform for removal of dye pollutants. Nano Res 4 (6):550–562

Sun P, Hui C, Azim Khan R, Du J, Zhang Q, Zhao YH (2015) Efficient removal of crystal violet using Fe_3O_4-coated biochar: the role of the Fe_3O_4 nanoparticles and modeling study their adsorption behaviour. Sci Rep 5:12638

Tan XF, Liu YG, Zeng G, Wang X, Hu X, Gu Y, Yang Z (2015) Application of biochar for the removal of pollutants from aqueous solutions. Chemosphere 125:70–85

Tan X-F, Liu Y-G, Gu Y-L, Xu Y, Zeng G-M, Hu X-J, Liu S-B, Wang X, Liu S-M, Li J (2016a) Biocharbased nano-composites for the decontamination of wastewater: a review. Bioresour Technol 212:318–333

Tan XF, Liu Y-G, Gu Y-L, Xu Y, Zeng G-M, Hu X-J, Liu S-B, Wang X, Liu S-M, Li J (2016b) Biochar-based nano-composites for the decontamination of wastewater: a review. Bioresour Technol 212:312–333

Tang J, Lv H, Gong Y, Huang Y (2015) Preparation and characterization of a novel graphene/biochar composite for aqueous phenanthrene and mercury removal. Bioresour Technol 196:355–363

Tian J, Sang YH, Yu GW, Jiang HD, Mu XN, Liu H (2013) Asymmetric supercapacitors based on graphene/MnO2 nanospheres and graphene/MoO3 nanosheets with high energy density. Adv Mater 25:5074–5083

Twardowski TE (2007) Introduction to nanocomposite materials: properties, processing, characterization. DEStech Publications. Pensylvania, 17601, USA

Upadhyay RK, Soin N, Roy SS (2014) Role of graphene/metal oxide composites as photocatalysts, adsorbents and disinfectants in water treatment: a review. RSC Adv 4:3823–3851

Vakili M, Rafatullah M, Salamatinia B, Abdullah AZ, Ibrahim MH, Tan KB, Gholami Z, Amouzgar P (2014) Application of chitosan and its derivatives as adsorbents for dye removal from water and wastewater: a review. Carbohyd Polym 113:115–130

Valsala TP, Roy SC, Shah JG, Gabriel J, Raj K, Venugopal V (2009) Removal of radioactive caesium from low level radioactive waste (LLW) streams using cobalt ferrocyanide impregnated organic anion exchanger. J Hazard Mater 166:1148–1153

Verbyla ME, Mihelcic JR (2015) A review of virus removal in wastewater treatment pond systems. Water Res 71:107–124

Vijayaraghavan K, Padmesh TVN, Palanivelu K, Velan M (2006) Biosorption of nickel(II) ions onto Sargassumwightii: application of two-parameter and three-parameter isotherm models. J Hazard Mater B 133:304–330

Wang S, Li H, Xu L (2006a) Application of zeolite MCM-22 for basic dye removal from wastewater. J Colloid Interface Sci 295:71–78

Wang S, Li H, Xie S, Liu S, Xu L (2006b) Physical and chemical regeneration of zeolitic adsorbents for dye removal in wastewater treatment. Chemosphere 65:82–87

Wang Y, Tang XW, Chen YM, Zhan LT, Li ZZ, Tang Q (2009) Adsorption behavior and mechanism of Cd(II) on loess soil from China. J Hazard Mater 172:30–37

Wang JP, Yang HC, Hsieh CT (2010) Adsorption of phenol and basic dye on carbon nanotubes/carbon fabric composites from aqueous solution. Sep Sci Technol 46:340–348

Wang S, Sun H, Ang HM, Tad MO (2013) Adsorptive remediation of environmental pollutants using novel graphene-based nanomaterials. Chem Eng J 226:336–347

Wang Z, Xing M, Fang W, Wu D (2016) One-step synthesis of magnetite core/zirconia shell nanocomposite for high efficiency removal of phosphate from water. Appl Surf Sci 366:67–77

Warren BE (1969) X-ray diffraction. Addison-Wesley, London

Wigginton N, Yeston J, Malakoff D (2012) WHO (World Health Organization), 2015. Drinking-water: Fact sheet No. 391. More treasure than trash. Science 337, 662–663. http://www.who.int/mediacentre/factsheets/fs391/en/

Wong Y (2004) Adsorption of acid dyes on chitosan—equilibrium isotherm analyses. Process. Biochemist 39:695–704

Wróbel D, Boguta A, Ion RM (2001) Mixtures of synthetic organic dyes in a photoelectrochemical cell. J Photochem Photobiol A Chem 138:7–22

Wu L, Liu Y, Zhang L, Zhao L (2014) A green-chemical synthetic route to fabricate a lamellar-structured Co/Co(OH)2 nanocomposite exhibiting a high removal ability for organic dye. Dalton Trans 43:5393–5400

Xie K, Shan CH, Qi JS, Qiao S, Zeng QS, Zhang LY (2015) Study of adsorptive removal of phenol by MOF-5. Desalin Water Treat 54:654–659

Xu M, Zhang Y, Zhang Z, Shen Y, Zhao M, Pan G et al (2011) Study on the adsorption of Ca^{2+}, Cd^{2+} and Pb^{2+} by magnetic Fe_3O_4 yeast treated with EDTA dianhydride. Chem Eng J 168(2):737–745

Yaoji T, Ma D, Zhu L (2014) Sorption behavior of methyl violet onto poly(acrylic acid-co-acrylamide)/kaolin hydrogel composite. Poly Plast Technol Eng 53:851–857

You Y, Vance GF, Sparks DL, Zhuang J, Jin Y (2003) Sorption of MS bacteriophage to layered double hydroxides: effects of reaction time, pH, and competing anions. J Environ Qual 32:2046–2053

Yu Y, Zhuang Y-Y, Wang Z-H (2001) Adsorption of water-soluble dye onto functionalized resin. J Colloid Interface Sci 242:288–293

Yu L, Hao G, Gu J, Zhou S, Zhang N, Jiang W (2015) Fe_3O_4/PS magnetic nanoparticles: synthesis, characterization and their application as sorbents of oil from waste water. J Magn Magn Mater 394:14–21

Yunus RF, Zheng YM, KGN N, Chen JP (2009) Electrochemical removal of rhodamine 6G by using RuO2coated Ti DSA. Ind Eng Chem Res 48:7466–7473

Yupan Y, Zifu L, Yi-Hung C, Mayiani S, Shikun C, Lei Z (2016) Reduction of nitrate in secondary effluent of wastewater treatment plants by FeO reductant and Pd- Cu/graphene catalyst. Water Air Soil Pollut 227:111

Yusuf M, Elfghi FM, Zaidi SA, Abdullah EC, Khan MA (2015) Applications of graphene and its derivatives as an adsorbent for heavy metal and dye removal: a systematic and comprehensive overview. RSC Adv 5:50392–50420

Zach-Maor A, Semiat R, Shemer H (2011) Synthesis, performance, and modeling of immobilized nano-sized magnetite layer for phosphate removal. J Colloid Interface Sci 357:440–446

Zadaka D, Nir S, Radian A, Mishael AG (2009) Atrazine removal from water by polycation-clay composites: effect of dissolved organic matter and comparison to activated carbon. Water Res 43:677–683

Zayed A, Badruddoza M, Shawon ZBZ, Daniel TWJ, Hidajat K, Uddin MS (2013) Fe_3O_4/cyclodextrin polymer nanocomposites for selective heavy metals removal from industrial wastewater. Carbohydr Polym 91:322–332

Zelmanov G, Semiat R (2008) Phenol oxidation kinetics in water solution using iron (3)-oxide-based nano-catalysts. Water Res, 42:3848–3856

Zhang X, Li A, Jiang Z, Jiang Q (2006) Adsorption of dyes and phenol from water on resin adsorbents: effect of adsorbate size and pore size distribution. J Hazard Mater 137:1115–1122

Zhang M, Lei D, Yin X, Chen L, Li Q, Wang Y, Wang T (2010) Magnetite/graphene composites: microwave irradiation synthesis and enhanced cycling and rate per-formances for lithium ion batteries. J Mater Chem 20:5538–5543

Zhang X, Zheng S, Lin Z, Zhang J (2012a) Removal of basic fuchsin dye by adsorption onto polyacrylamide/laponite nanocomposite hydrogels. Synth React Inorg Met-Org Nano-Met Chem 42:1273–1277

Zhang M, Gao B, Yao Y, Xue Y, Inyang M (2012b) Synthesis, characterization, and environmental implications of graphene-coated biochar. Sci Total Environ 435:567–572

Zhang M, Gao B, Yao Y, Inyang M (2013a) Phosphate removal ability of biochar/MgAl-LDH ultra-fine composites prepared by liquid-phase deposition. Chemosphere 92:1042–1047

Zhang M, Gao B, Varnoosfaderani S, Hebard A, Yao Y, Inyang M (2013b) Preparation and characterization of a novel magnetic biochar for arsenic removal. Bioresour Technol 130:457–462

Zhang X, Qian J, Pan B (2016) Fabrication of novel magnetic nanoparticles of multifunctionality for water decontamination. Environ Sci Technol 50:881–889

Zhao XM, Zhang BH, Ai KL, Zhang G, Cao LY, Liu XJ, Sun HM, Wang HS, Lu LH (2009) Monitoring catalytic degradation of dye molecules on silver-coated ZnO nanowire arrays by surface-enhanced Raman spectroscopy. Mater Chem 19:5547–5553

Zhou K, Zhang Q, Wang B, Liu J, Wen P et al (2014) The integrated utilization of typical clays in removal of organic dyes and polymer nanocomposites. J Clean Prod 81:281–289

Zhu C, Guo S, Fang Y, Dong S (2010a) Reducing sugar: new functional molecules for the green synthesis of graphene nanosheets. ACS Nano 4:2429–2437

Zhu HY, Jiang R, Xiao L, Zeng GM (2010b) Preparation, characterization, adsorption kinetics and thermodynamics of novel magnetic chitosan enwrapping nanosized γ-Fe2O3 and multi-walled carbon nanotubes with enhanced adsorption properties for methyl orange. Bioresour Technol 101:5063–5069

Zouboulis AI, Katsoyiannis IA (2002) Arsenic removal using iron oxide loaded alginate beads. Ind Eng Chem Res 41:6149–6155

Chapter 3
Nano-materials for Wastewater Treatment

Anjali Tyagi, Anshika Tyagi, Zahoor A. Mir, Sajad Ali, and Juhi Chaudhary

Abstract Heavy metal and ionic contamination in wastewater and surface water has been reported in many parts of the world and studied as a major global issue. Their exposure in a trace amount is hazardous for human health, thus; it is critical and challenging to remove undesirable metals from the water system. Today various methods have been placed for effective removal of heavy metals from the water like chemical precipitation, ion exchange, adsorption, membrane filtration and electrochemical technologies. The low-cost adsorbent has been studied as a substitute for costly current methods. The synthesis and characterization of nano-composite material for treating wastewater through adsorption using magnetic nano-sized ferric oxide (FeO) namely iron oxide have been discussed. It was then functionalized with graphene due to its large surface area. Due to the magnetic nature of this composite, it can be easily separated from the water under a magnetic field. Their chemical synthesis, characterization (XRD, SEM, TEM, and AAS), sorption behaviour of heavy metals [e.g. Pb(II), Cd(II), Cr(VI) and As (III)] from aqueous systems under varying experimental conditions has also been studied.

Keywords Adsorption; heavy metals; wastewater · Graphene · Granular ferric hydroxide · Lead · Chromium · Cadmium

---

A. Tyagi (✉)
CSIR-National Physical Laboratory, New Delhi, India

A. Tyagi · Z. A. Mir
NRCPB, IARI, New Delhi, India

S. Ali
CORD, University of Kashmir, Srinagar, Jammu and Kashmir, India

J. Chaudhary
College of Agricultural and Life Sciences, University of Florida, Gainesville, Florida

M. Oves et al. (eds.), *Modern Age Waste Water Problems*,
https://doi.org/10.1007/978-3-030-08283-3_3

3.1 Introduction

Water is very essential and used in many aspects of life. It is necessary for sustaining human, animal and plant life. Contaminated water mainly contains three types of contaminants say inorganic pollutants which include As, Cd, Pb, Cr, F, Hg, Se, NO_3^- etc., organic pollutants include pesticides, detergents, insecticides, etc. and biological contaminants include Fungi, algae, gram positive/gram-negative bacteria, etc. The sources of such contamination are either directly from the factory effluents and waste treatment plants or indirectly from the soil by agriculture or industrial runoff. So with this increasing industrial activity and wastewater from many industries such as chemical manufacturing, mining, battery manufacturing factories, etc. contains toxic heavy metals, which are not biodegradable and tend to accumulate in living organisms, and causing various diseases and disorders (75/440/EEC 1975; 80/778/EEC 1980; 98/83/EC 1998).

3.1.1 Sources of Contaminated Water May Include

Natural sources of contaminated water are those which are not made by people but instead occur naturally. Groundwater contamination is one of the main reason of waterborne related diseases in humans and is the main source of water for consumption purposes and its contaminated by nthropogenic activitiesSignificant sources of a natural disaster are Tsunamis, earthquake, hurricanes, floods, volcanoes and acid rain, etc. Natural sources are those, on which we do not have any control, it occurs naturally. Generally, the higher the distance between a source of contamination and a groundwater source, the more likely that natural process will reduce the impacts of contamination. Natural sources of pollution have relatively low significance in causing health and welfare effects because levels and the distances from sources and human population are usually massive. Moreover, major sources of natural pollution have episodic and transient character. (Richardson et al. 2007).

Anthropogenic contamination of the water may be defined as the chemical and biological contamination caused by the human activities. Its increase is associated with the rise in global population, industrialization, urbanization and intensive agriculture practices. In comparison to natural sources, anthropogenic sources of pollution are very serious, and the problem is still increasing due to the increase in human population and its increasing needs. It can be of various types:-.

3.1.1.1 Industrial Discharge

It can produce polluted water and waste. Before discharging the waste and effluents into the environment, some critical steps must be taken to ensure that any pollutants present are at a safe level. Industries that produce textiles, paper, and leather, generate effluents containing acids, fuels, oil and solvents. Ore and metal processing industries and those which uses metal in manufacturing can produce effluent that contains heavy metals. Many metals are highly toxic and carcinogenic. Even if they are present at low concentration in water, can pose a significant threat to human and ecosystem health. However, improper storage and release of chemicals on commercial and industrial sites can have adverse effects on water quality. To protect water sources, wastes must be treated appropriately before they are discharged to water-courses. (Byers 1995; Chambers and Winfiel 2000; Environment Canada 2001).

3.1.1.2 Municipal Discharge

Municipal wastewater refers to the contents of sewer systems. It can be made up of both sanitary sewage and stormwater, and can contain suspended solids, disease-causing pathogens, decaying organic wastes, nutrients and over 200 identified chemicals from industries, institutions, households and other sources. Municipal wastewater is released to lakes, rivers, and oceans and it must be cleaned before allowing any effluent to enter public waters. Most countries, municipalities, and communities have laws regulating effluents. (Regional Municipality of Waterloo 1994, 2000, 1996).

3.1.1.3 Heavy Metals

These are that whose density is exceeded 5 g/cm^3. Its presence even at traces level is believed to be a risk for human beings. It is still the challenging task to remove the undesirable metals from the water system. It can cause the severe health disorders i.e. reduced growth, organ damage, nervous system damage and in extreme may casuse death (Table 3.1).

3.1.1.4 Urban Pollution

Industry and human dwellings are also a source of potential contaminants. The most common contaminants are heavy metals, hydrocarbons from petroleum oils and solvents such as tri and tetrachloroethane found in ground water. It is evident these pollutants occur in small concentrations in drinking water and are enough to cause health effects, and some times these low molecular weight aromatic hydrocarbons will lead to sever odour problems in drinking water at concentrations of less than 30 mg/l.

Table 3.1 Common contaminants, their sources, and effects on human beings

Contaminant	Health effect	Sources
Arsenic	Skin damage, problems with circulatory systems, and may have increased risk of getting cancer.	Erosion of natural deposits, runoff from orchards, runoff from glass and electronics production wastes.
Cadmium	Kidney damage	Corrosion of galvanized pipes, erosion of natural deposits, discharge from metal refineries, runoff from waste batteries and paints (http://www.lenntech.com/periodic-chart-element/cd-en.html).
Fluoride	Bone disease	Water additive which promotes strong teeth, erosion of natural deposits, discharge from fertilizer and aluminum factories.
Lead	Infants and children: Delays in physical or intellectual progress; children might display slight shortfalls in attention span and grasping abilities.	Corrosion of household plumbing systems, erosion of natural deposits
	Adults: Kidney problems, high blood pressure	
Mercury	Kidney damage	Erosion of natural deposits; discharge from refineries and factories; runoff from landfills and croplands (http://www.epa.gov/ogwdw000/contaminants/dw_contamfs/mercury.html).
Chromium	Allergic dermatitis	Discharge from steel and pulp mills; erosion of natural deposits (http://www.epa.gov/OGWDW/contaminants/dw_contamfs/chromium.html).
Nitrate	It may sometimes be fatal to infants below 6 months and may cause some times death if untreated	Runoff from fertilizer use; leaking from septic tanks, sewage; erosion of natural deposits
	Symptoms such as breathing shortness and blue-baby syndrome	
Selenium	Hair or fingernail loss; numbness in fingers or toes; circulatory problems	Discharge from petroleum refineries; erosion of natural deposits; discharge from mines (http://en.wikipedia.org/wiki/selenium).

3.1.1.5 Agricultural Chemicals

Agriculture is a source of chemical contamination. In this case, the most important contaminant is nitrate, which can cause methemoglobinemia, or blue-baby syndrome, in infants less than 3 months and it is unpredictable at which level it has have the more effect, and some times the presence of microbial flora may cause infection is a signifant risk factor. WHO has projected a standard of 50 mg/l nitrate based pollutants n which the condition was seen below that concentration, although nitrite is also present and will be taken into consideration as it is 10 times as effective a methaemoglobinaemia agent as nitrate. The most important concern is the run-off of nutrients to surface waters, often combined with sewage discharges that lead to significant growths of cyanobacteria referred. The microbes produce wide range of toxins and it is likely that not all toxins have been identified yet. The treatments processes are limited and there is a possibility of undesirable concentration to be present in drinking water. Concerns are mainly directed at hepatotoxins such as the microcystins and cylindrospermopsin, and the neurotoxins such as saxitoxin (U.S. Environmental Protection Agency 2002, 2005) (Table 3.2).

The above table is showing the maximum permitted value of the major elements present in drinking water. These are the standard values given by the World Health Organisation (WHO) and environmental protection agency (EPA), and if the amount increases or decrease to the given maximum permissible values then it will be harmful to us. So after seeing the drawbacks of contaminated water, it is essential to purify water before its use as it is one of the basic needs for the existence of humanity. And the adequate disposal of heavy metals has been arousing worldwide concern in the last few decades. Till date, various treatment techniques and processes have been reported in literature survey for the removal of toxic contaminants from wastewater, such as adsorption, ion exchange, chemical precipitation, membrane-based filtration, photodegradation, evaporation, solvent extraction, reverse osmosis, oxidation, co-precipitation and so on.

Table 3.2 WHO & EPA permissible limits for major elements in drinking water (75/440/EEC 1975; 80/778/EEC 1980; 98/83/EC 1998)

Parameter's	Maximum permitted value (mg/l)
Aluminium	Less than 0.1–0.2
Arsenic	0.01
Cadmium	0.005
Fluorine	1.5
Lead	0.01
Mercury	0.001
Nitrate	50
Chromium	0.05
Sulphate	250
Chloride	Less than 200–300
Chlorine	0.3
Iron	0.3
Manganese	0.1
Zinc	Less than 0.1
Total hardness	Less than 100
Toluene	Less than 600

3.2 Removal Technologies

In the previous chapter, several types of common contaminants of water, a brief description of the sources of contaminated water, various kinds of heavy metals, sources and their effect on human beings, maximum permissible values of heavy metals in drinking water given by WHO and EPA were described. The course of this study was a focus on the various removal technologies, adsorbent materials, and their drawbacks, advantages of nano-sized material, proposed work all are outlined below in this chapter.

3.2.1 Various Techniques Used for the Purification of Contaminated Water

3.2.1.1 Filtration

It is the simplest technique and used for the removal of solid impurities from a liquid and collection of a desired solid from the solution from which it was precipitated or crystallized. It is generally used for the separation of large/big particles which we can see with our necked eyes. It is the classical separation technique.

It is mainly of three types:

Simple or Gravity

It is the most common method of filtration and is used to take out an insoluble solid residue from a solution. The residue which is left in the filter paper can be the required product or impurity or an additive such as a drying agent. Generally, the choice of filter paper is to depend on the size of the solid particle present in the solution like if the particle size is in the nano range than we preferred to use the watt man filter paper otherwise for large sized particles we use ordinary filter paper. A conventional or fluted filter paper is folded and set in a filter funnel which is then placed in the neck of a conical flask or supported in a clamp or ring stand. The solution is then carefully and slowly filtered by pouring it into the filter funnel by taking care not to fill the funnel above the edge of the filter paper (http://www.chem.ucalgary.ca/courses/351/laboratory/filtration.pdf) (Fig. 3.1).

Hot Filtration

Sometimes during a gravity/simple filtration, crystal starts to develop in the filter funnel, and it may block the pores of the filter paper and stop the filtration. This problem can be avoided by using a hot filtration method where the whole filtration

Fig. 3.1 Gravity filtration using fluted paper

apparatus (conical flask) is heated so that to avoid the sudden cooling of the solution. This hot filtration process is carried out by using a fluted filter paper and a stemless filter funnel. Sometimes due to the presence of hot vapors, the pressure build-up in the flask, and this will slow down or even stop the filtering action. But the fluted filter paper reduces this problem. The hot solution should be filtered quickly through the fluted paper in a stemless filter funnel into a conical flask.

To prevent pressure build-up a small piece of wire may be introduced between the funnel and conical flask. To prevent the crystal formation in the filter funnel the conical flask containing the solvent and funnel should be heated on a steam bath, and the process should be carried out on the steam bath. (http://www.chem.ucalgary.ca/courses/351/laboratory/filtration.pdf) (Fig. 3.2).

Vacuum Filtration

In a vacuum filtration, the solution which is to be filtered is passed through the filter paper by applying a vacuum to a Buchner flask. Vacuum filtration is generally the fast and effective way of filtering. The crystals are collected by swirling the mixture of the solid and liquid and then pouring quickly it into the filtration apparatus. This typically contains a Buchner funnel which is fitted with the appropriate size of filter paper, a clamped filter flask with conical filter adapter, and a vacuum applied to the side arm of the filter flask (http://www.chem.ucalgary.ca/courses/351/laboratory/filtration.pdf) (Fig. 3.3).

Fig. 3.2 Hot filtration

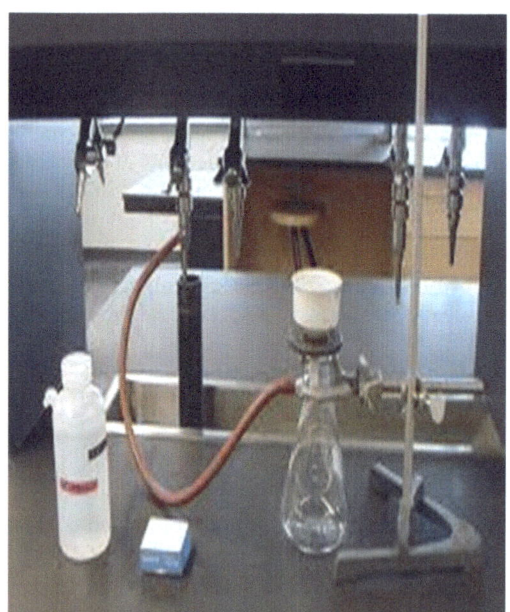

Fig. 3.3 Vacuum filtration using Buchner funnel

3.2.1.2 Ion-Exchange Technique

It is the modern separation technique. The term ion exchange generally means the exchange of ions of like sigh between a solution and a solid highly insoluble in it. The solid known as ion exchanger carries exchangeable cations and anions. When the exchanger is in contact with an electrolyte, these ions can be exchanged for a stoichiometrically equivalent amount of other ions of the same sigh. Carriers of exchangeable cations are known as cation exchangers and carriers of exchangeable anions as anion exchangers. Specific materials are capable of both cation and anion exchange. These are known as amphoteric exchangers (Ion Exchange, by F. Helfferich, McGraw Hill Book Company).

A cation exchange reaction is:

$$2NaX + CaCl_2(aq) \leftrightarrow CaX_2 + 2NaCl(aq)$$

A anion exchange reaction is:

$$2XCl + Na_2SO_4(aq) \leftrightarrow X_2SO_4 + 2NaCl(aq)$$

Where, X represents a structural unit of the ion exchanger.

In the first process, a solution containing dissolved $CaCl_2$, say something like hard water, is treated with a solid exchanger, NaX, containing exchangeable Na^+ ions. The exchanger removes the Ca^{2+} ions from the solution and replaces them with Na^+. Thus, a cation exchanger in Na^+ form is converted to Ca^{2+} form.

The ion-exchange resin can be divided into two:-

(a) *Cation-exchange resin*: – These are a high molecular weight cross-linked polymer containing a functional group such as phenolic, sulfonic, carboxylic, etc. sulfonic acid includes phenol formaldehyde, styrene divinyl benzene. Carboxylic acid includes chloromethylate styrene and phosphorous containing acrylic acid.
(b) *Anion-exchange resin:* – These are also high molecular weight cross-linked polymer containing basic groups are such as amino groups or substituted amino or quaternary ammonium.

Example: – Dowex-50, it is quaternary ammonium polystyrene (Ion Exchange Resins, by R. Kunin, John Wiley and sons).

3.2.1.3 Properties of Ion-Exchanger

1. They should be insoluble in water and an organic solvent.
2. They should contain charged groups that will exchange reversibly with another ion 0.in liquid Phase in which there is no permanent change in the structure.
3. They are porous and polymeric in nature.
4. The physical and chemical properties of ion-exchange resins are determined by the degree of cross-linking.
5. The swollen exchanger must be denser than water. Highly cross-linked resins are more brittle, harder and porous than the ion exchanger which contains light cross-linking agents (Ion Exchange in Analytical chemistry, by O. Samuelson, John Wiley and Sons) (Fig. 3.4).

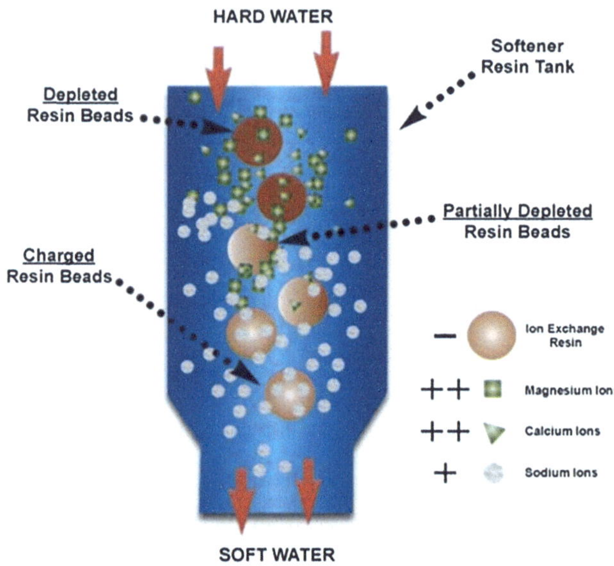

Fig. 3.4 Ion-exchange

3.2.1.4 Reverse Osmosis

It is the modern and most common home treatment method for contaminated drinking water. It is probably best known for its use in turning seawater into drinking water. It can reduce the amounts of organics, inorganic (As, Cr, Al, Cd, Pb, Hg, K, Nitrate, etc.), bacteria, particulates (Asbestos, Protozoan cysts, Cryptosporidium) and pesticides (Lindane, Heptachlor, Endrin) that can be found in contaminated drinking water. It is used to reduce the level of total dissolved solids and suspended particles within water (Dvorak and Skipton 2008).

RO is based on the principle of osmosis. In osmosis, a membrane separates two solutions containing different amount of dissolved chemicals. The membrane allows some compounds like water to pass through it but does not allow the large compounds to pass through it. This membrane is called the semi-permeable membrane. Pressure is applied to pass the pure water through the semi-permeable membrane from the dilute to the more concentrated solution. This pressure is called the osmotic pressure, and the process is called the osmosis. The hydrostatic pressure can vary from 2 MPa to around 6 MPa depending upon the salt content of the feed mixture. The natural tendency is for water to move from the dilute to the concentrated solution through a membrane until chemicals reach the equal concentrations on both sides of the membrane. But in reverse osmosis, the pressure is applied in the reverse direction say to the concentrated side of the membrane. This forces the osmotic pressure in the reverse direction so that the pure water is forced from the

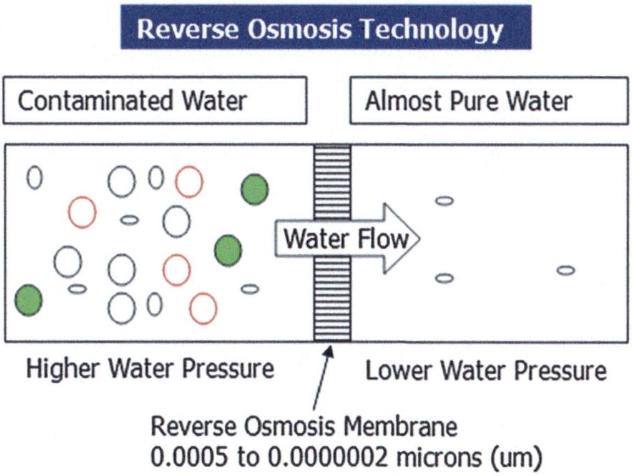

Fig. 3.5 Reverse osmosis process

concentrated side to the dilute side. Treated water is collected in a storage container. The rejected contaminants on the concentrated side of the membrane are washed away as wastewater (Handbook of Industrial membranes 1995; Winston Ho et al. 1992) (Fig. 3.5).

3.2.1.5 Ultrafiltration

In this technique, the solvent along with micro solutes pass through the membrane, and the membranes retain macro solutes. This process is similar to sieving, and the driving force is the hydrostatic pressure across the membrane.

The process is used to fractionate the solutes in a solution based on their size or molecular weight difference. Micro solutes whose effective sizes are smaller than the pore size of the membranes pass along with the solvent whereas macro solutes whose effective sizes are larger than the pore size of the membrane are retained. The driving force used in ultra-filtration processes is of the order of 500 kPa or so. The pore sizes of the membrane ranging from around 30 A° to about 200 A°. This technique is used for the separation of colloidal particles, oils, and macromolecules from effluent water as well as from surface waters (Winston Ho et al. 1992).

3.2.2 Various Adsorbent Materials Used

3.2.2.1 Activated Alumina

It is an adsorbent used in the packed bed form to remove fluoride, arsenic, selenium and natural organic matter from water. It is a granulated form of aluminium oxide. This material has a very high surface area of about 200 m^2/g and is highly porous. It is the chemical and physical process in which the ions present in the solution are moved out on the oxide surface. In this process the water is continuously passed through one or more activated alumina beds, loading rate ranges from 3 to 5 gpm/ft^2. At a regular interval of time, the activated alumina media is backwashed to remove any solids that have to gather in the system. The activated alumina should be regenerated with a strong base after occupying all the available sites (Wang et al. 2000).

3.2.3 Waste Disposal

The waste material that is produced by the regeneration of passed activated alumina is corrosive and contains a high concentration of the contaminants which is removed by the whole process. This waste material is regarded as a hazardous waste material (Industrial Wastewater Management, Treatment, and Disposal 2008).

3.2.4 Maintenance

The activated alumina (adsorbent) requires high maintenance. There may be bacterial growth in the pours or canister of the adsorbent due to continuous repetition of the same bed. So the bed should be replaced after the defined period. It should be cleaned with an appropriate regenerates like $Al_2(SO_4)_3$, specific acids, caustics that will extend the media (bed) life. Approximately every 1–3 years, the activate alumina need to be replaced (Industrial Wastewater Management, Treatment, and Disposal 2008).

3.2.5 Zeolite

It is the naturally occurring crystalline hydrated aluminosilicate alkaline and earth alkaline mineral. It belongs to the "tectosilicates" class of mineral (Badillo-Almaraz et al. 2003). The original form of zeolite formula is $M_{m/z}$ [$mAlO_2.nSiO_2$].qH_2O. Its structure is consists of an interlocking framework of three-dimensional tetrahedrons of $[SiO_4]^-$ and $[AlO_4]^-$. Generally, the natural zeolites are formed by alteration of glass rich volcanic rocks with fresh water in playa lakes or seawater. The ratio of (Si + Al)/O must be equal to 1:2, in order to be a zeolite (Barer 1987a). The size of the aluminium ion is so small that it occupies the center of the tetrahedron of four oxygen atoms. The replacement of isomorphous Si^{4+} by Al^{3+} produces a negative charge in the lattice, and these negative charges are balanced by the exchangeable cations like sodium, potassium or calcium. These cations are exchangeable by other cations that are present in the solution such as lead, cadmium, zinc, and manganese (Breck 1964). There is a large vacant space or cage in the structure of zeolite that also allows space for large molecules like ammonia, water, carbonate ions and nitrate ions. The cavity in the zeolite is of about 0.1–2 nm in diameter. Zeolite exchangeable ions are harmless and make them suitable for removing heavy metals ion from the contaminated water. It is the aluminosilicate, but other tetrahedral atoms such as phosphorus, gallium, boron, germanium also exist in the framework as well (Barer 1987b; Brunner 1993; Zeng et al. 2010).

3.2.6 Activated Carbon

Activated carbon has been used as an adsorbent for centuries. It is a material used to filter the harmful chemicals from contaminated water. It is composed of black granules of coal, wood, nutshells or other carbon-rich materials. The heavy metals stick to the surface of granules as the contaminated water pass through the carbon bed and are easily removed from the water. In spite of heavy metals, it can also treat the radon, fuel oil, dioxins, solvents, industrial chemicals and other radioactive materials. Generally, adsorption occurs when the attractive force at the carbon surface overcome the attractive force of the liquid. Its treatment generally consists of one or more column or tanks filled with granules activated carbon. Then the contaminated water is passed over the tank from top to bottom, but it flows in the upward direction. During process, contaminants are stick to the inner or outer side of the GAC, and this is how it purifies the contaminated water (Dvorak and Skipton 2013).

3.2.7 Maintenance

The GAC bed should be replaced once the contaminant occupies all the available sites or pours of the activated carbon and no longer contaminants sticks to them. The passed activated carbon is replaced by the fresh GAC. To regenerating the GAC, it is sent to the offsite where it is heated to a very high temperature to destroy the contaminants.

3.2.8 Graphene Oxide

Nowadays graphene oxide has attracted much more attention in the past several years due to its excellent electronic, thermal, optical, mechanical and adsorptive properties (Novoselov et al. 2004). Recently it attracts the researcher's attention as an adsorbent material for removing the heavy metals and dyes from the aqueous solution (Deng et al. 2010; Ramesha et al. 2011; Yang et al. 2011). Graphene is a 2-D, one atom thick carbon. All the carbons are sp^2 hybridized. Due to its significant properties like the high surface area of about 2600 m^2/g, large surface to volume ratio, high room temperature carrier mobility, conductance quantization, the presence of functional groups like –COOH, –OH and epoxy groups form a complex with the metal ions. It can adsorb number of heavy metals from the contaminated water. So its adsorbing power is greater than the other adsorbing materials mentioned above (Amann et al. 1998).

But all these techniques have some disadvantages. Its required high skilled operator for chemical dosing, removal of the microflocs formed is often difficult and critical for the process efficiency, higher reagent, and energy requirement, precipitation is often ineffective if metals are complexes or if they are present as anions, precipitated metals often form small particles that do not settle readily, generation of toxic sludge and other waste product that require disposal or treatment, usually not feasible with high levels of total dissolved solid and process is expensive. Normal *filtration* technique is not able to remove the nano-sized particles from the solution; it can remove only the large/big particles. *Ion exchange* technique cannot handle the concentrated metal solution as the matrix gets easily fouled by organics and other solids in the wastewater. It is non-selective and highly sensitive to the pH of the solution (Kurniawan et al. 2006). In *chemical precipitation*, a large number of chemicals are required to reduce the metal to obtain the acceptable level. Other drawbacks are its excessive sludge production that requires further treatments, slow metal precipitation, poor settling, and the aggregation of the metal precipitates and the long-term environmental impacts of sludge disposal (Aziz et al. 2008). In *Electro-winning technology* the corrosion could become the significant limiting factor, where electrodes would frequently have to be replaced (Liu et al. 2003). In *reverse osmosis,* wastage of electricity and water, membrane fouling hindered it from wider applications in wastewater treatment. For this, numerous adsorbents/catalysts, materials like activated carbon (AC) and zirconium coated activated carbon, iron hydroxide, iron (II) and iron (III) oxides, sand and zero-valent iron, hardened paste of Portland cement, iron oxide polymers, biological systems, activated alumina have been developed for the removal of such hazardous chemicals from contaminated water to make it portable and for household utilities, but all these adsorbent required more maintenance (Abu Qdaisa and Moussa 2004).

So among all these techniques, *adsorption* offers flexibility in design, operation, and it will generate high quality treated effluent. In addition because of their reversible nature, the adsorbents can be regenerated by suitable desorption process for multiple uses, and many desorption processes are of low maintenance cost, high efficiency and ease of operation. Therefore, the adsorption technique has come to the forefront as one of the major technique for the removal of heavy metals from wastewater.

Another most important investigated material is nanomaterials. Nano-materials had gained special attention since last decade because the materials of such kind possess unique properties than the bulk materials. Like different nano materials, single and multi-metal or doped metal oxides are also subject of much interest since that materials possess a high surface-to-volume ratio, enhanced the magnetic property, special catalytic properties, etc.

Nanofiltration is the most advanced nano-enabled technology for water treatment and is already running in the market. This technology is most promising and is also eco-friendly, the more advancement of this technology is under process.. The most common cation in water affecting human and animal health is NH^{4+}. In drinking water, ammonia removal is very important to prevent oxygen depletion, and algae bloom and due to its extreme toxicity to most fish species. It can be replaced with biologically acceptable cations, like Na^+, K^+ or Ca^{2+} in the zeolite. To bring this technology into use the research is underway, and it holds promise of making new materials and devices that takes the adventage of elusive phenomena understood at length scales, because of their high affinity.

3.3 Toxic Heavy Metal

3.3.1 Cadmium

Cadmium is a naturally occurring element found in the earth crusts at the concentration of 0.1–0.5 ppm. The concentration of cadmium in natural surface water and groundwater is usually <1 μg/l. Natural emission of cadmium to the environment resulting from a volcanic eruption, forests fire, generation of sea salt aerosol, or other natural phenomena. Human activities are the main source that is responsible for the production of toxic cadmium, include tobacco smoking, mining, smelting, refining of non-ferrous metals, fossil fuel combustion and electric and electronic waste. In its purest form, cadmium is a soft silvery-white metal in the earth crust; its density is 8.64 g/cm^3, the melting point is 320.9 °C, the boiling point is 765 °C at 100 kPa (IARC 1976). Cadmium is generally soluble in dilute nitric acid and concentrated sulphuric acid. In the environment, it exists only in one oxidation state, i.e. (+2) and does not undergo oxidation and reduction reactions. In groundwater and surface water, it mainly exist as the hydrated ion or as ionic complexes with other organic or inorganic substances. It also exist in combination with other elements in the environment like cadmium chloride, cadmium sulfide, and cadmium oxides. It has no smell and taste. Cadmium metal is mainly used as anticorrosive, electroplated onto the steel. Cadmium sulfide and selenide are commonly used as pigments in plastics. Its compounds are mainly used in electric batteries, electronic components and nuclear reactors (Ros and Slooff 1987; Ware 1989).

Exposure of cadmium in the human being may be due to breathing cadmium-contaminated air. If someone works for a battery manufacturer or working in metal soldering or welding, then the exposure of cadmium is higher in the workplace. Its exposure can also occur by eating foods containing low levels of cadmium. The most common source of exposure to cadmium is mainly through eating food, especially shellfish, liver, and kidney meats.

The cadmiun present in soil is absored by roots, and the the fishes intake it from the water, but this is not of much concern. Cadmium is also found in tobacco plants and it is hence present in cigrates and the people who smoke have more cadmium present in their bodies than non-smokers. The the industries which use coal and fossil fuels also release cadmiun in higher concentrations and hence that is mixed with the air and inhaled by humans (ATSDR 1989).

Cadmium may also be released to the environment from zinc, lead or copper smelters. If you are near in these areas or working, then your exposure to cadmium may be higher than the average person. The primary and most serious adverse health effect of long-term exposure to cadmium are kidney dysfunctions, lung cancer, and prostate cancer. It may cause local skin or eye irritation and can affect long-term effects if inhaled or ingested. Many cases of Itai-Itai disease and low-molecular-weight proteinuria have been reported among people living in contaminated areas in Japan and exposed to cadmium via food and drinking-water. The daily intake of cadmium in the most heavily contaminated areas amounted to 600–2000 μg/day; in other less heavily contaminated areas, daily intakes of 100–390 μg/day have been found.

3.3.1.1 Exposure Limits for Cadmium

Under the OSHA cadmium standard, there are three exposure limits include first is the action-level, second is the permissible exposure limits (PEL), and the third is the SECAL (Separate engineering control air limit). The action level for workplace exposure to cadmium is 2.5 μg per cubic meter of air (2.5 $\mu g/m^3$). The PEL is a time-weighted average concentration that must not be exceeded during any 8 h work shift of a 40 h work week. The standard sets a PEL of 5 μg of cadmium per cubic meter of air (5 $\mu g/m^3$) for all cadmium compounds, dust, and fumes. The SECAL for cadmium is 15 $\mu g/m^3$ or 50 $\mu g/m^3$, depending on the processes involved (Agency for Toxic Substances and Disease Registry (ATSDR) 2008; OSHA (occupational safety and health administration) 2004).

3.3.2 Lead

Lead is a toxic heavy metal (atomic number 82) that has low exposure levels and has sharp and very bad or serious effects on human health. It belongs to the IVA group of the periodic table. The only possible oxidation state of lead is +2 means it is stable at that state only. It is pale silvery grey in colour when freshly cut, but on exposure to air, it becomes darkened. It is dense, readily fusible, low melting point, heavy, malleable and a poor conductor of electricity. It is very soft enough that it can be scratched with a fingernail. It can be used in the pure elemental form or combined with other elements to form lead compounds. Inorganic lead compounds are used in the paints, glasses, plastics, rubber compounds, and pigments, etc. (Liu et al. 2006). Lead is a multi-organ system toxicant that can cause the neurological, haematological, gastrointestinal, cardiovascular and renal systems (Agahian et al. 1990; Health and safety guidelines 2011). The type and extremity of effects depend on the level, duration and timing of the exposure. Natural abundance of lead is very small mainly as lead sulide. However, the widespread occurrence of lead in the environment is due to human activities like mining, smelting, refining, production of lead-acid batteries and paints, ceramics, jewelry making, water pipes, soldering, etc. (Viswanathan et al. 1991).

3.4 Tolerable Intake Value Given by WHO

The study conducted in 2010 by joint Food and Agriculture Organization of the United Nations (FAO)/WHO and (JECFA) projected that the earlier established provisional tolerable weekly intake (PTWI) of 25ug/kg body weight is no longer measured as health protective and hence it was withdrawn. As it was predicted that the dose response studies did not provide any indications of health related effects by lead, hence the committee decided not to establish a new PTWI that would be health protective. The dose response analysis should be used as reference to classify the magnitude of effects linked with identified levels of dietic lead exposure in diverse populations.

The tolerable, intake value of lead given by WHO is 10 μg/l and in the air its value is 0.5 $\mu g/m^3$. (Annual average) (George Foundation 1999; Romieu et al. 1995).

Lead exposure mainly occurs through inhalation of air and dust and ingestion of foodstuffs, water, etc. it is especially dangerous for children's as they absorb 4–5 times as much lead as adults. Infants, young children's (below 5 years of age) and pregnant women are more sensitive to the adverse effects of lead. The harmful effects on childrens is more than adults due to (1) the consumption of lead per unit average body weight. (2) The blood-brain barrier is not yet fully developed. (3) The chances of ingestion of dust are more in children than adult (Nriagu et al. 1996). The most critical effect of lead is on the developing nervous system of the children. The insidious effect of 50ug/l of lead in blood shows on Intelligence Quotient which may increase by increasing level of lead in the blood (Agahian et al. 1990). Exposure of lead on pregnant women can cause miscarriage, stillbirth, premature birth and low birth weight, as well as minor malformation. Lead exposure can cause serious health problem to a human being like it may cause gastrointestinal disturbance (vomiting, nausea, anorexia, and abdominal pain), hepatic and renal damage, hypertension and neurological effects (malaria, drowsiness) that may lead to death. It is also evidence that long-term exposure to lead may lead to the development of cancer (Health and safety guidelines 2011). In men, the reproductive effects of lead include decreased sperm count and increase number of abnormal sperms. In the environment, lead found in the form of particle-bound and has low mobility and bioavailability; though highly soluble ionic forms also exist mainly in the marine environment. In the environment, it exists as a lead compound. The speciation of lead differs in the freshwater and seawater. In the freshwater, it mainly found in divalent form $(Pb)^{2+}$ under acidic condition. And in seawater, it is in the chloride form, and the primary species are $PbCl_3^-$, $PbCO_3$, $PbCl_2$, $PbCl^+$ and $Pb(OH)^+$.

3.5 Synthesizing Oxides of Iron

Keeping the advantages and disadvantages of the above methods and material, synthesizing oxides of iron by wet chemical route and their utilization for water purification is important. Iron is one of the most widespread transition elements in the earth. It is environmentally friendly. There are total 16 oxides, hydroxides and oxyhydroxides are known, but the intensively used for the removal of heavy metals from the wastewater are magnetite (Fe_3O_4), hematite (α-Fe_2O_3), ferric hydroxide [$Fe(OH)_3$], ferrous hydroxide [$Fe(OH)_2$], maghemite (γ-Fe_2O_3). Also iron has a strong magnetic property due to which it can easily absorb the heavy metals from the contaminated water. It is non-toxic in nature, easy to dispose of and its disposal is further is used for the making of bricks. Even if it is dissolved in the water, it works as an essential element for human beings. It has high adsorption capacity and very slightly soluble in water.

Graphene, a two dimensional (2D) one atom thick nanomaterial consisting sp^2 – hybridised carbon atom has attracted great interest due to its unique properties, including high specific surface area of 2600 $m^2\ g^{-1}$, highly porous and presence of functional group like –O, –OH, –COOH which form complexes with the metal ions, are used to remove the heavy metal ions. So on combining the graphene with the iron oxides, makes the properties of adsorbing material twice time greater than the individual material (Fig. 3.6).

Mechanism of synthesized material:

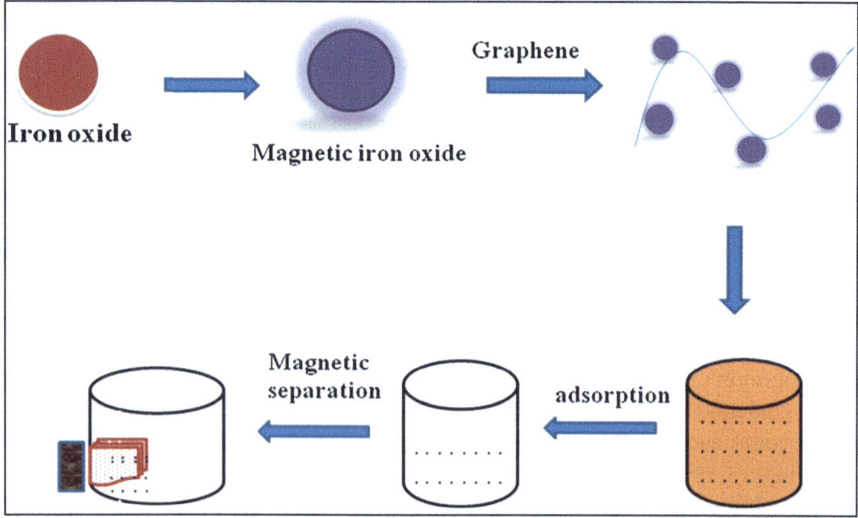

Fig. 3.6 Mechanism of graphene doped iron oxide

3.6 Conclusion

For the purification of water contaminated by toxic metallic species like Cd and Pb, a simple, fast, low cost and non-toxic material have been presented. The material can be synthesized in bulk without the help of any costly equipment. The material is of non-hazardous nature pH is the major deciding factor for quality of water, the proposed material do not alter even after adding and after removal of toxic contaminants like Cd and Pb. So the proposed material can be used safely for water purification. But the proposed material required deep study in laboratory placed because all the experiment has been carried out on lab scale. So to check the real utility of the material natural water purification is required which contain several organic, inorganic and biological species. But the material will be beneficial for bulk purification of water like in army troops, drain, ponds, and river, etc.

References

75/440/EEC (1975) EC Directive concerning the quality required of surface water intended for the abstraction of drinking water in the Member States. Off J Eur Commun Legis 194(25/7):26–31

80/778/EEC (1980) EC Directive relating to the quality of water intended for human consumption. Off J Eur Commun Legis 229(30/8):11–29

98/83/EC (1998) EC Directive on the quality of water intended for human consumption. Off J Eur Commun Legis 330(5/12):32–54

Abu Qdaisa H, Moussab H (2004) Removal of heavy metals from wastewater by membrane processes: a comparative study. Desalination 164:105–110

Agahian B, Lee JS, Nelson JH et al (1990) Arsenic levels in fingernails as a biological indicator of exposure to arsenic. Am Ind Hyg Assoc J 51(12):646–651

Agency for Toxic Substances and Disease Registry (ATSDR) (1989) Public health statement for cadmium. U.S. Department of Health and Human Services, Atlanta

Agency for Toxic Substances and Disease Registry (ATSDR) (2008) Toxicological profile for cadmium (Draft for Public Comment). U.S. Department of Public Health and Human Services, Public Health Service, Atlanta

Amann R, Lemmer H, Wagner M (1998) Monitoring the community structure of wastewater treatment plants: a comparison of old and new techniques. FEMS Microbiol Ecol 25:205–215

Aziz HA, Adlan MN, Ariffin KS (2008) Heavy metals (Cd, Pd, Zn, Ni, Cu and Cr(III)) removal from water in Malaysia: post treatment by high quality limestone. Bioresour Technol 99:1578–1583

Badillo-Almaraz V, Trocellier P, Davila-Rangel I (2003) Adsorption of aqueous Zn(II) species on synthetic zeolites. Nucl Inst Methods Phys Res B 210:424–428

Barer RM (1987a) Zeolites and clay minerals as sorbent and molecular sieves. Academic, New York

Barer RM (1987b) Zeolites and clay minerals as sorbent and molecular sieves. Academic, New York

Breck DW (1964) Crystalline molecular sieves. J Chem Educ 41:678

Brunner GO (February 1993) Quantitative zeolite topology, can help to recognize erroneous structures and to plan syntheses. Zeolites 13(2):88–91

Byers A (ed) (1995) Reader's Digest Atlas of Canada. The Reader's Digest Association (Canada) Ltd, Montreal

Chambers C, Winfield M (2000) Mining's many faces: environmental mining law and policy in Canada. Canadian Institute for Environmental Law and Policy, Toronto

Deng X, Lu L, Li H, Luo F (2010) The adsorption properties of Pb(II) and Cd(II) on functionalized graphene prepared by electrolysis method. J Hazard Mater 183:923–930

Bruce ID, Sharon OS (2008) Drinking water treatment: reverse osmosis, University of Nebraska-Lincoln Extension, and Institute of agriculture and Natural Resources

Bruce ID, Sharon OS (2013) Drinking water treatment: activated carbon filtration. University of Nebraska-Lincoln Extension, and Institute of Agriculture and Natural Resources, Lincoln

Environment Canada (2001) Threats to sources of drinking water and aquatic ecosystem health in Canada. National Water Research Institute, Burlington

George Foundation (1999) Project lead-free: a study of lead poisoning in major Indian cities. In: Proceedings of the International Conference on Lead Poisoning, Bangalore, India, 8–10 February Bangalore, The George Foundation. 79–86

Handbook of Industrial membranes (1995) By Keith Scott, Elsevier Advanced Technology, Oxford

Health and safety guidelines (2011) Occupational health and safety branch ministry of labour. ISBN 978-1-4435-6227-0 (PDF)

http://en.wikipedia.org/wiki/selenium

http://www.chem.ucalgary.ca/courses/351/laboratory/filtration.pdf

http://www.epa.gov/OGWDW/contaminants/dw_contamfs/chromium.html

http://www.epa.gov/ogwdw000/contaminants/dw_contamfs/mercury.html

http://www.lenntech.com/periodic-chart-element/cd-en.html

IARC (1976) Cadmium, nickel, some epoxides, miscellaneous industrial chemicals and general considerations on volatile anaesthetics. In: IARC Monographs on the Evaluation of the Carcinogenic Risk of Chemicals to Man, vol 11. International Agency for Research on Cancer, Lyon, pp 39–74

Industrial Wastewater Management, Treatment, and Disposal (2008) Water Environment Federation. McGraw-Hill, New York

Ion Exchange, by F. Helfferich, McGraw Hill Book Company

Ion Exchange in Analytical chemistry, by O. Samuelson, John Wiley and Sons

Ion Exchange Resins, by R. Kunin, John Wiley and sons

Kurniawan TA, Chan GYS, Lo WH, Babel S (2006) Physico-chemical treatment technique for wastewater laden with heavy metals. Chem Eng J 118:83–98

Liu XD, Tokura S, Nishi N, Sarkairi N (2003) A novel method for immobilization of chitosan onto non-porous glass beads through a 1, 3-thiazolidine linker. Polymer 44:1021–1026

Liu CW, Liang CP, Huang FM, Hsueh YM (2006) Assessing the human health risk from exposure of inorganic arsenic through oyster (Crassostrea gigas) consumption in Taiwan. Sci Total Environ 361:57–66. https://doi.org/10.1016/j.scitotenv.2005.06.005. PMID: 16122780

Novoselov KS, Geim AK, Morozov SV, Jiang D, Zhang Y, Dubonos SV, Grigorieva IV, Firsov AA (2004) Electric field effect in atomically thin carbon films. Science 306:666–669

Nriagu JO, Blankson ML, Ocran K (1996) Childhood lead poisoning in Africa: a growing public health problem. Sci Total Environ 181(2):93–100

OSHA (occupational safety and health administration) (2004) Cadmium, OSHA 3136-06R 2004

Ramesha GK, Vijayakumar A, Muralidhara HB, Sampath S (2011) Graphene and graphene oxide as effective adsorbents towards anionic and cationic dyes. J Colloid Interface Sci 361:270–277

Regional Municipality of Waterloo (1994) Water resources protection strategy. Regional Municipality of Waterloo, Kitchener

Regional Municipality of Waterloo (1996) Take care of your land and the land will take care of your water: best management practices to ensure profitable production and the continued quality of your drinking water. Regional Municipality of Waterloo, Kitchener

Regional Municipality of Waterloo (2000) Facts on Tap. Issues 1–5, October 1999 – August 2000. Regional Municipality of Waterloo, Waterloo

Richardson D, Plewa J, Wagner D, Schoeny R, Demarini M (2007) Occurrence, genotoxicity, and carcinogenicity of regulated and emerging disinfection by-products in drinking water: a review and roadmap for research. Mutat Res 636(1–3):178–242

Romieu I et al (1995) Environmental urban lead exposure and blood lead levels in children of Mexico City. Environ Health Perspect 103(11):1036–1040

Ros JPM, Slooff W eds. (1987) Integrated criteria document. Cadmium. Bilthoven, National Institute of Public Health and Environmental Protection (Report No. 758476004)

U.S. Environmental Protection Agency National Water Quality Inventory: Report to Congress (2002) Reporting cycle: findings, rivers and streams, and lakes, ponds and reservoirs

U.S. Environmental Protection Agency (2005) Protecting water quality from agricultural runoff. March 2005

Viswanathan P et al (1991) Biological monitoring and lead and cadmium. In: Krishna Murti CR, Viswanathan P (eds) Toxic metals in the Indian environment. Tata McGraw-Hill Publishing Company, New Delhi, pp 212–235

Wang L, Chen A, Fields K (2000) Arsenic removal from drinking water by ion exchange and activated alumina plants. EPA/600/R-00/088, October 2000

Ware GW (ed) (1989) Cadmium. US Environmental Protection Agency Office of drinking water health advisories. Rev Environ Contam Toxicol 107:25–37

Membrane handbook (1992) By W.S. Winston Ho and Kamlesh K. Sirkar, Van Nostrand Reinhold, New York

Yang ST, Chen S, Chang Y, Cao A, Liu Y, Wang H (2011) Removal of methylene blue from aqueous solution by graphene oxide. J Colloid Interface Sci 359:24–29

Zeng Y et al (2010) Adsorption of Cr (VI) on hexadecylpyridium bromide (HDPB) modified natural zeolites. J Micropor Mesopor Mater 130(1–3):83–91

Chapter 4
Chitosan Based Nanocomposites as Efficient Adsorbents for Water Treatment

Nafees Ahmad, Saima Sultana, Mohammad Zain Khan, and Suhail Sabir

Abstract Excessive growth and rapid development put a huge burden on the environment and the large fraction of water bodies getting polluted through industries. Though one third part of the earth contain water, but still there is a crises of water for drinking and other purposes. These water contaminants, contain many harmful and toxic metals, dyes, drugs, pesticides etc. Polluted water can be treated by numerous methods to make it fit for the drinking purpose. Chitosan being the most abundant biodegradable and cationic polymer can be effectively used for the treatment of wastewater for the removal of toxic and hazardous contaminants, heavy metals and other impurities present in water. Chitosan with characteristic functional groups such as amino and hydroxyl group can also be useful for the treatment of water as an adsorbent. Chitosan being the environmental friendly, cost effective and non toxic in nature is also applied for the purpose of treatment of water through adsorption, chelation, precipitation, ion exchange techniques etc. Nanoparticle doped with chitosan in the form of bionanocomposites has been a very successful tool for the removal of pollutants from the water. Wastewater contains different types of metal ions, toxic substance,and different pH i.e. alkaline as well as acidic nature. Chitosan in the acidic media is very helpful as the amino functional group can easily be protonated in acidic media to bind with the anionic part of the organic pollutants.

Keywords Chitosan · Nanocomposites · pH · Metal · Adsorbents

4.1 Introduction

Water is the most important renewable resource on the earth for human being as well as for the animals and aquatic species. Rapid industrialization and huge demands of the resources leads to discharge of large volume of waste product into the water

N. Ahmad (✉) · S. Sultana · M. Z. Khan · S. Sabir
Environmental Research Laboratory, Department of Chemistry, Aligarh Muslim University, Aligarh, Uttar Pradesh, India

M. Oves et al. (eds.), *Modern Age Waste Water Problems*,
https://doi.org/10.1007/978-3-030-08283-3_4

bodies. Waste product contains toxic metals, hazardous product like dye from the textile, leather, food and cosmetic industries, cholorophenols from paper industries, (Bingbing et al. 2017; Neeta et al. 2017; Qin et al. 2007) drugs and fertilizers. There are various techniques which are used these days for the removal of the pollutants from the water such as precipitation and coagulation, ion exchange filtration, reverse osmosis, adsorption, (Fenglian and Wang 2011) incineration, electrofloatation and electrochemical treatment (Mondal 2008; Swami and Buddi 2006; Lefebvre and Moletta 2006). Low cost adsorbents and synthetic polymer which have the antibacterial and anti fungal properties and flocculation efficiency (Haritma and Ruhi 2016) have already been used for this purposes however there are few shortcomings related to this type of technology as the synthetic polymer are non biodegradable in nature which once used cannot be degraded and create additional burden and adverse effect on the environment. With the issue of major environmental concern there is need to develop a new technology which can be used for the treatment of the wastewater and after fulfilling the need of the water treatment can easily be degraded by the action of the food borne fungi and bacterial action (Nafees et al. 2017; Sultana et al. 2007) and can dramatically reduce the environmental burden and many diseases. However major part of the surface water in addition to the chemically organic and inorganic wastewater are affected by the bacteria, fungi and algae that makes the water unfit for the drinking and daily needs. For this purpose a biodegradable polymer chitosan can be effectively employed for the treatment of the wastewater containingorganic waste, heavy metals and toxic substances. The characteristic properties of the chitosan such as high content of primary amino groups, non toxicity, its bacterial and fungistatic effect, protein separations, and itschelating property with many semiconducting and transitional metal ions make it suitable for water treatment. Chitosan is a deactylated form of chitin (Seong et al. 1999) and hence it has some very unique properties. The amino group along with hydroxyl group allows chitosan to be highly reactive for the purpose of removal of pollutants from water. Chitosan can be further doped with the various kinds of synthesized nanoparticles in order to improve the treatment efficiency (Duong et al. 2018; Sultana et al. 2007) for the treatment of water such as Chitosan/CNT, Chitosan/copper nanoparticle, Chitosan/Ag nanoparticle (Rania et al. 2017), Chitosan/alginate nanocomposites (Gokila et al. 2017), Chitosan/SnO_2 (Gupta et al. 2017), Chitosan/ZnO (Farzana et al. 2015). These nanocomposites due to the complexion properties of chitosan are also employed for the removal of microorganism for the wastewater. Chitosan can be effectively applied as an additive for water processing in filtration process as the adsorbent property causes the fine sediment particles to bind together and removed with the sediment during filtration. In the filteration process chitosan along with the sand is very much effective that can removes up to 99% of the turbidity and used for detoxifying water (Nithya et al. 2014). The chelating property of the chitosan make it feasible to purify the water through complexion and absorb impurities associated with the water such as negatively charged toxins from dyes, oil, grease, and other toxic substances. Chitosan in the acidic media with its cationic group used for coagulation and flocculation. The main factor that is very much helpful in coagulation and flocculation is the degree of acetylation and high nitrogen content (Zemmouria et al. 2013) to make water suitable for chemical and biological purposes. The extent of coagulation and flocculation can be investigated by the solubility and cationic charge of chitosan. Coagulation techniques are also used for activated sludge in food and beverage processing industries. Treated water on the basis of their appearance must be colorless, odorless, and tasteless and quality parameters such as pH, total dissolve solids (TDS), turbidity, and different concentration of the metal ions must be present in water to be used. Apart from water purification properties chitosan based nanocomposites also possess better anti cancerous properties when doped with nanoparticle and also can be used for different biomedical applications such as tissue engineering, bone regeneration, drug delivery etc.

4.2 Structure, Sources and Preparation of Chitosan Nanocomposites

Chitosan, a high molecular weight biopolymer is a derived product of chitin through deacetylation and is the second most abundant biopolymer. Chitosan is a linear polysaccharides composed of α-1, 4-linked 2-amino 2-deoxy-α-D glucose are presented in Fig. 4.1.

It is found basically in the shell of shrimps,crustaceans and crabs, cartilage of squids, outer shell of the insects (Kim et al. 1997; No et al. 2000). Biologically, chitosan can be synthesized by numerous living organisms, such as arthropods, fungi, yeast and other biological sources and industrially it can be derived either from deacetylation of chitin in the presence of strong alkali such as NaOH as presented below in Fig.4.2 or enzymatic hydrolysis in the presence of chitin

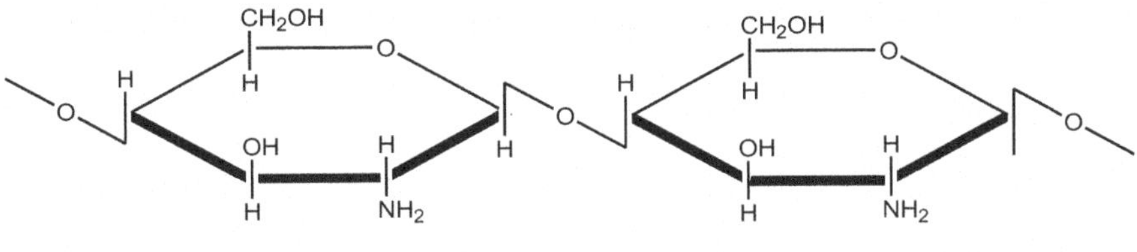

Fig. 4.1 Structure of Chitosan

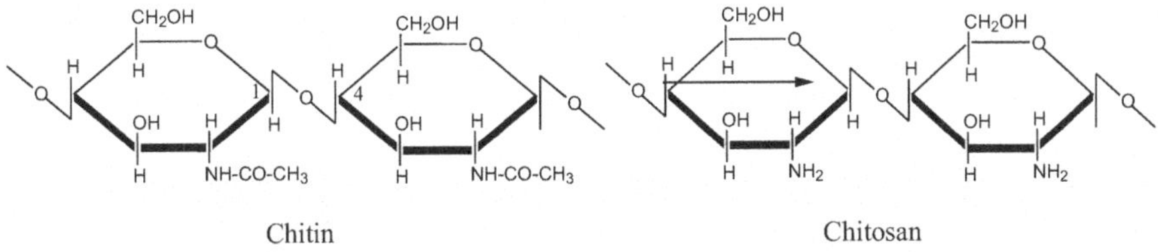

Fig. 4.2 Deacetylation process of chitosan from chitin

deacetylase. Chitosan can show molecular weight range from 5 to 1000 kg mol^{-1}. In the solid state Chitosan is a semicrystaline polymer form that exists in different allomorphs depending upon the degree of acetylation (Chivrac et al. 2009). Chitosan based nanocomposites can be synthesized by doping it to nanoparticle by different techniques such as simple precipitation method, hydrothermal method, sol gel method etc. Numbers of Chitosan based nanocomposites have been reported for the treatment of water such as Polypyrrole/Chitosan, Polyaniline/Chitosan/Ag nanocomposites are synthesized by Sultana et al. 2007 for the photocatalytic degradation of the dyes. Chitosan with the other metal and semiconductor nanoparticle can be synthesized for the effective removal of the dye from the water as well as for the removal of microorganism.

4.3 Chitosan Based Nanocomposites for the Wastewater Treatment

Chitosan with its remarkable properties can be used as good adsorbent, and in order to improve its efficiency it can be modified by grafting, cross linking, fictionalization to form new composites. Many reagents such as formaldehyde, glutaraldehyde, ethylene glycol and glycerol (Zheng et al. 2011) have been used to improve its performance. The modification of cross linking process improves it's properties, which can be used for the removal of water contaminants. Water treatment consist mainly three main stages which removes the heavy as well as small particle from the water and all the impurities which assimilates with the water and further through chemical process removes its microorganism e.g. (bacteria and fungi). The three processes are:

- Primary: Mechanical and physical method for the removal of solid particles, immiscible liquids, suspended substance from the effluents
- Secondary: Biological Method
- Tertiary: Treatment of the sludge formed

There are many materials which are known to be used as coagulants /flocculants i. e. inorganic and organic flocculants such as mineral additives, hydrolyzed metal salt, polyelectrolyte, organic flocculants including cationic and anionic polyelectrolyte, non ionic polymer, and hydrophobic polymer. The major pollutants with water such organic dyes, drugs, pesticides, heavy metals, fertilizers, microorganisms can be removed by chitosan based nanocomposites (Fig. 4.3).

4.3.1 Removal of Organic Pollutants

Organic pollutants such as dyes, drugs, fertilizers etc. are the major pollutants of the water pollution that are discharged from textile, paper and leather industries and

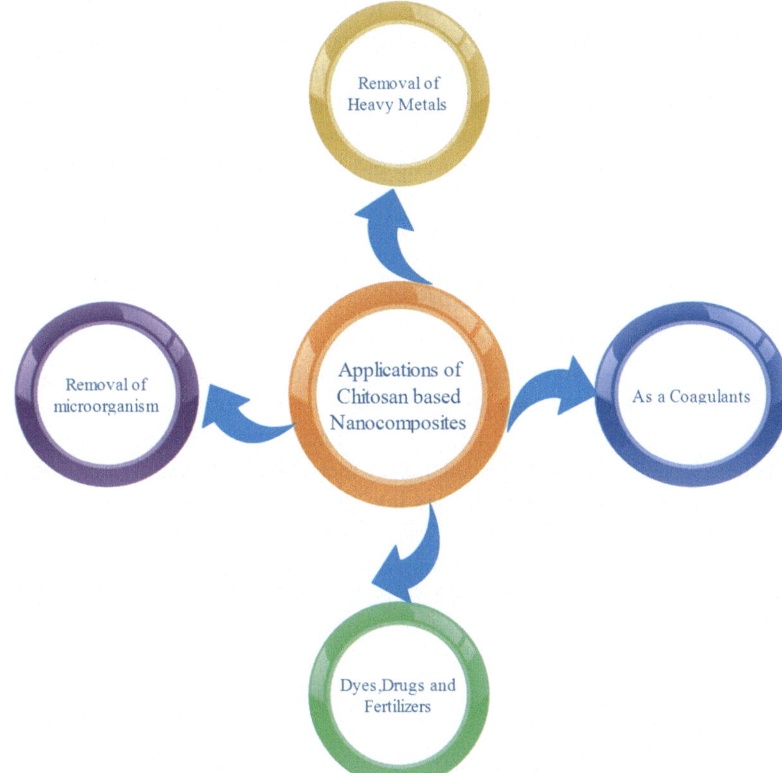

Fig. 4.3 Applications of chitosan based nanocomposites for the water treatment

major part from the dye manufacturing industries. Removal of organic pollutants from the wastewater by the sedimentation and adsorption techniques (Dashairya et al. 2018) has been the most promising way for last few decades. Dyes discharges into water are quite toxic and carcinogenic in nature which affects the aquatic species as well as human health and cause many diseaseslike cancer, inflammation, dermatitis etc. (Bhatangar and Silanpaa 2009). So before discharging the organic waste into the water bodies there is need for the treatment of these waste products. Chitosan and its derivatives can successfully be utilized for this purpose. Dyes such as Rhodamin B can be successfully degraded using chitosan/Ag nanocomposites by Sultana et al. 2007 Reactive Red 222(RR222) were degraded by Wu et al. 2001 using chitosan.RR222, RB222,RY145 are successfully degraded to the final product by using chitosan in the form of flake type and highly swollen beads type (Wu et al. 2001) and found that chitosan beads are more effective in removal of dye thanchitosan flakes. Removal of the dye from the water is quite effective in the acidic media pH less than 5 ($pH < 5$) due to the presence of the $-NH_2$ group in acidic media (Chiou et al. 2003) which leads to theprotonation of NH_2 to NH_3^+ and electrostatic attraction between the chitosan NH_3^+ and dye anion leads to the new byproductand results in the removal of dye from the water. Chitosan in the form of flakes, beads can be used as an adsorbent. However apart from dyes other organic pollutants can also be removed from the water. In treatment of water chlorine is used for the disinfectant thatresults into the formation of some by products i.e. trichloromethane and chlorophenol which can be removed by chitosan due to present of NH_2 group which grab the halide group in water (Haritma and Ruhi 2016).

4.3.2 Removal of Heavy Metals

Chitosan can be shaped into different form such as beads, flakes, powder and thick film so as to be used for different purposes (Portes et al. 2009) as the surface area of different forms of chitosan vary according to their shape, thus chelating property towards heavy metals can be enhanced. Heavy metals such as lead, mercury, arsenic, cadmium discharged from the industries into the water are toxic and carcinogenic leading to many diseases such nausea, vomiting,diarrhea and chronic exposure may damage body organs sometimes causing cancer (Hosseinzadeh and Ramin 2018). The main problems for these heavy metals are that they are quite toxic and persistent in nature and can't be easily removed from the water (Harmita and Ruhi 2016). Metal ions i.e. Mg^{2+}, Ca^{2+}, k^{+}, Na^{+} and iron are present in water in a limited amount which should be acceptable for the drinking purpose (Saha et al. 2008) however extensive amount of these metal ions can leads to many diseases i.e. hemolysis, hepatotoxicity and nephrotoxicity, carcinogenesis, mutagenesis etc (Ozer and Ozer 2004). Chitosan due to its characteristic functional group of NH_2 and OH having lone pair of electrons and can bind to the metals by donating the lone pair to the metal and form chelate to the metals. Heavy compounds form chelates very easily and remove metals from the water (Varma et al. 2004; Kumar et al. 2017). For the removal of the heavy metals numerous composites are well known for the wastewater treatment such as chitosan ceramic alumina composites can be used as an adsorbent due to the amphoteric character of aluminium hydroxide for the removal of heavy metals. Removal of As(III), As(V) (Veera et al. 2008a, b), Cr(VI)(Veera et al. 2003) Cu(II), Ni(II) (Veera et al. 2008a, b) has been done by previous groups. Chitosan with the other composites such as magnetite (Miquel et al. 2007) cotton fiber, sand (Wan et al. 2007), cellulose (Saito and Isogai 2005), polyvinyl alcohol (Wang et al. 2007), polyvinyl chloride etc.,Chitosan/ Polyvinyl chloride have very efficient adsorbing property for the removal of heavy metals due to high surface area, good physical and chemical stabilities in concentrated acidic and basic medium. Chitosan Zeolite composites due to good thermal and chemical stability have been applied for the separation of the ions from the water. It has been observed from the above study that in the aqueous condition most adsorption occurs for the removal of heavy metals.

4.3.3 Removal of Microorganism from Water

Nanoparticles are best parameter for the antifungal and antibacterial treatment. Chitosan which are basically positively charged used in the reverse osmosis process, filtration in which microorganism are trapped by the chitosan adsorbents (Qin et al. 2006) by binding on the microbial cell. There are many chitosan nanocomposites which are used for the antibacterial and antifungal treatment from the water i.e. FeO doped chitosan (Amin et al. 2014), Cu doped chitosan nanocomposites, Ag doped chitosan (Sultana et al. 2007) nanocomposites are used for the antibacterial action. Many disease are caused by the existence of microorganism such as bacterial and fungal growth in water for drinking as well as other purposes even on the removal of heavy metals, heavy pollutants, rags, suspended colloidal particles etc. There is basic need for the removal of bacteria. Thousands of pathogenic microorganism reported to cause water pollution such as bacteria (E Coli., Shigella and V. cholera), viruses (Hepatitis A,polio virus, Rota virus) and parasites (E. Hiystolytica, Hook warm) etc. have been treated by various groups (Ali and Gupta 2007, 2012). The antibacterial treatment of the chitosan based nanocompositesis due to its cationic nature and the interaction between and the negatively charged cell microbes and positively charged amine group of the chitosan provoking internal osmotic imbalance that inhibits the growth of the microorganism (Tripathi et al. 2011; Hadwiger et al. 2016; Shahidi et al. 1999). Chitosan nanocomposites when used for the treatment of the wastewater must be in acidic media i.e. pH less than 7 because of the protonation of the amino group on the chitosan which results in the degradation of the polymeric chain. However there are many factors which depend on the solubility of the chitosan such as concentration, pH, species of the bacteria etc. Chitosan due to its antibacterial and antioxidative properties can also be used for removal of impurities other than microorganism from the wastewater. The anti bacterial and anti oxidative properties of the chitosan can be enhanced by doping it with several metals, semiconductor nanoparticles, such as copper, ZnO,$ZrO_{2,}$ Ag etc. Silver/Chitosan nanocomposites among all the other chitosan nanocomposites have excellent antibacterial properties due to its complexing property have been applied for the treatment of water (Masheane et al. 2016).

4.3.4 As a Coagulant

Chitosan based nanocomposites used for the coagulation process to improve the settling property of the pollutants. Previous studies shows the use of Chitosan based nanocomposites used as a flocculants for the treatment of water, complexing by membrane filtration (Krajewska. 2005) and adsorption (Gerente et al. 2007; Crini and Badot 2008; Guibal and Roussy 2007). The unique properties of chitosan which make it suitable for flocculants and coagulants are nontoxicity, non corrosive nature, biodegradability, complexing property and easy handling (Varma et al. 2004).

Chitosan along with its composite derivatives being water soluble material and easily complexing with the pollutants can be suitably applied for the water treatment. It can binds with the metal ions as well as dye by adsorption process. It can also form liquid state to solid state in the form of gels, beads, membranes etc. High cationic charge density, long polymer chain and precipitation make it effective coagulants for the removal of the water contaminants i.e. mineral, organic matters, dissolved solids etc. (Rinaudo 2006; Kurita 2006). Chitosan along with its composites enhance the settling ability of the pollutants,thus very easily removed from the water (Cheng et al. 2005). However the adverse effect of chitosan is that it can only be applied for limited pH value, and limited concentrations, as excessive use of chitosan can leads to dispersion of the medium. Solubility is the very important factor for the treatment of the water and chitosan due to its hydrophilic nature and amino group can easily be protonated in acidic media and becomes completely soluble. Solubility of the chitosan mainly depends on the degree of deacetylation and molecular weight. Some other intrinsic factors such as crystallinity, purity, hydrophilicity, charge density are taken into consideration for chitosan to be used as a coagulant (Renault et al. 2009). There are number of effluents which are already treated by chitosan as a coagulants such as inorganic suspension (bentonite, kaolinite) (Roussy et al. 2005; Roussy et al. 2004; Huang and Chen 1996; Divakaran and Pillai 2001), Bacterial suspension (Xie et al. 2002; Strand et al. 2001), Effluents containing humic substance and dyes (Bratskaya et al. 2002; Guibal 2004), Pulp and paper (Rodrigues et al. 2008; Wang et al. 2007; Chen et al. 2006) Metals ion (Wu et al. 2008; Assaad et al. 2007; Gamage and Shahidi 2007) Phenol derivatives (Wada et al. 1995).

4.4 Properties of Chitosan and Chitosan Nanocomposites

Chitosan the second most abundant biopolymer and naturally occurring polysaccharides found acidic in nature. However, other polysaccharides are basic in nature i.e. cellulose, agar, pectin etc. Chitosan has many phenomenal characteristics due to its electrolytic behavior, solubility as chitosan is soluble in many solvents like hexaflouro-isopropanal, hexaflouroacetate, choloro-alcohol and forms conjugate with the aqueous solution, formation of oxysalts, metal chelation and ability to form films (Fig. 4.4).

4.4.1 Biological Properties

There are numerous properties of chitosan due to which chitosan nanocomposites find many applications in water treatment, industrial application as well as in biomedical applications. Some important properties of the chitosan are as follows:

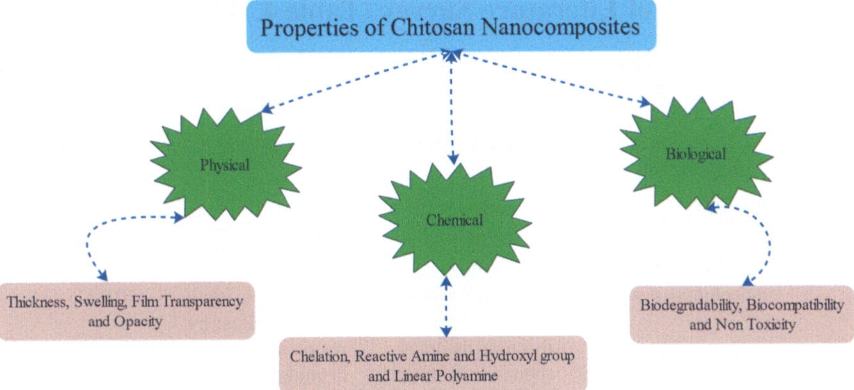

Fig. 4.4 Different properties of chitosan based nanocomposites

- Biodegradability, biocompatibility and Non toxicity
- Anticancerous and antitumor
- Helps in bone regeneration
- Regenerative effect on gum tissue
- Anticholestermic
- Central nervous system depressant and immunoadjuvants
- Hemostatic and fungistatic

4.4.2 *Chemical Properties*

- Chelation
- Reactive amine and Hydroxyl group
- Linear polyamine

4.4.3 *Physical Properties*

- Thickness
- Swelling
- Film transparency and Opacity
- Mechanical strength

4.5 Summery of Literature Survey (Table 4.1)

Table. 4.1 Literature survey on the Chitosan based bionanocomposites for the removal of contaminants from the waste water

S. No.	Types of Nanocomposites	Methods of synthesis	Applications	References
1.	Chitosan/Palladium nanocomposites	Nanocomposites of chitosan/Palladium were synthesized through chemical reduction method	Removal of 4-nitrophenol,a toxic water pollutant by photocatalytic degradation	Dhanavel et al. (2018)
2.	Zirconium tungstate/ Chitosan bio-nanocomposite	Zirconium tungstate (ZrW) nanocomposites were synthesized via hydrothermal method.	Mineralization through photocatalytic degradationof ZrW as compared to C/ ZrW	Vanamudan et al. (2018)
3.	Chitosan-cross linked κ-carrageenan bionanocomposites	The magnetic adsorbents were prepared via in situ coprecipitation of Fe^{2+}/Fe^{3+} ions in the presence of κ-carrageenan	Bionanocomposites were assessed as adsorbents to remove methylene blue from its aqueous solution through Photocatalytic degradation	Mahdavinia and Mosallanezhad (2016)
4.	Chitosan and crosslinked chitosan/ MMT nanocomposite	Nanocomposite of chitosan and clay-chitosan were prepared by ionic gelation technique	Adsorption of arsenate ions from water	Gogoi et al. (2016)
5.	Cellulose and chitosan nanoparticles/ composites	Nanocomposites are prepared by ionic gelation method	Removal of Heavy metals and dyes through adsorbent property	Olivera et al. (2016)
6.	Chitosan and 5% Bentonite/Chitosan nanocomposites	The nano 5%Bentonite/chitosan were prepared by precipitation method	Removal of chromium ions (Cr(VI)) from wastewater through Adsorption behaviour	Moussout et al. (2018)
7.	Chitosan/fluorescein-copper	Chitosan/fluorescein (Ch/f) film and hydrogel nanocomposites were prepared by direct reaction with Cu^{2+} ions	Removal of Methylene blue employing N_2H_4 as reducing agent by Photocatalytic activity	Saldias et al. (2018)

(continued)

Table. 4.1 (continued)

S. No.	Types of Nanocomposites	Methods of synthesis	Applications	References
8.	Chitosan-SnO_2 nanocomposites	Nanocomposites were synthesized by simple precipitation followed by sonication method.	Removal of Rhodamine B and Methyl Orange by Photocatalytic process in the presence of UV light	Gupta et al. (2017)
9.	Chitosan and Alginate nanocomposites	Alginate nanoparticles were prepared by ionic gelation method and Chitosan nanoparticle was prepared by simple precipitation method.	Removal of chromium (VI) from water waste by adsorption	Gokila et al. (2017)
10.	Chitosan /Zeolite Y/ Nano ZrO_2 nanocomposite	The Chitosan/Zeolite Y/Nano ZrO_2 nanocomposite was prepared by silica dissolution method	Removal of nitrate from the aqueous solution by Adsorbtion	Teimouri et al. (2016)
11.	ZnO/Chitosan core-shell nanocomposite	ZnO/Chitosan core-shell nanocomposite (ZOCS) was fabricated by direct precipitation method	Removal of toxic metals from water.	Saad et al. (2018).
12.	Chitosan-bound pyridinedicarboxylate Ni(II) and Fe(III) complex	Nanocomposites were synthesized by adsorption method	Decyanidation of waste water	Adewuyi et al. (2016)
13.	Chitosan/polyaniline-polypyrrole	Hybrid Nanomaterial	Adsorptionof metal and antimicrobial activity	Kumar et al. (2017)

4.6 Conclusion and Future Perspective

Problem of water is global environmental concern. With rapid industrialization, demand of fresh water has been increasing day by day. Nanomaterials have already been the most promising tools for the remediation of the wastewater due to its unique size dependant properties in various forms such as membranes photocatalyst, surface coatings etc. However there are certain drawbacks such as instability and agglomeration properties of the nanoparticles. So nanoparticles doped with any biopolymer such chitosan, chitin etc. can be efficiently applied for the wastewater remediation. From the literature survey and the study it has been found that chitosan based nanocomposites can be effectively used for remediation of wastewater. Chitosan with its different form have been used for the removal of organic pollutants, removal of heavy metals, removal of microorganism and used as an coagulants, still there are certain problems which inhibit it from many applications such as many nanoparticles doped or adsorbed with chitosan have adverse effect of the human health as well as environment. So there must be a thorough evaluation of the toxicity and other properties of the nanoparticles before doping with chitosan. In order to make an efficient and cost effective water remediation, evaluation of the materials is the most promising factor. Apart from the nanoparticle, chitosan with reactive hydroxyl group, amine functionalized group can be easily protonated in acidic media thus helping in the removal of dyes with the anionic ligands and the microorganism with the negatively charged shells. Chitosan when doped with the nanoparticles become more effective in the activity whether it is removal of the pollutants from water i.e. dyes, pesticides, drugs, fertilizers, heavy metals etc. or the removal of bacteria and fungi from the water. Chitosan because of its unique properties such as biodegradability, biocompatibility, mucoadhessiveness and antibacterial activity have been used for many industrial applications. Chitosan based nanocomposites due to its low mechanical strength, high degree of swelling, low solubility and thermal stability make it limited for the application, but a lot of work based on chitosan derivatives is under progress to improve its property and make the maximum use of chitosan based nanocomposites for the wastewater remediation and biomedical application in future.

Acknowledgements Authors are thankful to all those who have already worked in the field treatment of waste water using chitosan based nanocomposites and also thankful to those who have been working in this field. Authors acknowledge Department of Chemistry, Aligarh Muslim University, Aligarh for providing necessary facilities. SS also thanked UPCST for providing Young Scientist fellowship.

References

Adewuyi S, Jacob JM, Olaleye O, Abdulraheem TO, Tayo JA, Oladoyinbo F (2016) Chitosan-bound pyridinedicarboxylate Ni(II) and Fe(III) complex biopolymer films as waste water decyanidation agents. Carbohydr Polym 151:1235–1239

Ali I (2012) New generation adsorbents for water treatment. Chem Rev 112:5073–5091

Ali I, Gupta VK (2007) Advances in water treatment by adsorption technology. Nat Protoc 1:2661–2667

Amin MT, Alazba AA, Manzoor U (2014) A review of removal of pollutants from water/wastewater using different types of nanomaterials. Adv Mater Sci Eng 2014:1–24

Assaad E, Azzouz A, Nistor D, Ursu AV, Sajin T, Miron DN (2007) Metal removal through synergic coagulation–flocculation using an optimised chitosan-montmorillonite system. Appl Clay Sci 37:258–274

Bhatnagar A, Sillanpaa A (2009) Applications of chitin- and chitosan-derivatives for the detoxification of water and wastewater – a short review. Adv Colloid Interface Sci 152(2009):26–38

Bingbing L, Zhou F, Huang K, Wang Y, Mei S, Zhou Y, Jing T (2017) Environmentally friendly chitosan/PEI-grafted magnetic gelatin for the highly effective removal of heavy metals from drinking water. Sci Rep 7:43082

Bratskaya SY, Avramenko VA, Sukhoverkhov SV, Schwarz S (2002) Flocculation of humic substances and their derivatives with chitosan. Colloid J 64:756–761

Chen X, Sun HL, Pan JH (2006) Decolorization of dyeing wastewater with use of chitosan materials. Ocean Sci J 41:221–226

Cheng WP, Chi FH, Yu RF, Lee YC (2005) Using chitosan as a coagulant in recovery of organic matters from the mash and lauter wastewater of brewery. J Polym Environ 13:383–388

Chiou MS, Kuo WS, Li HY (2003) Removal of reactive dye from waste water by adsorption using ECH cross-linked Chitosan beads as medium. J Hazard Mater A 38:2621–2631

Chivrac F, Pollet E, Averous L (2009) Progress in nano-biocomposites based on polysaccharides and nanoclays. Mater Sci Eng R 67:1–17

Crini G, Badot PM (2008) Application of chitosan, a natural aminopolysaccharide, for dye removal from aqueous solutions by adsorption processes using batch studies: a review of recent literature. Prog Polym Sci 33:399–447

Dashairya L, Sharma M, Basu S, Saha P (2018) Enhanced dye degradation using hydrothermally synthesized nanostructured Sb2S3/rGO under visible light irradiation. J Alloys Compd 735:234–245

Dhanavel S, Manivannan N, Gupta VK, Narayanan V, Stephen A (2018) Preparation and characterization of cross-linked chitosan/palladium nanocomposites for catalytic and antibacterial activity. J Mol Liq 257:32–41

Divakaran R, Pillai VNS (2001) Flocculation of kaolinite suspensions in water by chitosan. Water Res 35:3904–3908

Duong HV, Chau L, Dang N, Vanterpool F, Sanchez MS, Lizundia E, Tran HT, Nguyen VL, Nguyen TD (2018) Biocompatible chitosan-functionalized upconverting nanocomposites. ACS Omega 3:86–95

Farzana M, Meenakshi H, Sankaran V (2015) Facile synthesis of Chitosan/ZnO composite for the photodegradation of Rhodamine B dye. J Chitin Chitosan Sci 3:21–31

Fenglian F, Wang Q (2011) Removal of heavy metal ions from wastewater. J Environ Mang 92:407–418

Gamage A, Shahidi F (2007) Use of chitosan for the removal of metal ion contaminants and proteins from water. Food Chem 104:989–996

Gerente C, Lee VKC, Le Cloirec P, McKay G (2007) Application of Chitosan for the removal of metals from wastewaters by adsorption – mechanisms and model review. Crit Rev Environ Sci Technol 37:41–127

Gogoi P, Ashim J, Rashmi T, Devi R, Das B, Maji TK (2016) A comparative study on sorption of arsenate ions from water by crosslinked chitosan and crosslinked chitosan/MMT nanocomposite. J Environ Chem Eng 4(4):4248–4257

Gokila S, Gomathi T, Sudha PN, Anil S (2017) Removal of the heavy metal ion chromiuim(VI) using Chitosan and Alginate nanocomposites. Int J Biol Macromol 104:1459–1468

Guibal E (2004) Interactions of metal ions with chitosan-based sorbents: a review. Sep Purif Technol 38:43–74

Guibal E, Roussy J (2007) Coagulation and flocculation of dye-containing solutions using a biopolymer (chitosan). React Funct Polym 67:33–42

Gupta VK et al (2017) Degradation of azo dyes under different wavelengths of UV light with chitosan-SnO2 nanocomposites. J Mol Liq 232:423–430

Hadwiger LA, Kendra D, Fristensky BW, Wagoner W (2016) Chitosan both activates genes in plants and inhibits RNA synthesis in fungi. In: Chitin in nature and technology. Springer, Boston, pp 209–214

Haritma C, Ruhi G (2016) Eco friendly chitosan: an efficient material for water purification. Pharma Innovation J 5(1):92–95

Hosseinzadeh H, Ramin S (2018) Effective removal of copper from aqueous solutions by modified magnetic chitosan/graphene oxide nanocomposites. Int J Biol Macromol 113(1):859–868

Huang C, Chen Y (1996) Coagulation of colloidal particles in water by chitosan. J Chem Technol Biotechnol 66:227–232

Kim CY, Choi HM, Cho HT (1997) Effect of deacetylation on sorption of dyes and chromium on chitin. J Appl Polym Sci 63:725–736

Krajewska B (2005) Membrane-based processes performed with use of chitin/chitosan materials. Sep Purif Technol 41:305–312

Kumar R, Oves M, Almeelbi T, Al-Makishah NH, Barakat MA (2017) Hybrid chitosan/polyaniline-polypyrrole biomaterial for enhanced adsorption and antimicrobial activity. J Colloid Interface Sci 490:488–496

Kurita K (2006) Chitin and chitosan: functional biopolymers from marine crustaceans. Mar Biotechnol 8:203–206

Lefebvre O, Moletta R (2006) Treatment of organic pollution in industrial saline wastewater: a literature review. Water Res 40:3671–3682

Mahdavinia GR, Mosallanezhad A (2016) Facile and green rout to prepare magnetic and chitosan-crosslinked κ-carrageenan bionanocomposites for removal of methylene blue. J Water Pro Eng 10:143–155

Masheane M, Nthunya N, Malinga S, Nxumalo E, Barnard T, Mhlanga S (2016) Antimicrobial properties of chitosan-alumina/f-MWCNT nanocomposites. J Nanotechnol 2016:1–8

Miquel R, Souad A, Lara D, Javier G, Joan R, Jordi B (2007) Interaction of uranium with in situ anoxically generated magnetite on steel. J Hazard Mater 147:726–731

Mondal S (2008) Methods of dye removal from dye house effluent – an overview. Environ Eng Sci 25:383–396

Moussout H et al (2018) Performances of local chitosan and its nanocomposite 5%Bentonite/Chitosan in the removal of chromium ions Cr(VI) from wastewater. Int J Biol Macromol 108:1063–1073

Nafees A, Saima S, Ameer A, Suhail S, Mohammad ZK (2017) Novel bio-nanocomposite materials for enhanced biodegradability and photocatalytic activity. New J Chem 41:10198

Neeta P, Shukla SK, Singh NB (2017) Water purification by polymer nanocomposites: an overview. Nanocomposites 3(2):47–66

Nithya JK, Prabhu S, Jeganathan K (2014) Chitosan based nanocomposite materials as photocatalyst – a review. Mater Sci Forum 781:79–94

No HK, Lee K, Meyers SP (2000) Correlation between physical chemical characters binding capacity of Chitosan products. J Food Sci 65:1134–1137

Olivera S, Basavarajaiah H, Krishna M, Vijay V, Guna K, Gopalakrishna K, Kumar Y (2016) Potential applications of cellulose and chitosan nanoparticles/composites in wastewater treatment: a review. Carbohydr Polym 153:600–618

Ozer D, Ozer A (2004) The adsorption of copper (II) ions on to dehydrated wheat bran (DWB): determination of the equilibrium and thermodynamic parameters. Process Biochem 39:2183–2191

Portes E, Gardrat C, Castellan A, Coma V (2009) Environmentally friendly films based on chitosan and tetrahydrocurcuminoid derivatives exhibiting antibacterial and antioxidative properties. Carbohydr Polym 76:578–584

Qin C, Li H, Xiao Q, Liu Y, Zhu J, Du Y (2006) Watersolubility of chitosan and its antimicrobial activity. Carbohydr Polym 63:367–374

Qin Y, Cai L, Feng D, Shi B, Liu J, Zhang W, Shen Y (2007) Combined use of chitosan and alginate in the treatment of wastewater. J Appl Polym Sci 104(6):3581–3587

Rania EM, Alsabagh AM, Nasr SA, Zaki MM (2017) Multifunctional nanocomposites of chitosan, silver nanoparticles, copper nanoparticles and carbon nanotubes for water treatment: antimicrobial characteristics. Int J Biol Macromol 97:264–269

Renault F, Sancey B, Badot PM, Crini G (2009) Chitosan for coagulation/flocculation processes – an eco-friendly approach. Eur Polym J 45:1337–1348

Rinaudo M (2006) Chitin and chitosan: properties and applications. Prog Polym Sci 31:603–632

Rodrigues AC, Boroski M, Shimada NS, Garcia JC, Nozaki J, Hioka N (2008) Treatment of paper pulp mill wastewater by coagulation – flocculation followed by heterogenous photocatalysis. J Photochem Photobiol A Chem 194:1–10

Roussy J, Vooren VM, Guibal E (2004) Chitosan for the coagulation and flocculation of mineral colloids. J Dispers Sci Technol 25:663–677

Roussy J, Vooren VM, Dempsey BA, Guibal E (2005) Influence of chitosan characteristics on the coagulation and the flocculation of bentonite suspensions. Water Res 39:3247–3258

Saad AHA, Azzam A, El-Wakeel ST, Mostafa B, El-Latif MB (2018) Removal of toxic metal ions from wastewater using ZnO@Chitosan core-shell nanocomposite. Environ Nanotechnol Monit Manag 9:67–75

Saha P, Datta S, Sanyal SK (2008) Study on the effect of different metals on soil liner medium. Indian Sci Cruiser 22:50–56

Saito T, Isogai A (2005) Ion-exchange behavior and carboxylate groups in fibrous cellulose oxidized by TEMPO-mediated system. Carbohydr Polym 61:183–190

Saldias C et al (2018) In situ preparation of film and hydrogel bio-nanocomposites of chitosan/fluorescein-copper with catalytic activity. Carbohydr Polym 180:200–208

Seong HS, Kim JP, Ko SW (1999) Preparing chito-oligosaccharides as antimicrobial agents for cotton. Text Res J 69(7):483–488

Shahidi F, Arachchi JKV, Jeon YJ (1999) Food applications of chitin and chitosans. Trends Food Sci Technol 10:37–51

Strand SP, Vandvik MS, Vårum KM, Ostgaard K (2001) Screening of chitosan and conditions for bacterial flocculation. Biomacromolecules 2:121–133

Sultana S, Ahmad N, Faisal SM, Owais M, Sabir S (2007) IET Nanobiotechnol 2017:1–8

Swami D, Buddhi D (2006) Removal of contaminants from industrial wastewater through various non-conventional technologies: a review. Int J Environ Pollut 27:324–346

Teimouri A et al (2016) Chitosan/Zeolite Y/Nano ZrO_2 nanocomposite as an adsorbent for the removal of nitrate from the aqueous solution. Int J Biol Macromol 93:254–266

Tripathi S, Mehrotrap GK, Dutta PK (2011) Preparation and antimicrobial activity of chitosan-silver oxide nanocomposite film via solution casting method. Bull Mater Sci 34:29–35

Vanamudan A, Padmaja MS, Pamidimukkala S (2018) Nanostructured zirconium tungstate and its bionanocomposite with chitosan: wet peroxide photocatalytic degradation of dyes. J Taiwan Inst Chem Eng 85:74–82

Varma AJ, Deshpande SV, Kennedy JF (2004) Metal complexation by chitosan and its derivatives. Carbohydr Polym 55:77–93

Veera MB, Krishnaiah A, Jonathan LT, Edgar DS (2003) Removal of hexavalent chromium from wastewater using a new composite chitosan biosorbent. Environ Sci Technol 37:4449–4456

Veera MB, Krishnaiah A, Ann JR, Edgar DS (2008a) Removal of copper (II) and nickel (II) ions from aqueous solutions by a composite chitosan biosorbent. Sep Sci Technol 43:1365–1381

Veera MB, Krishnaiah A, Jonathan LT, Edgar DS, Richard H (2008b) Removal of arsenic (III) and arsenic (V) from aqueous medium using chitosan-coated biosorbent. Water Res 42:633–642

Wada S, Ichikawa H, Tatsumi K (1995) Removal of phenols and aromatic amines from wastewater by a combination treatment with tyrosinase and a coagulant. Biotechnol Bioeng 45:304–309

Wan MW, Kan CC, Lin CH, Buenda DR, Wu CH (2007) Adsorption of copper (II) by chitosan immobilized on sand. Chia-Nan Annu Bull 33:96–106

Wang JP, Chen YZ, Ge XW, Yu HQ (2007) Optimization of coagulation– flocculation process for a paper-recycling wastewater treatment using response surface methodology. Colloids Surf A Physicochem Eng Aspects 302:204–210

Wu FC, Tseng RL, Juang RS (2001) Enhanced abilities of highly swollen chitosan beads for color removal and tyrosinase immobilization. J Hazard Mater B81:167

Wu ZB, Ni WM, Guan BH (2008) Application of chitosan as flocculant for coprecipitation of Mn (II) and suspended solids from dual-alkali FGD regenerating process. J Hazard Mater 152:757–764

Xie W, Xu P, Wang W, Liu Q (2002) Preparation and antibacterial activity of a water-soluble chitosan derivative. Carbohydr Polym 50:35–40

Zemmouria H, Drouiche M, Sayeh A, Lounici H, Mameri N (2013) Chitosan application for treatment of beni-amrane's water Dam. Energy Procedia 36:558–564

Zheng C, Beach ES, Anastas PT (2011) Modification of chitosan films with environmentally benign reagents for increased water resistance. Taylor Francis 4:35–40

Chapter 5
Graphene and Its Composites: Applications in Environmental Remediation

Uzma Haseen, Khalid Umar, Hilal Ahmad, Tabassum Parveen, and Mohamad Nasir Mohamad Ibrahim

Abstract The research on graphene has attracted a lot of interest in the scientific world, especially in material science. The scaled-up and reliable production of graphene composites explores their utility with wider applications. In this chapter, we focus on the description of various methods to synthesize graphene-based composites, especially those with organic polymers, inorganic nanostructures, carbon nanotubes, etc. The possibility of conjugation of graphene with different materials to form composite has received particular attention concerning the design of stable materials particularly adsorbents for the advancement of the water treatment process. This chapter describes the preparation of graphene and its composites, efforts, and trends in environmental water remediation highlighting the strategies for the optimization of composite properties.

Keywords Graphene · Toxicity · Metal ions · Wastewater

5.1 Introduction

The keen observation and characterization of a mechanically exfoliated graphene monolayer by Novoselov in 2004 (Novoselov et al. 2004) has stimulated the advancement of graphene research in both the scientific and engineering communities. Graphene, a 2D layered honey comb like structure, consist of sp^2 hybridized carbon offering broad opportunities in different areas of nanotechnology. Although

U. Haseen
Department of Chemistry, Aligarh Muslim University, Aligarh, India

K. Umar · M. N. Mohamad Ibrahim
School of Chemical Sciences, Universiti Sains Malaysia, Pulau Pinang, Malaysia

H. Ahmad (✉)
Centre for Nanoscience and Nanotechnology, Jamia Millia Islamia (A Central University), New Delhi, India

T. Parveen
Department of Civil Engineering, Indian Institute of Technology, Roorkee, India

M. Oves et al. (eds.), *Modern Age Waste Water Problems*,
https://doi.org/10.1007/978-3-030-08283-3_5

the concept of graphene has been around in the early 1990s, it was initially thought to be too thermodynamically unstable to exist under ambient conditions (Dreyer et al. 2010). Graphene due to their unique mechanical strength, high theoretical specific surface area (2360 m^2 g^{-1}), and excellent optical and electrochemical properties develop an interest in various applications (Shen et al. 2012; Molina et al. 2016). This discovery opened up a whole selection of properties on graphene. The point overlap between the valence and conduction band at the Dirac point resulted in graphene being a material with zero-bandgap (Jorgensen et al. 2016).

The ambipolar electric field effect where the charge carriers could be electron or holes gave rise to its high charge carrier concentration and mobilities. High optical transparency was also inherent to graphene, with the absorption of approximately 2.3% towards visible light. To furthermore explore the properties of graphene in various kinds of applications, convenient and reliable synthetic routes have been developed to prepare graphene and its composites, ranging from the bottom-up epitaxial growth to the top-down exfoliation of graphite using oxidation, intercalation, and/or chemical functionalization. Specifically, economical manufacturing of chemically exfoliated graphene oxide (GO) and graphene sheets has been realized which possess numerous reactive oxygen-containing groups for further functionalization and tuning properties of graphene and GO sheets. One of the most popular approaches to graphene-based nanomaterials is to use graphene oxide (GO), due to its lower production costs. With these added advantages, it is desirable to take benefits of the valuable properties of graphene and its derivatives in composites, through the consolidation of various kinds of functional materials. To date, graphene-based composites have been successfully made with inorganic nanostructures, organic crystals, polymers, metal–organic frameworks, biomaterials and carbon nanotubes (Georgakilas et al. 2012).

Even though graphene holds great promise for various application, the physical handling of graphene sheets is challenging since graphene is not soluble in most solvents, only soluble in solvents exhibiting surface tension close to 40–50 mJ m^{-2}, such as benzyl benzoate, N,N-dimethylacetamide, etc. In aqueous media frequent problems of graphene sheets is to form irreversible agglomerates or restacking to form graphite *via* pi-pi stacking and van der Waals interactions (Islam et al. 2014), Thus requires extensive research on the development of such major issues before use in real applications.

Chemical functionalization of graphene is one of the many solutions to address the above challenges of graphene. Functionalization through synthetic chemistry methods allow for the preparation of p- and n-doped graphene based on the selection of electron donating or electron withdrawing complexes covalently bonded to the graphene carbon network. Similarly, functionalization with inorganic nanoparticles, organic monomers, metal-organic framework, and carbon nanotubes leads to the formation of graphene-based composites (Gandhi et al. 2016). Furthermore, by varying the groups which are covalently bonded on the graphene carbon network, modifications of the graphene solubility in both the organic and aqueous media could be easily achieved.

5.2 Graphene Polymer Composites

The graphene-polymer composites based on the 3D spatial orientation of the groups and the type of binding interactions between the graphene and polymer can be classified into three types. (i) Adsorption of surfactant hemimicelles on the surfaces of the graphene and graphene oxide causes its dispersion in surfactant micelles in an aqueous media. (ii) The self-assembly of anionic sulfonate surfactant on graphene surface with the opposite charges metal cations (like Sn^{2+}) and the transition into the lamella mesophase towards the formation of SnO_2 graphene nanocomposite, where hydrophobic graphene sheets are sandwiched in the hydrophobic domains of the anionic surfactant. (iii) Layered metal oxide-graphene composites are the most common type and composed of alternating layers of metal oxide nanocrystals and graphene stacks after crystallization of metal oxide and removal of the surfactant.

Solution mixing is the most straight forward method for the preparation of polymer composites. The solvent compatibility of the polymer and the filler is critical in achieving good dispersibility. Due to the residual oxygen-containing functional groups, graphene oxide can be directly mixed with water-soluble polymers, such as polyvinyl alcohol, at various concentrations. Sonication and ultrasonication process had been employed to produces metastable dispersions of graphene, which are further mixed with the polymer solution such as poly(methyl methacrylate), polyaniline, polycaprolactone and polyurethane to create composites. High-speed shearing combined with ice-cooling has also been applied to well mix graphene-based fillers and the polymer matrices. In both the above-mentioned approaches re-stacking, aggregation, and folding of the graphene sheets are unavoidable hence, significantly reduces the specific surface area of the 2D fillers. Thus, surface functionalization of graphene before solution mixing must be attempted to provide them good dispersibility in aqueous media.

Moreover, instead of being used as fillers to enhance the properties of polymers, graphene derivatives can be applied as 2D templates for polymer decoration via covalent and non-covalent functionalization. The solubility of graphene can also be improved by polymer immobilization onto graphene surface. Polymer coating introduces new functional groups which lay to new hybrid sheet. Usually, carboxylic groups of graphene oxide (GO) sheets react with the functional groups of target polymers. Esterification of the carboxylic groups in GO with the hydroxyl groups in PVA has been demonstrated in the synthesis of GO–PVA composite sheets. The carboxylic groups on GO were also involved in the carbodiimide-catalyzed amide formation process to bind with the six-armed polyethylene glycol (PEG)-amine stars. However, the carboxylic groups are mainly confined to the periphery of the GO sheets. In addition, the grafting of certain polymers requires the presence of non-oxygenated functional groups on GO, such as amine and chloride groups. Therefore, alternative strategies with additional chemical reactions have to be developed to modify GO/graphene surfaces with desired functional groups before the grafting of polymers.

For example, after 4-bromophenyl groups were coupled on the graphene surface through the diazonium reaction, a fluorene–thiophene–benzothiadazole polymer was covalently grafted to the graphene surface through the Suzuki reaction via the 4-bromophenyl groups. The pre-bonded diazonium group on graphene can also act as the initiator for atomic transfer radical polymerization (ATRP), based on which, polystyrene has been successfully grafted onto the graphene sheet with controlled density. Additionally, acyl-chlorinated GO sheets can further react and connect to triphenylamine-based polyazomethine (TPAPAM) and MeOH-terminated P3HT, and the APTES-modified GO surface can be further bonded with maleic anhydride-grafted polyethylene (MA-g-PE).

5.3 Graphene Composites for Water Decontamination

Water can be purified in multiple ways, i.e. desalination, filtration, osmosis, adsorption, disinfection, and sedimentation, however, adsorption holds many advantages over other methods. Adsorption is the surface phenomenon where pollutants are adsorbed on the surface of an adsorbent via physical forces. Adsorption depends on many factors such as temperature, pH, the concentration of pollutants, contact time, particle size, temperature, and nature of the adsorbate and adsorbent. Carbon-based materials such as activated carbon and carbon nanotubes have been used for water purification. The large-scale production of functionalized graphene at low cost should result in good adsorbents for water purification. This is due to the two-dimensional layer structure, large surface area, pore volume and presence of surface functional groups in these materials (Islam et al. 2014). The inorganic nanoparticles present in 2D graphene nanocomposites prevent graphene aggregation, as such a high surface area and pore volume can be maintained.

Graphene materials have been applied as absorbents for the removal of organic pollutants, such as dyes, antibiotics, hydrocarbons, crude oil, pesticides, and natural organic matter (Liu et al. 2012a, b). The mechanism of interaction between nanomaterials and organic compounds is dependent on their structural properties (e.g., molecular conformation, dipole moment, the presence of functional groups). Hence, the adsorption capacity of the same molecule might be different whether the absorbent materials are composed of GO or pristine graphene sheets. Similarly, the presence or absence of functional groups (–NH2, –OH, –COOH) in the adsorbate structure will determine the mechanism and efficiency of the adsorption process (Liu et al. 2010). Electrostatic interaction is prevalent when the adsorbate has charged functional groups while the adsorbents preserve their charged surface. For instance, the adsorption of cationic dyes such as methylene blue and ethyl violet by GO over a wide pH range (6–10) is mediated through electrostatic interactions between exfoliated GO and the dye molecules.

Conversely, the adsorption of anionic dyes (rhodamine B and orange G) by GO was not favorable at the same pH range. As the carboxyl groups in both materials were negatively charged, a subsequent electrostatic repulsion was possibly created between GO sheets and the anionic dyes molecules. The effective removal of cationic dyes using GO sheets via electrostatic interactions has been reported so far (Perreault et al. 2015). The adsorption of cationic dyes by pi-pi interactions is goverened by the delocalized electron of graphene. In addition, it has been considered that under neutral pH conditions, the formation of hydrogen bonds could be an important factor in the molecular interaction between a cationic dye (which contains –NH2 groups) and GO monolayers. In another study, the adsorption of cationic methylene blue by GO exfoliated layers was extensively improved when samples with higher oxidation degree were used. With increasing oxidation degree, the mechanism of sorption was presumed to change from parallel pi–pi stacking to vertical electrostatic interactions (Liu et al. 2010).

5.4 Graphene-Based Photocatalytic Composites for Water Decontamination

When Graphene used in photocatalysis, it was observed that they can tune the band gap energy of semiconductors which present in their composites. Due to high electron mobility in graphene, it suppresses the rapid recombination of electron-hole pairs, thus leading to an enhanced photocatalytic activity (Liu et al. 2010). The formation of composites between semiconductor particles and graphene sheets can, therefore, contribute to extending the photocatalytic activity of conventional photocatalysts, such as TiO_2, ZnO etc. by decreasing the frequency of electron–hole pair recombination (Umar et al. 2013, 2015). Graphene-based photocatalyst are prepared by anchoring photoactive nanostructures on graphene. Prior reviews on this subject described in detail the methods used to prepare graphene nanocomposites for photocatalytic purposes. Additional studies have also concluded that graphene–TiO_2 photocatalyst were more efficient in the degradation of organic dyes in comparison to bare TiO_2 (Lightcap et al. 2010). The increased adsorption of the organic dyes on graphene and the excellent ability to transfer electrons were correlated with the exceptional photocatalytic performance of graphene-related photocatalysts. Another important factor that may be associated with the improvement in photocatalytic activity for graphene-based photocatalyst is their increased surface area.

In addition to dye degradation, graphene-based photocatalysts have also shown increased efficiency for the degradation of hydrocarbon derivatives. As an example, graphene–CdS nanocomposites, prepared by self-assembling positively charged CdS nanostructures with negatively charged GO sheets, were applied as photocatalysts for selective reduction of nitroaromatic compounds. Additional studies also demonstrated the enhanced degradation of pesticides, methanol, and endocrine disruptors (phenol, bisphenol, and atrazine) by graphene-hybrid photocatalysts. All these studies consistently reported that graphene sheets played a crucial role in the enhancement of the photocatalytic ability of pristine semiconductor particles (e.g., TiO_2, Ag nanoparticles, and CdS). Even though the mechanism of degradation has been associated with the electron-accepting capacity of graphene and its ability to prevent hole-pair recombination, Zhang has proposed an alternative mechanism to explain the role of graphene in the selective oxidation of alcohols and alkenes by graphene–ZnS nanocomposites (Zhang et al. 2010). To prove their proposed mechanism, experiments were conducted under visible light irradiation, where ZnS is not able to be photoexcited. Rather than providing an electron conductive platform as suggested by most studies in the literature, graphene sheets were found to act as a macromolecular "photosensitizer". In other words, upon visible light irradiation, photo-induced electrons from the graphene itself could be shuttled into the conductance band of ZnS nanoparticles. As a primary consequence, the presence of graphene imparts to ZnS particles photocatalytic activity under visible light.

5.5 Conclusion and Perspective

During the past few years, appreciable efforts have been focused toward the preparation of graphene-based nanocomposites, and the unique properties of these materials have been demonstrated for a variety of applications; catalysts, adsorbents, supercapacitors, and fuel cell batteries. It is indubitable that making the use of graphene-based nanocomposites in either of these technologies is a new venture, indicating that future research efforts will be abundant. Moreover, how to retain the advantageous properties of graphene in their sucessive composite is the next challenge. It is significant that GO sheets are unquestionable effective precursors for bulk production of hybrid materials. However, an obvious prediction is that a more convincing route can be develop and employ via direct oxidation of pristine graphene in soft manners to avoid higher degree of functional groups produced at the surface.

Acknowledgment The authors are thankful to department of science and technology and science and engineering research board of India for providing basic facilities under the project NPDF 000995. The authors also gratefully acknowledged the post doctoral financial support (USM/PPSK/FPD(BW) 1/06 (2018) and a research grant (304/PKIMIA/6316174) by School of Chemical science, Universiti Sains Malaysia.

References

Dreyer DR, Ruoff RS, Bielawski CW (2010) From conception to realization: an historial account of graphene and some perspectives for its future. Angew Chem Int Ed 49:9336–9344

Gandhi MR, Vasudevan S, Shibayama A, Yamada M (2016) Graphene and graphene-based composites: a rising star in water purification-a comprehensive overview. ChemistrySelect 1:4358–4385

Georgakilas V, Otyepka M, Bourlinos AB (2012) Functionalization of graphene: covalent and non-covalent approaches, derivatives and applications. Chem Rev 112:6156–6214

Islam A, Ahmad H, Zaidi N, Kumar S (2014) Graphene oxide sheets immobilized polystyrene for column preconcentration and sensitive determination of lead by flame atomic absorption spectrometry. ACS Appl Mater Interfaces 6:13257–13265

Jorgensen JH, Cabo AG, Balog R (2016) Symmetry-driven band gap engineering in hydrogen functionalized graphene. ACS Nano 10:10798–10807

Lightcap IV, Kosel TH, Kamat PV (2010) Anchoring semiconductor and metal nanoparticles on a two-dimensional catalyst mat. Storing and shuttling electrons with reduced graphene oxide. Nano Lett 10:577–583

Liu J, Bai H, Wang Y, Liu Z, Zhang X, Sun DD (2010) Self-assembling TiO_2 nanorods on large graphene oxide sheets at a two-phase interface and their anti-recombination in photocatalytic applications. Adv Funct Mater 20:4175–4181

Liu F, Chung S, Oh G, Seo TS (2012a) Three-dimensional graphene oxide nanostructure for fast and efficient water-soluble dye removal. ACS Appl Mater Interfaces 4:922–927

Liu Q, Shi J, Jiang G (2012b) Application of graphene in analytical sample preparation trends. Anal Chem 37:1–11

Molina J, Cases F, Moretto LM (2016) Graphene-based materials for the electrochemical determination of hazardous ions. Anal Chim Acta 946:9–39

Novoselov KS, Geim AK, Morozov SV, Jiang D, Zhang Y, Dubonos SV, Grigorieva IV, Firsov AA (2004) Electric field effect in atomically thin carbon films. Science 306:666–669

Perreault F, Faria AF, Elimelech M (2015) Environmental applications of graphene-based nanomaterials. Chem Soc Rev 44:5861–5896

Shen J, Zhu Y, Yang X, Li C (2012) Graphene quantum dots: emergent nanolights for bioimaging, sensors, catalysis and photovoltaic devices. Chem Commun 48:3686–3699

Umar K, Haque MM, Muneer M, Harada T, Matsumura M (2013) Mo, Mn and La doped TiO_2: synthesis, characterization and photocatalytic activity for the decolourization of three different chromophoric dyes. J Alloys Compd 578:431–437

Umar K, Aris A, Parveen T, Jaafar J, Majid ZA, Reddy AVB, Talib J (2015) Synthesis, characterization of Mo and Mn doped Zno and their photocatalytic activity for the decolorization of two different chromophoric dyes. Appl Catal A Gen 505:507–515

Zhang LS, Jiang LY, Yan HJ (2010) Mono dispersed SnO_2 nanoparticles on both sides of single layer graphene sheets as anode materials in Li-ion batteries. J Mater Chem 20:5462–5546

Chapter 6
Fabrication of Polyaniline Supported Nanocomposites and Their Sensing Application for Detection of Environmental Pollutants

Mohammad Shahadat, Mohammad Oves, Abid Hussain Shalla, Shaikh Ziauddin Ahammad, S. Wazed Ali, and T. R. Sreekrishnan

Abstract Nanoscale composite materials have played a significant role in sensing of gases, owing to their high surface area, higher mechanical strength with efficient chemical activity as well as cost effective nature. The present chapter deals with synthesis and characterization of Polyaniline (PANI) based nanocomposites ion-exchanger and nanomaterials in addition to their sensing applications in various fields. These nanomaterials have been explored on the basis of advanced techniques of characterizations. Besides the sensing materials, on the basis of ion uptake capacity, these nanocomposite ion-exchange materials can also be used for the treatment of meal ions from industrial wastewaters. This chapter mainly focuses on the synthesis of PANI based nanocomposites and their applications as gas sensors and biosensors. The PANI nanomaterials demonstrated impressive results and outstanding sensing behaviour. It has been found that PANI based nanocomposite materials are not only used for the detection of toxic gases, but, these materials

M. Shahadat (✉)
Department of Biochemical Engineering and Biotechnology, Indian Institute of Technology Delhi, New Delhi, India

Department of Textile Technology, Indian Institute of Technology Delhi, New Delhi, India

M. Oves
Centre of Excellence in Environmental Studies, King Abdul Aziz University, Jeddah, Saudi Arabia

A. H. Shalla
Department of Chemistry, Islamic University of Science & Technology (IUST), Pulwama, Jammu and Kashmir, India

S. Z. Ahammad · T. R. Sreekrishnan
Department of Biochemical Engineering and Biotechnology, Indian Institute of Technology Delhi, New Delhi, India
e-mail: sree@dbeb.iitd.ac.in

S. Wazed Ali
Department of Textile Technology, Indian Institute of Technology Delhi, New Delhi, India

M. Oves et al. (eds.), *Modern Age Waste Water Problems*,
https://doi.org/10.1007/978-3-030-08283-3_6

also facilitated immobilization of bioreceptors (e.g., enzymes, antigen–antibodies, and nucleic acids, etc.) for the exposure of biological agents through a combination of biochemical and electrochemical reactions. In future, PANI based nanocomposite materials are expected to open new approaches for demonstrating their outstanding applications in diverse fields.

Keywords Synthesis · Characterization · Polyaniline · Sensing behavior · Biosensors · Metal ions

6.1 Introduction

Recently, polyaniline (PANI) based organic–inorganic conducting nanocomposite materials have played important role in the field of nanotechnology. These nanocomposites with combinations of two or more components have received more and more attention, owing to their interesting physical properties as well as many potential applications in various areas (Gangopadhyay and De 2000). Conducting polymers are a new class of materials with essential applications in a different growing advance technologies, such as energy storage devices and chemical sensors (Novák et al. 1997; Santhanam and Gupta 1993; Feng and MacDiarmid 1999; Ayad et al. 2009; Singh et al. 2008; Sutar et al. 2007; Khan et al. 2010a; Jain et al. 2003; Fuke et al. 2008, 2009). In the development of nanocomposite materials, organic polymers such as polyaniline, polythiophene, polyacrylonitrile, etc. have been extensively used. Because of their significant proton dopability, these materials have the prominent features of variable electrical properties, low cost, ease of synthesis, excellent redox recyclability and thermal as well as chemical stability (Nabi et al. 2011a; Shahadat et al. 2015).

Among all conducting polymers, PANI has attracted much attention (Genies et al. 1990; Syed and Dinesan 1991; Huang et al. 1986; Shahadat et al. 2012) on account of its ability, under certain conditions to exhibit a high level of electrical conductivity (Adams et al. 1996; Shahadat et al. 2017). PANI demonstrates electrical conductivity because of their conjugated pi-bond system. These conjugated double bonds permit transfer of electron mobility throughout the molecule due to the delocalization of electron. Besides this, it is also considered as an environmentally stable and highly tunable conducting polymer which can be obtained in the form of bulk powder, cast films, or fibers. In conjunction with feasibility of low-cost, large-scale production, makes PANI an ideal candidate for various applications in different fields and therefore, can be successfully employed as a sorbent (Vatutsina et al. 2007) ion-exchanger (Nabi et al. 2010, 2011b, c; Bushra et al. 2014), catalyst (Arrad and Sasson 1989), ion selective electrode (Khan 2006) and sensor (Raman et al. 1996), conducting material (Shahadat et al. 2012) and also used in the host guest chemistry (Khan et al. 2011).

It is also observed that polymer is believed to be deprotonated by ammonia, which results change in conductivity of nanocomposite material (Khan et al. 2011). On the basis of alteration of conductivity, PANI based nanocomposite materials can be used towards sensing behaviour of ammonia. Determination of ammonia is essential for industrial, agricultural and medical fields together with environmental monitoring areas because ammonia is one of the critical industrial exhaust gas having high toxicity. In the scientific literature, a number of articles have been published that deals with the application of nanocomposite materials as a gas sensor. A number of metal oxides conjugated with PANI based ion-exchange materials have been demonstrated their sensing behaviour towards gases (Docquier and Candel 2002; Hu et al. 2002; Riegel et al. 2002; Ampuero and Bosset 2003; Dubbe 2003). The present book chapter reposts synthesis and characterization of PANI based nanocomposite ion-exchange materials and their sensing behaviour for the detection of gases.

6.2 PANI Based Nanocomposite Ion-Exchange Materials and Their Sensing Applications

The nanocomposite of polyaniline–titanium(IV)phosphate (PANI–TiP) was synthesized using a very simple sol–gel chemical route (Khan et al. 2011). The nanocomposite material was characterized on the basis of Fourier transform infrared spectroscopy (FTIR), thermogravimetric analysis (TGA), and scanning electron microscopy (SEM) analysis, the latter two confirmed smooth spherical morphology covered a particle size of ~25–45 nm. The material was examined to detect sensing at room temperature (25 ± 2 °C). Sensing behaviour measurement confirmed that nanocomposite of PANI–TiP demonstrated good reversible sensitivity towards ammonia (3–6%). Besides the doping of p-toluene sulphonic acid (p-TSA), a nanocomposite of PANI–TiP was also doped with hydrochloric acid (HCl) which demonstrated higher sensing response of p-TSA than HCl together with detection limit $\leq$1% ammonia. The higher electrical resistivity of the PANI-TiP nanocomposite may be attributed to the dopant p-toluene sulphonates consumed by NH_4^+ ions which make a weak charge complex of p-toluene ammonium sulphate ($C_7H_7SO_3 - NH_4^+$). In addition, it increases the resistivity of the nanocomposite, owing to the resistance in the mobility of the charge carriers along the backbone of the polymer. The weak charge complex starts to dissociate immediately as the sensor kept in the air and it regains its resistivity after a certain specific period. Hydrochloric acid also has the potential to form complex, however, ammonia molecule actively interacts with chloride ions which was not easily dissociated by putting the sensor in air (Koul and Chandra 2005). Proposed mechanism for complex formation of ammonia ($C_7H_7SO_3 - NH_4^+$) with PANI-TiP nanocomposite is shown in Scheme 6.1.

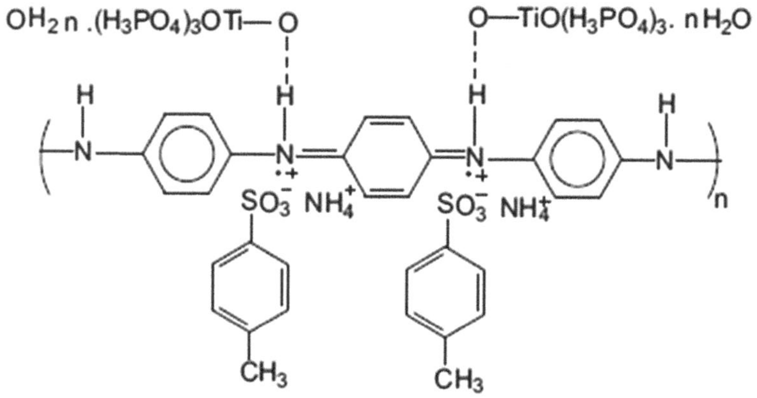

Scheme 6.1 Sensing behaviour of PANI-TiP nanocomposite ($C_7H_7SO_3$ – NH_4^+) towards ammonia. (Reprinted with permission, Khan et al. 2011)

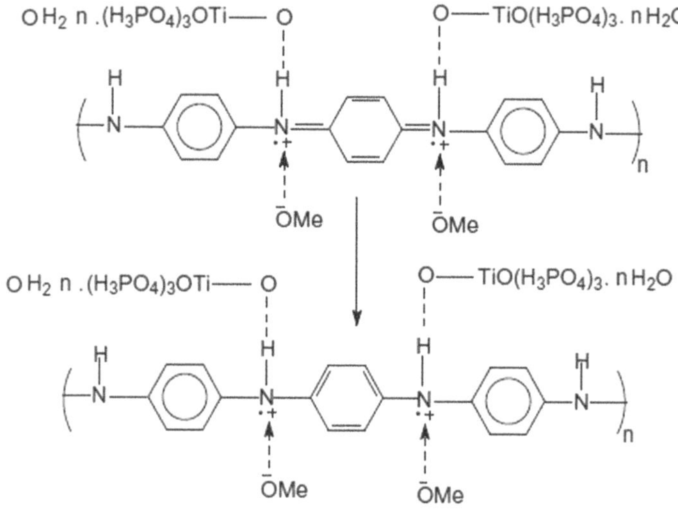

Fig. 6.1 Interaction of methanol with imine nitrogen of PANI. (Reprinted with permission, Khan et al. 2010b)

Same nanocomposite material (PANI–TiP) was synthesized to investigate the IEC, DC electrical conductivity and sensing behaviour towards alcohol vapors (Khan et al. 2010b). It exhibited good ion-exchange capacity and isothermal stability in terms of DC electrical conductivity retention at ambient conditions below 100 °C. Sensing characteristics of PANI–TiP nanocomposite towards alcohol was found to be higher than that of native PANI. The interaction between nanocomposite (PANI–TiP) and methanol molecule is shown in Fig. 6.1.

Nanofiber of V_2O_5@polyaniline was fabricated by in-situ oxidative polymerization of aniline in the presence of V_2O_5 and surfactant (cetyltrimethylammonium

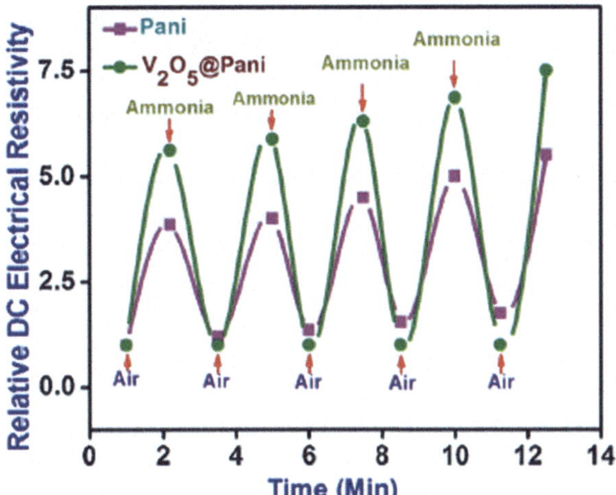

Fig. 6.2 Variation in the resistivity of pTSA-doped PANI and V2O5 composite nanofibers upon intermittent exposure of ammonia (0.1 M). (Reprinted with permission, Hasan et al. 2015)

bromide) (Hasan et al. 2015). Besides SEM, TEM, TGA-DTA and XRD analyses, the nanocomposite materials were also characterized by Raman spectroscopy, X-ray photoelectron spectroscopy. The DC electrical conductivity of V_2O_5@Pani composite nanofibers measured at 25 ± °C as 1.33 S cm^{-1}, which was found significantly higher than that the nanocomposite of HCl-doped PANI (Ansari et al. 2013). Under ambient conditions, the materials (pTSA-doped Pani and V_2O_5@Pani composite nanofiber) were examined to determine sensing response of ammonia vapour at low concentration (0.01 M). It was found that V_2O_5@pani nanofiber sensor demonstrates higher sensing response as compare to pTSA-doped PANI nanofiber sensor (which was fabricated in the presence of the surfactant; CTAB). Since, V_2O_5 (Ansari and Mohammad 2011) and PANI (Raj et al. 2010) independently exhibited ammonia sensing behaviour, however, V_2O_5@Pani nanocomposite showed higher sensing response owing to the synergistic/additional effect of both V_2O_5 and PANI as well as polymerization of aniline on V_2O_5 may increase the surface area of PANI, which result improvement in sensing response. A comparative study related to the sensing behaviour of both nanocomposites is shown in Fig. 6.2.

Chemical oxidative polymerization of 3-methythiophene (3MTh) in the presence of titanium(IV)molybdophosphate (TMP) has been carried out to synthesized poly (3-methythiophene)-titanium(IV)molybdophosphate (P3MTh–TMP) cation exchange nanocomposites (Khan and Baig 2013a). The characterization results confirmed a strong interaction between P3MTh and TMP particles which result higher thermal stability of P3MTh–TMP nanocomposites than pure P3MTh. The nanocomposite material showed good ion-exchange capacity, electrical conductivity, and isothermal stability under ambient condition below 100 °C. The results of sensing behaviour indicated enhancement in resistivity of nanocomposites on exposure to ammonia at room temperature (25 °C) and a linear relationship between the responses and the concentration of ammonia was achieved.

Electropolymerisation method was used for the fabrication of graphenated polyaniline/tungsten oxide (PANI/WO_3/GR/GCE) nanocomposite sensor on a graphene-modified glassy carbon electrode (Tovide et al. 2014). The electrode of PANI/WO_3/GR/GCE nanocomposite was examined for sensing behaviour of phenanthrene using cyclic voltammetry. It was found that PANI/WO_3/GR/GCE nanocomposite exhibited higher sensitivity towards phenanthrene and covered a dynamic linear range (1.0–6.0 pM) with a detection limit of 0.123 pM. In addition, it showed excellent reproducibility and long-term stability as well as lower detection in sensitivity than the WHO permissible limit (1.12 nM phenanthrene in wastewater. Advanced techniques namely Cyclic voltammetric (CVs), square wave voltammetric (SWVs), X-ray diffraction (XRD). Atomic force microscope (AFM) Raman spectra, Electrochemical impedance spectra (EIS) measurement and HRTEM and FTIR analyses were employed for the characterization of the nanocomposite. On the basis of good sensing response towards the detection of phenanthrene, PANI/WO_3/GR/GCE nanocomposite can be applied in a reactor for the degradation of PAHs with minimal energy requirement.

Chemically oxidative polymerization method was employed to prepare Polypyrrole-zirconium(IV)selenoiodate (PPy/ZSI) cation exchange nanocomposite (Khan et al. 2013a). The XRD pattern of the PPy, ZSI confirmed amorphous nature and SEM analysis indicated its globular surface. On the basis of EDX analysis the tentative structure of inorganic precipitate (ZSI) was found to be as $[(ZrO)_4(OH)(HSeO)_3(IO_3)_2]_n$. The ion-exchange capacity (IEC) and electrical conductivity of the nanocomposite was found to be 2.49 meq g^{-1} and 0.436 Scm^{-1} respectively. In terms of thermal stability, the nanocomposite demonstrated significant isothermal DC electrical conductivity retention up to 130 °C under ambient condition. The material was tested to determine the sensing behaviour of formaldehyde vapors at room temperature (25 $\pm$ 2 °C). It was found that the resistivity of PPy/ZSI increases with increasing percent concentration of formaldehyde at room temperature. Thus, PPy/ZSI based sensor showed good reversible response towards formaldehyde vapors in the range of 5–7%. In addition, the percentage composition of formaldehyde varpour was also measured in terms of mole fractions which was found to be 0.099%, 0.142% and 0.211% for 5, 7 and 10%, respectively. A schematic presentation related to the reversible interaction between PPy/ZSI and formaldehyde is shown in Fig. 6.3. This study confirms that PPy/ZSI based nanocomposite can be used as a smart sensor for the detection of formaldehyde.

Another electrically conducting Poly-o-toluidineetitanium(IV)phosphate (POT-TiP) cation exchange nanocomposite was fabricated by the mixing of Poly-o-toluidine and inorganic precipitate (titanium(IV)phosphate) (Khan and Baig 2013b). TEM analysis of POT-TiP nanocomposite confirmed its spherical morphology having an average particle size of 25–30 nm. The nanoparticles of TiP were shown in the form of dark spots dispersed in the matrix of poly-o-toluidine. The XRD pattern of POT-TiP showed its amorphous nature due to the dispersion of inorganic precipitate in the matrix of Poly-o-toluidine. In terms of conductivity, nanocomposite cation exchange material was examined to check humidity responses in the air. With increasing percentage of relative humidity, the resistivity of the

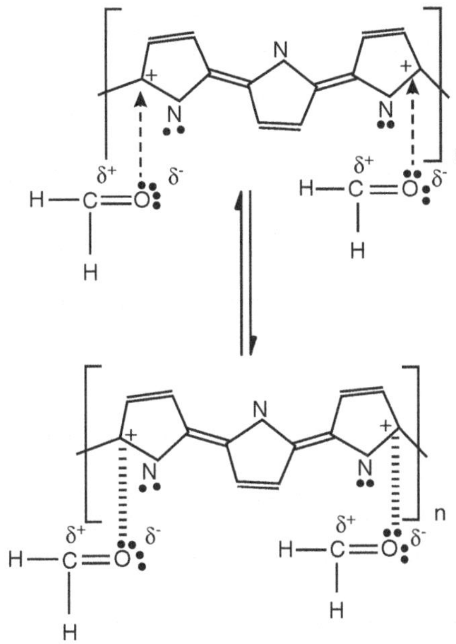

Fig. 6.3 A mechanistic representations for reversible interaction between PPy/ZSI and formaldehyde. (Reprinted with permission)

nanocomposite was found to be decreased. Improvement in conductivity or decline in resistivity with the enhancement of humidity can be attributed to the mobility of the dopant ion in the polymer matrix (Singla et al. 2007) which were loosely attracted to the chain of poly-o-toluidine by very weak forces (Van der walls forces of attraction). It was also found that at low humidity, the mobility of dopant was restricted due to dryness which curled polymer chain. In other words, at high humidity content, polymer chains become aligned with respect to each other. Decline in resistivity with increasing humidity content confirmed the adsorption of the water molecules, therefore, PANI chain of the nanocomposite act as a p-type semiconductor (the lone pair e increased, i.e., the concentration of hole$^-$ from the conducting complex toward the TiP water molecules). The partial charge transfer process of conducting species with that of water molecules decreased the resistivity of nanocomposite as shown in Fig. 6.4. The nanocomposite material demonstrated good humidity response along with ion-exchange capacity and electrical conductivity. In terms of DC electrical conductivity retention, POT-TiP based nanocomposite was found stable under ambient conditions below 90 °C. Significant results of sensing behaviour towards humidity established that the POT-TiP can be employed as a good sensing material for detection of humidity.

To develop a new humidity sensor, poly-o-anisisine based poly-o-anisidine-Sn (IV) arsenophosphate (POA-SAP) nanocomposite has been synthesized by the sol-gel method (Khan et al. 2013b). Besides sensing behaviour, the ion-exchange property of the material was also determined to use it as a cation exchange material. Physico-chemical properties were also measured in terms of diffusion coefficient (D°), the energy of activation (E_a) and standard entropy change (ΔS°). The humidity

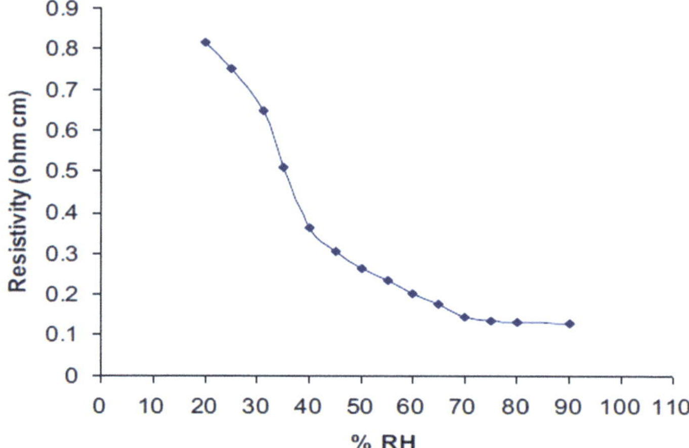

Fig. 6.4 Humidity response of Poly-o-toluidine-TiP cation exchange nanocomposite for 20–90% RH. (Reprinted with permission, Khan and Baig 2013b)

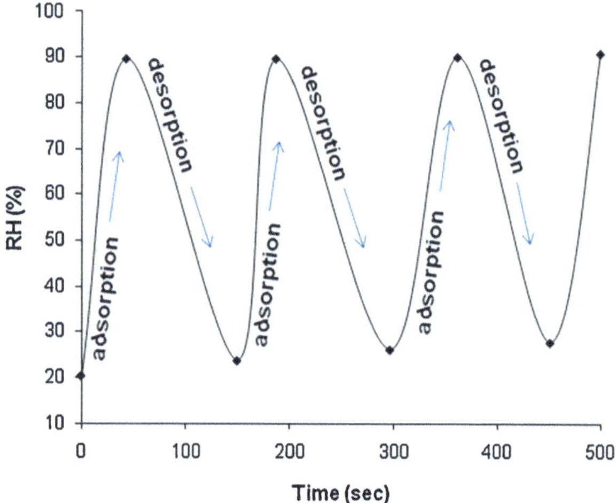

Fig. 6.5 Recovery response of humidity sensor (poly-o-anisidine Sn(VI) arsenophosphate nano-composite cation exchanger). (Reprinted with permission, Khan et al. 2013b)

sensing (adsorption) and recovery (desorption) response of the nanocomposite was measured in the form of electrical conductivity with respect to time at 25 ± 2 °C (as shown in Fig. 6.5). The time response for continuous humidification was found 60 s (from 20 to 90% RH) and recovery time was achieved in 140 s. The long recovery time pointed out towards slow surface desorption process of water molecules at ambient temperature. Therefore, POA-SAP nanocomposite material showed an advantage for humidity sensing applications owing to its quick response and recovering characteristics.

Khan et al., fabricated PANI based electrically conductive polyaniline-titanium (IV)molybdophosphate (PANI-TMP) cation exchange nanocomposite for detection

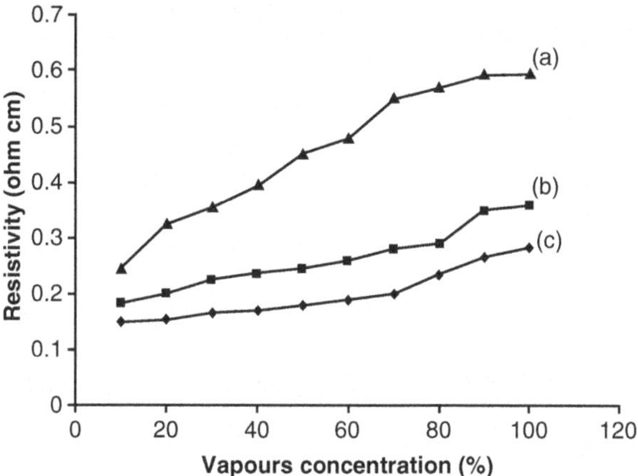

Fig. 6.6 Resistivity response of PANI-TMP nanocomposite towards (**a**) methanol (**b**) ethanol and (**c**) 1-propanol at different concentrations. (Reprinted with permission, Khan et al. 2013a)

of sensitivity towards the series of aliphatic alcohols (methanol, ethanol, and 1-propanol) (Khan et al. 2013a). The nanocomposite was prepared by the mixing of PANI with an inorganic precipitate of titanium (IV)molybdophosphate (TMP). Characterization was of the nanocomposite was performed on the basis of and FTIR, SEM, TEM, XRD, TGA, and UV-Vis. Comparative FTIR spectra of methanol exposed to PANI-TMP confirmed shifting of the quinoid peak from 1579 to 1547 cm^{-1} which attributed to the interaction of methanol with imine nitrogen, thereby causing the reducing effect. Therefore, the effective positive charge on imine nitrogen was reduced by the molecules of methanol. SEM analysis of PANI-TMP nanocomposite demonstrated its granular structure, and TEM image of PANI-TMP nanocomposite revealed spherical morphology which covered particle size in the range of 20–30 nm. The XRD pattern of PANI and TMP showed crystalline nature, however, after binding TMP with PANI, the morphology of nanocomposite became amorphous. TGA analysis of PANI-TMP nanocomposite confirmed its thermally stability than pure PANI and this nanocomposite can be used cover a wide range of temperature. In terms of resistivity, nanocomposite material was tested to examine its applicability as a sensor for aliphatic alcohols (methanol, ethanol, and 1-propanol). The resistivity response of the nanocomposite towards various concentrations of different alcohols has been determined (as shown in Fig. 6.6).

The reversible resistivity response of PANI-TMP for methanol, ethanol, and 1-propanol was also observed at 25 ± 2 °C. It is commonly known that PANI act as a p-type semiconductor; therefore, the exposure of electron-donating gases to PANI result decline in conductivity (Dimitriev 2003). The PANI-TMP sensor showed good reversible response towards methanol vapours compared to ethanol and 1-propanol vapours. These reversible resistivity responses were found to be due to the adsorption and desorption of the alcohol vapours. The reversible response for different concentration of methanol was also examined. It was found that the resistivity response increases with increasing methanol concentration (as shown in Fig. 6.7). The response time for continuous exposure of alcohol vapour was 20 s and

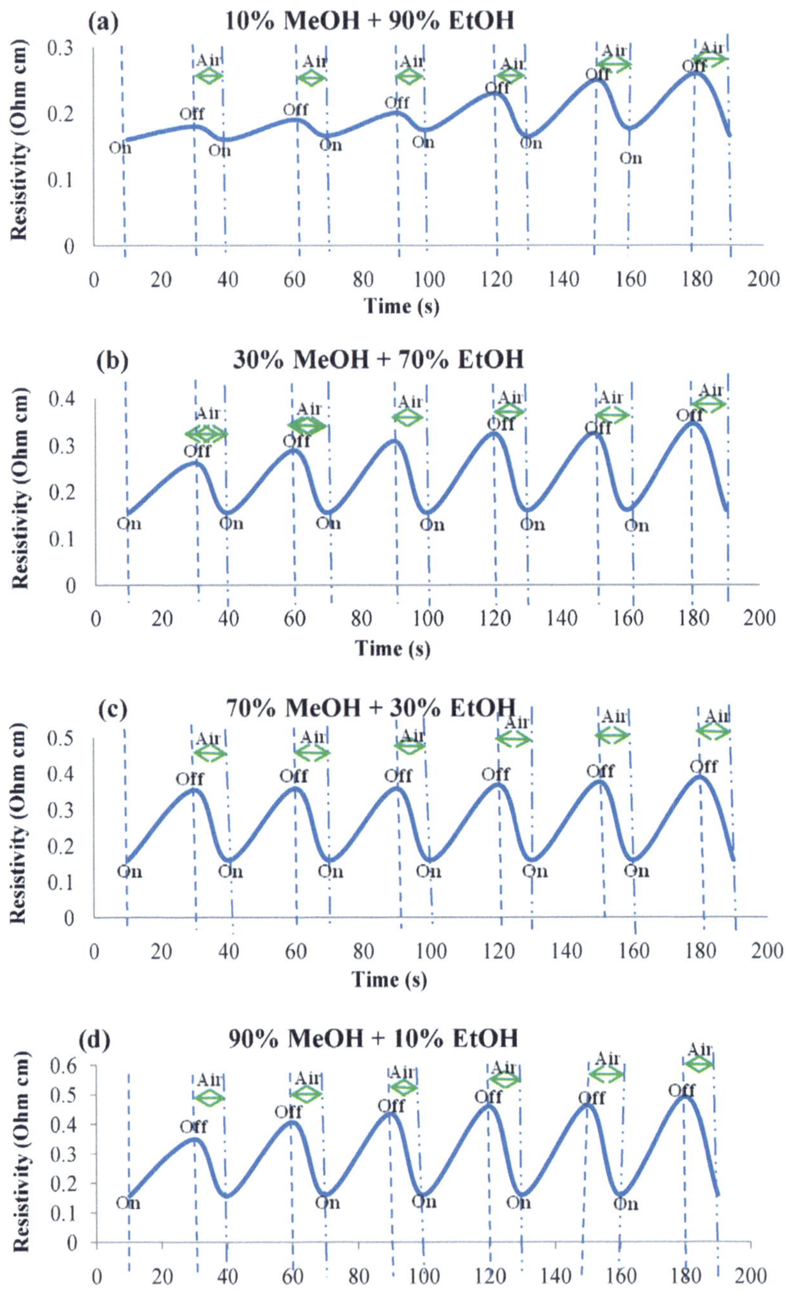

Fig. 6.7 Response transient curves of PANI-TMP nanocomposite towards different concentrations of methanol and ethanol mixture. (Reprinted with permission)

the recovery time out of the gas chamber was 10 s. The alteration in current–voltage data recorder confirmed that ethanol and 1-propanol vapours are less sensitive towards the PANI-TMP which may be due to the low polarity nature of ethanol and 1-propanol (Athawale and Kulkarni 2000). Owing to the small size and more polar nature of methanol molecules it can be efficiently interacted with nanocomposite material than other alcohols molecules (ethanol and 1-propanol). The response at each concentration justified that methanol molecules were found more sensitive and selective for PANI-TMP nanocomposite. PANI based nanocomposite demonstrated good resistivity response towards alcohol vapours; therefore, it can be used as a good sensing material for the detection of methanol vapours at room temperature.

Transition metal-supported (Cu^{2+}, Ni^{2+}, Co^{2+}, Fe^{2+}, and Mn^{2+}) polyaniline–zeolite (PANI-ZSM-5) nanocomposite ion-exchange material was synthesized and characterization was conducted by FTIR, XRD, TGA and SEM analyses (Kaur and Srivastava 2015). Sensing behaviour of PANI-Nano-ZSM-5 was examined for the detection of epinephrine, paracetamol, and folic acid. Among all synthesized materials, Cu^{2+}-PANI-ZSM-5 demonstrated the highest electro-catalytic activity together with excellent stability, sensitivity, and selectivity. Under equilibrium, epinephrine, paracetamol and folic acid covered a wide linear range (10 nM–600 M, 15 nM–800 M and 13 nM–700 M, respectively). The limit of detection for epinephrine, paracetamol, and folic acid was found as 4, 8, and 5 nM, respectively. On the basis of significant performance, the Cu^{2+}-PANI-ZSM-5 sensor can be applied for the determination of epinephrine, paracetamol, and folic acid in the commercial pharmaceutical preparations.

Another HCl doped multi-walled carbon nanotube (MWCNT)/polyaniline (PANI) nanocomposites based Pani@MWCNT nanocomposite ion-exchange sensor was prepared in the presence of surfactant (cetyl-trimethylammonium bromide) using in-situ oxidative polymerization of aniline (Ansari et al. 2014). The stability of the sensor was determined based on electrical conductivity retention under isothermal and cyclic aging conditions (as shown in Fig. 6.8).

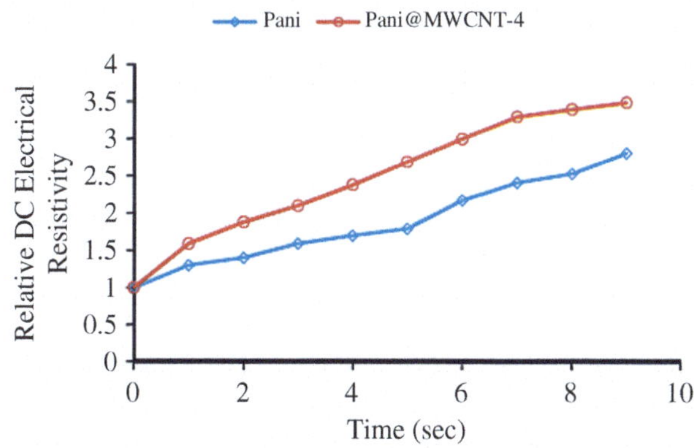

Fig. 6.8 Effect on the resistivity of HCl doped Pani and Pani@MWCNT-4 nanocomposite on exposure to ammonia (0.1 N) with respect time of exposure. (Reprinted with permission, Ansari et al. 2014)

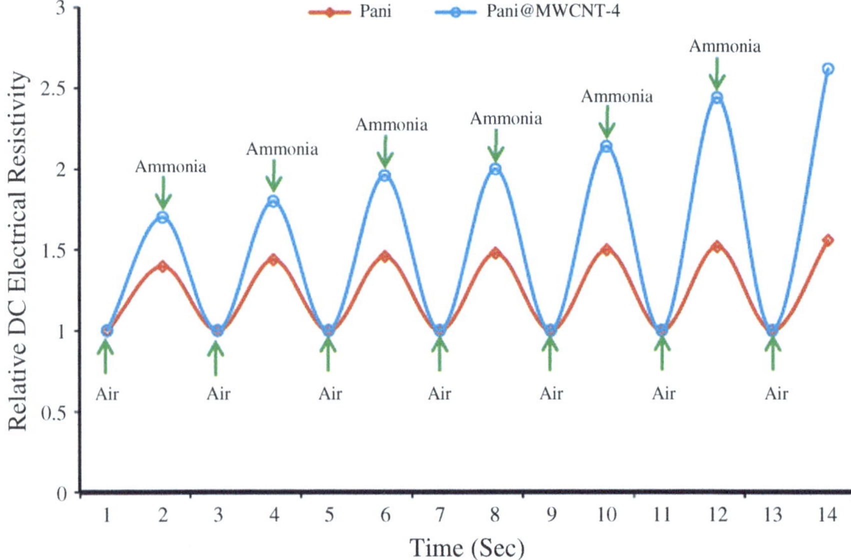

Fig. 6.9 Variation in resistivity of HCl doped (**a**) Pani and (**b**) Pani@MWCNT-4 nanocomposite on intermittent exposure to ammonia (0.1 N). (Reprinted with permission, Ansari et al. 2014)

On the basis of good isothermal and cyclic stability, nanocomposite sensor was tested for detection of ammonia sensing behaviour (Ansari et al. 2014). Relative resistivity of nanocomposite increased linearly on the exposure of ammonia and the resistivity did not reach to its initial value on exposure under ambient condition (Fig. 6.9). Excellent sensing response of Pani@MWCNT-4 can be attributed to the high surface area of PANI coated over MWCNTs. A proposed mechanism related to sensing behaviour of nanocomposite for the detection of ammonia vapor is shown in Fig. 6.10. Abudllah et al., also synthesized PANI/MWCNTs based nanocomposite for trace level detection of ammonia (NH_3) gas (Abdulla et al. 2015). The gas sensor behaviour of C-MWCNT and PANI/MWCNTs nanocomposite for detection of NH_3 gas at trace level (2–10 ppm) under ambient conditions were determined, and their performances were compared. It was examined that in terms of response and recovery time, PANI/MWCNTs nanocomposite based sensor showed significant sensitivity towards NH_3 gas in few seconds (6 s), which confirmed potential application of PANI/MWCNTs for detection of NH_3 in trace level.

A novel layer-by-layer (LBL) deposition technology was used for the fabrication of layered double hydroxide (LDH)/conductive polymer multilayer films by the alternate assembly of exfoliated ZnAl-LDH nanosheets and polyaniline (PANI) on silicon wafer substrates (Xu et al. 2013a). UV–vis absorption spectroscopy measured the regular growth of the (LDH/PANI)n multilayer films upon increasing deposition cycles. Uniform surface and thickness (2 nm per bilayer) of the nanocomposite was determined by SEM and AFM analyses. Multilayer films of $(LDH/PANI)_n$ nanosheets demonstrated a highly selective response to ammonia at

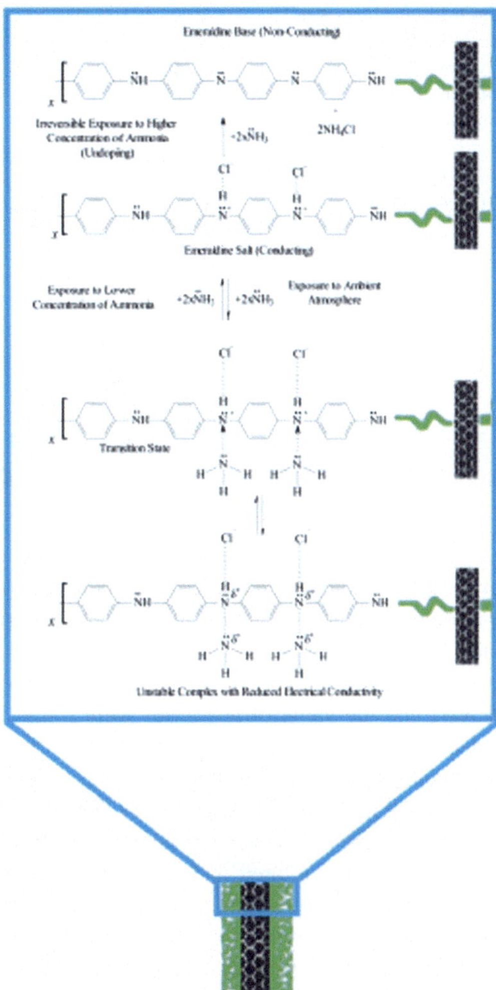

Fig. 6.10 Proposed mechanism for the ammonia vapor sensing for Pani@MWCNT nanocomposites. (Reprinted with permission)

room temperature (25 ± 2 °C) due to the existence of LDH in nanosheets. It can be seen that all the (ZnAl-LDH/PANI)$_n$ multilayer films have a reversible response to NH_3 at room temperature. The experiment was conducted by exposing the multi-layer films for 90 s in the air, and the air was replaced by 1000 ppm ammonia and kept it for 450 s before the environment was switched back to air. The response values to ammonia (1000 ppm) increased from 14 to 16 by increasing bilayers numbers (from 12 to 30). The response and recovery time of (ZnAl-LDH/PANI)$_{12}$ at room temperature were found to be 120 and 150 s, respectively, however, for (ZnAl-LDH/PANI)$_{30}$, though the response increases from 14 of (ZnAl-LDH/PANI)$_{12}$ multilayer film to 16, the response/recovery are prolonged to 150/240 s (as shown in Fig. 6.11).

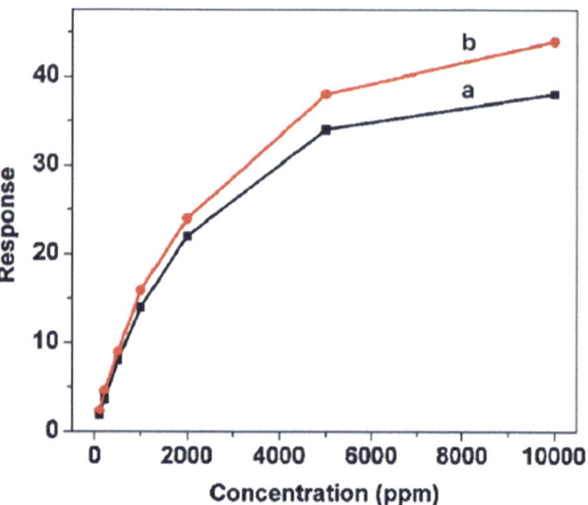

Fig. 6.11 Responses of (ZnAl-LDH/PANI)n multilayer film to different concentration of ammonia at room temperature ((**a**) n = 12 and (**b**) n = 30). (Reprinted with permission, Xu et al. 2013a)

The response of gas sensors can be changed by altering the concentration of ammonia at a constant temperature. In addition, multilayer films can also be detected ammonia gas below100 ppm, and the corresponding values were found to be as 1.8 and 2.4 for (ZnAl-LDH/PANI)$_{12}$ and (ZnAl-LDH/PANI)$_{30}$, respectively. The responses of multilayer films surprisingly improved with increasing concentration of NH_3 and the value of responses were found to be 3.6, 8, 14, 22, 34 and 38 for (ZnAl-LDH/PANI)$_{12}$ multilayer film to 200, 500, 1000, 2000, 5000 and 10,000 ppm ammonia, and the corresponding values for (ZnAl-LDH/PANI)$_{30}$ multilayer film are 4.6, 9, 16, 24, 38 and 44, respectively, Thus, LDH nanosheets play an essential role to improve the gas sensing properties of PANI because ZnAl-LDH nanosheets provide a rigid, confined and stable microenvironment. The uniform orientation and a suppressed aggregation of nanosheets may increase the void for reaction with ammonia molecule, and the ammonia sensing response primarily enhanced. To determine selective behaviour of (ZnAl-LDH/PANI)$_{12}$ and (ZnAl-LDH/PANI)$_{30}$ multilayer film with respect to different gases as (NO_2, CO, H_2, CH_4, C_2H_2 and C_2H_5OH) upto 10,000 ppm, (which is ten times concentration of ammonia), their responses were compared with the response of the sensor toward ammonia (Fig. 6.12). It was observed that both multilayer films demonstrated high response towards ammonia, however, low responses were found for NO_2, CO, H_2, and no significant response achieved for CH_4, C_2H_2, and C_2H_5OH even with increasing concentration (ten times) of ammonia which indicated that the multilayerfilms of (ZnAl-LDH/PANI)$_n$ could be employed as good candidates for detection of ammonia.

Another polyaniline–titanium dioxide based PANI/TiO_2 nanocomposite was synthesized in the presence of colloidal TiO_2 using in-situ chemical oxidation polymerization approach (Tai et al. 2007). Prepared PANI/TiO_2 nanocomposite was mounted on a silicon substrate covered with interdigital electrodes to fabricate a gas sensor via the self-assembly method. Surface morphology of nanocomposite

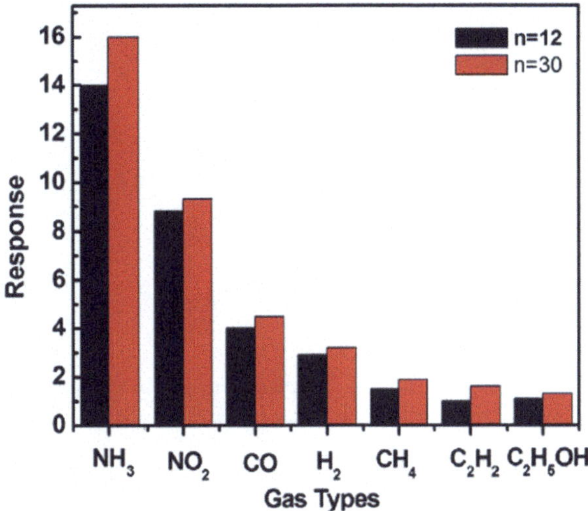

Fig. 6.12 Gas responses of $(ZnAl\text{-}LDH/PANI)_n$ multilayerfilms to 1000 ppm of ammonia and 10,000 ppm of NO_2, H_2, CO, CH_4, C_2H_2 and ethanol at room temperature. (Reprinted with permission, Xu et al. 2013a)

was examined by SEM analysis as shown in Fig. 6.13. It is evident from SEM image that a number of pores on the surface of nanocomposite were found which seem to contribute short response time and good reversibility of the sensors. Therefore, diffusion of gas occurs more readily in porous structures, and the reaction between gas molecules and the thin film occurs rapidly (Matsuguchi et al. 2002). However, the size of pores on the PANI/TiO_2 thin film (a) was much larger than the pure PANI film (b), which significantly improved the diffusion owing to its larger exposure area and penetration depth for gas molecules. Therefore, the response values of the PANI/TiO_2 thin film sensor increased due to its high surface-to-volume ratio to a degree (Sadek et al. 2006).

The material was used to examine gas-responses of NH_3 and CO. It was found that in terms of response, reproducibility and stability, a nanocomposite of PANI/TiO_2 demonstrated sensing behaviour towards NH_3 that was superior to CO gas. It was also found that as compared with NH_3 and CO gases, humidity was found less active on the resistance of the PANI/TiO_2 thin film. On the basis of SEM analysis, the microstructure of the PANI/TiO_2 nanocomposite demonstrated good agreement with the sensor performance. It was revealed that the difference gas sensing properties of pure PANI and PANI/TiO_2 thin films was occurred due to the alternation in surface the morphology.

In-situ oxidation polymerization of m-aminophenol was carried out in the presence of Carboxyl-functionalized multi-walled carbon nanotube (c-MWCNT) to synthesis PmAP/c-MWCNT nanocomposite (Verma et al. 2015). Interaction between the conjugated PmAP chain and the π-bonded surface of c-MWCNT was established by FT-IR and Raman spectroscopy. The surface morphology of the PmAP/c-MWCNT nanocomposite film was characterized by SEM and TEM analyses which demonstrated globular (spherical) microstructure of the PmAP deposited on the surface of c-MWCNT. The globular PmAP particles are consisting of

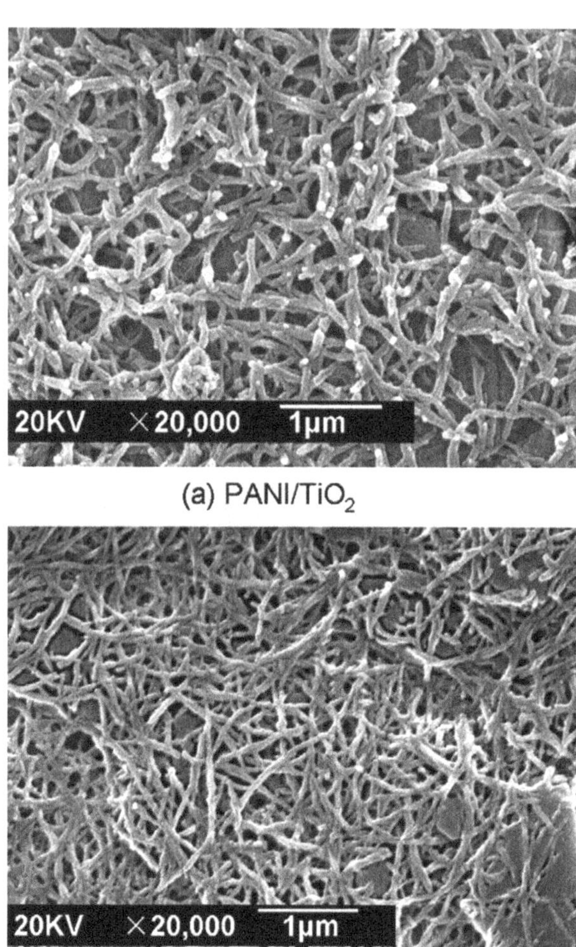

Fig. 6.13 SEM images of PANI/TiO_2 (**a**) and PANI (**b**) thin films. (Reprinted with permission, Tai et al. 2007)

considerable number of polymer chains. The XRD pattern of pure PmAP exhibits a broad amorphous diffraction peak which confirmed the amorphous nature of PmAP owing to periodicity parallel to polymer chain (Fig. 6.14). In addition, the PmAP/c-MWCNT nanocomposites demonstrated several sharp diffraction peaks that symbolized the crystalline nature of the PmAP. On increasing concentration of c-MWCNTs, the electrical resistance of the PmAP is significantly decreased to 11.6, 1.9, 0.25, and 0.06 kΩ cm for the PmAP/c-MWCNT nanocomposite containing 0.5, 1.0, 1.5 and 2.0 wt% c-MWCNT, respectively. The increase of conductivity can be attributed to the strong π–π* electron/H-bonding interactions between the c-MWCNTs and PmAP chains, leading to the formation of highly ordered crystalline nature of PmAP matrix that in turn decreased the resistance of the nanocomposite.

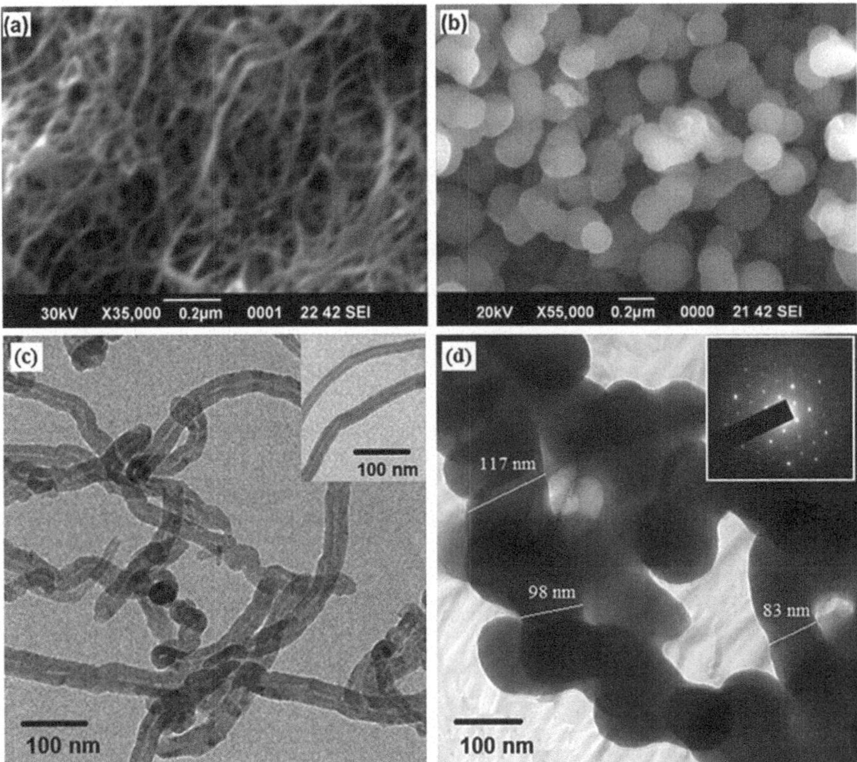

Fig. 6.14 SEM (**a** and **b**) and TEM images of carboxyl-functionalized MWCNTs and 2 wt% c-MWCNT contained PmAP/c-MWCNT nanocomposite. Inset of the figure (**d**) represents electron diffraction pattern of the nanocomposite. (Reprinted with permission, Verma et al. 2015)

The incorporation of c-MWCNTs with the matrix of PmAP matrix substantially improve the alcohol sensing potential of PmAP at relatively lower vapor concentration consequently the synergic effects of the components. Increasing concentration of c-MWCNTs enhanced the dynamic responses of different PmAP/c-MWCNT nanocomposites toward methanol and ethanol vapor at 90 ppm. The absorption of alcohol molecules successfully altered the electrical conduction of the PmAP/c-MWCNT. The extent of alcohol absorption and subsequent change in electrical resistance were dependent on the incorporation of c-MWCNT in the matrix of PmAP/c-MWCNT.

Oxidative polymerization of aniline was carried out in the presence of finely divided In_2O_3 to synthesize a PANI/In_2O_3 nanocomposite (Sadek et al. 2006). Prepared nanocomposite material was used to develop a surface acoustic wave (SAW) sensor was developed for the examination of sensing behaviour of different gases. The dynamic response to a sequence of various gases (H_2, CO, and NO_2) by varying concentrations in the synthetic air was determined (as shown in Figs. 6.15, 6.16 and 6.17). In synthetic air, the measured sensor responses were found

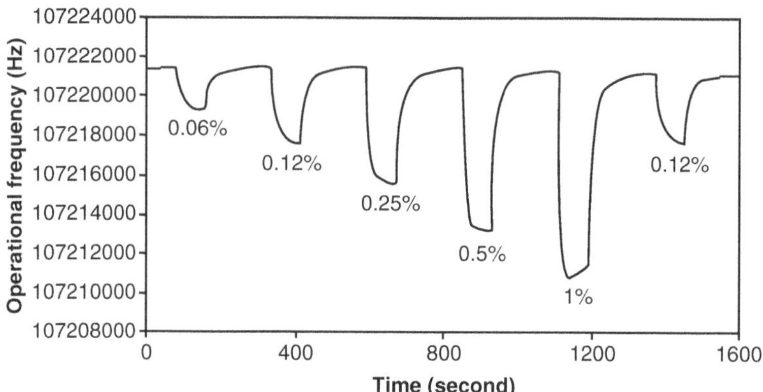

Fig. 6.15 Dynamic response of the SAW sensor towards different concentrations of H_2 at room temperature. (Reprinted with permission, Sadek et al. 2006)

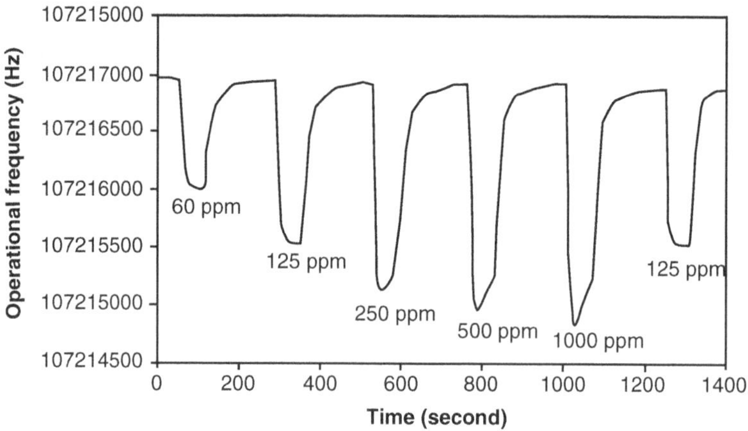

Fig. 6.16 Dynamic response of the SAW sensor towards different levels of CO at room temperature. (Reprinted with permission, Sadek et al. 2006)

approximately 11.0, 2.0 and 2.5 kHz towards 1% H_2, 500 ppm CO and 2.12 ppm NO_2, respectively. In addition, 90% response times; 30, 24 and 30 s was achieved for H_2, CO, and NO_2, respectively and the corresponding 90% recovery times were 40, 36 and 65 s. It was also found that, as compared to high concentration, the slower response was achieved at low density for all gases (H_2, CO, and NO_2). Thus, a high gas exposure time is required to get a stable sensing response (frequency shift) at low concentration. Reproducibility test was also performed to examine repeatability of nanocomposite by applying a second pulse of 0.12% H_2, 125 ppm CO and 510 ppb NO_2 into the sensor chamber which indicated that except for NO_2, PANI/In_2O_3 nanofibre based sensor has the potential to generate repeatable responses of the same magnitude with good baseline stability for H_2 and CO gases.

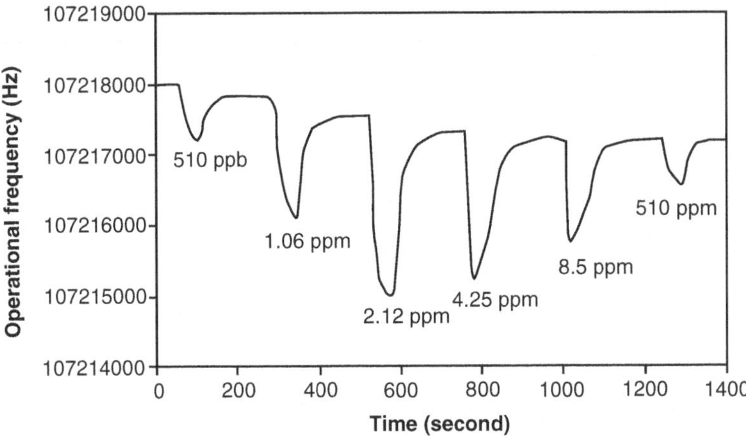

Fig. 6.17 Dynamic response of the SAW sensor towards different concentrations of NO_2 at room temperature. (Reprinted with permission, Sadek et al. 2006)

Reproducibility was observed as indicated when the second pulse of 0.12% H_2, 125 ppm CO and 510 ppb NO_2 was introduced into the sensor chamber. The PANI/In_2O_3 nanofibre based sensor produces repeatable responses of the same magnitude with good baseline stability for H_2 and CO gases, however, not for NO_2. Repeatability of the nanocomposite was confirmed by testing the sensor continuously over a period of 5 days. It was found that from the dynamic response curve of gases (frequency shift) increases almost linearly by increasing concentration of H_2, however, the frequency shifts for CO and NO_2 gases covered a nonlinear trend. For NO_2 gas, sensor response was not repeatable, and baseline seems to drift downward (at 4.25 and 8.5 ppm NO_2 concentration). The sensor response magnitude was found lower at 2.12 ppm which confirmed that high level of NO_2 gas is poisoning for the nanocomposite film. Thus, on the basis of fast response and recovery times with excellent repeatability, PANI/In_2O_3 nanocomposite could be used as a good sensing material for detecting H_2, CO and NO_2 gases at room temperature.

Ultrathin films of conducting polymers; Polyhexylthiophene (PHTh), poly(ethylene dioxythiophene) (PEDT), PHTh–PEDT copolymer, sulfonated polyaniline, polyaniline (PANI)–SnO_2, polypyrrole (PPy)–SnO_2, PEDT-SnO_2, PHTh–SnO_2 and copolymer (HTh-EDT)–SnO_2 were fabricated to examine sensing behaviour of gases (Ram et al. 2005a). The physical properties of films were investigated in terms of UV and FTIR analyses before and after the NO_2 gas treatment. The sensitivity of gases was measured in terms of resistance of PHTh, PEDT, PHTh–PEDT (as shown in Fig. 6.18). The value of resistance decreases exponentially upon exposure to NO_2 which established dopant nature of gas for conducting polymer. The films showed excellent reproducibility due to its thin nature. Among all synthesized nanocomposites and its copolymer P(HTh-PEDT), metal oxide nanocomposites films demonstrated the detection potential of NO_2 gas with high sensitivity.

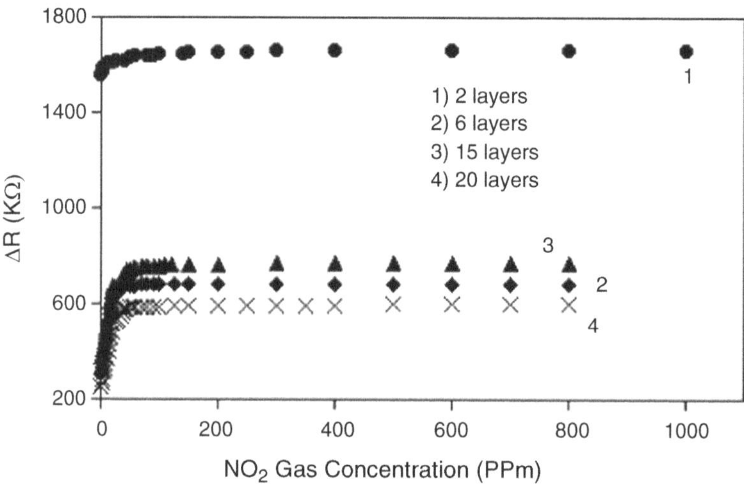

Fig. 6.18 Resistance change vs. NO_2 gas concentration along with a number of layers. (Reprinted with permission, Ram et al. 2005a)

Novel PANI based PANI/Mn_3O_4 composite material was fabricated to determine its sensing behaviour towards relative humidity. The material was doped with perchloric acid, sulphuric acid, *ortho*-phosphoric acid, acetic acid and acrylic acid to examine alteration in sensitivity with respect to relative humidity (Singla et al. 2007). On increasing relative humidity, the resistance of each polyaniline/Mn_3O_4 increased. The response of each composite was found to be almost linear with the percentage of relative humidity. However, the composites doped with organic acids showed more sensitivity than those with inorganic dopants. Under the identical conditions, these samples demonstrated reverse behaviour, and the value of resistance decreased with increasing relative humidity. This anomalous behaviour of the acid doped composite material may be due to lowering p-type nature of the polymer chains with the enhancement in humidity. Therefore, owing to the improvement in the resistance of composite materials doped with organic acids, it can act as a humidity sensor.

Titanium and polytetrafluoroethylene (Ti–PTFE) nanocomposite thin films were synthesized on glass substrates by the combination of dc and rf magnetron sputtering (Rujisamphan et al. 2016). Prepared Ti–PTFE composites below the percolation threshold (i.e., 27% metal volume filling (F)) produced Ti clusters with the average sizes of 7 $\pm$ 2 nm. The percolation threshold (F = 62%) of nanocomosite increased with increasing the % of Ti content, the connecting regions of Ti were formed within the polymer matrix. The electrical conductivity of nanocomposite changed rapidly from insulator-like to metal-like properties. The Ti–PTFE composites prepared near the percolation threshold showed sensitivity in terms of electrical response towards different volatile organic compounds (VOCs). The sensitivities of Ti–PTFE device to acetone and ethanol as a function of VOCs concentration are shown in Fig. 6.19. It was observed that the sensitivity of acetone and ethanol increased with increasing

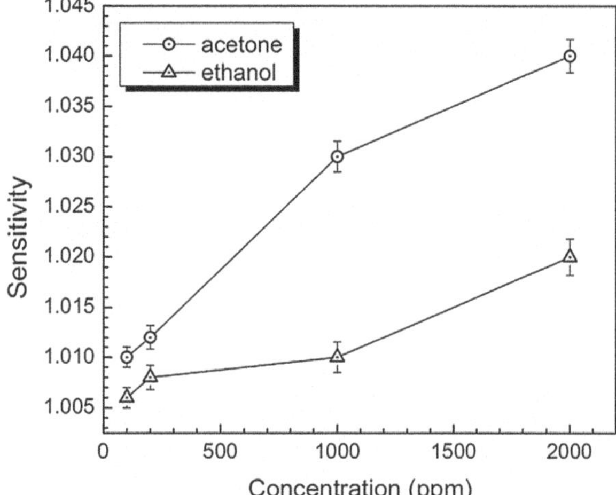

Fig. 6.19 Sensitivity of Ti–PTFE composite films for VOCs (acetone and ethanol). (Reprinted with permission, Rujisamphan et al. 2016)

VOCs concentration leading to higher changes in resistance. The sensing behaviour of nanocomposite towards acetone vapour was found to be higher than that of ethanol which may be due to different degrees of swelling on the polymer matrix. Thus, a number of factors including chemical bonding of the polymer, degree of crosslink, the concentration of metal nanoparticles including chemical reaction at the surface affect the degree of swelling. Other physico-chemical property namely; chemisorption on the TiF/TiO_2 clusters on the surface and a change in a composite structure might play significant role in the swelling mechanism. The sensitivities towards acetone vapor were found to be in the range of 1.01–1.04 corresponding to concentrations between 100 ppm and 2000 ppm. The XPS analysis confirmed the existence of chemical bonds (C C, C CF, C F and CF_2 and TiF) on the surface of TiO_2. Surface morphology was evaluated by AFM analysis demonstrated RMS surface roughness of nanocomposite as 13.3 nm. The particle size of Ti clusters dispersed in the PTFE matrix was found in the range of 10–30 nm (using TEM analysis).

Sensing behaviour towards the water, methanol, toluene, were investigated by synthesized carbon nanotube (CNT) grafted with poly(ε-caprolactone) (PCL) by layer by layer spray from PCL–CNT solutions (Castro et al. 2009). Grafting of ε-caprolactone on the CNT surface through in-situ ring opening polymerisation was performed usingNMR analysis after solvent extraction of ungrafted chains. The mechanism of grafting between the hydroxyl groups of the CNT surface and the aluminium-based catalyst leads to the formation of an intermediate compound. It has a potential to initiate a ring-opening polymerisation of **ε**-CL for the synthesis of the CPC in which PCL chains were grafted on CNT via covalent bonding (as shown in Fig. 6.20).

AFM analysis evaluated CNT coating and dispersion level. It was observed that commercial PCL led to less conductive CPC as compared to synthesized PCL which may be due to the low molar mass PLCs chains which can more effectively interact

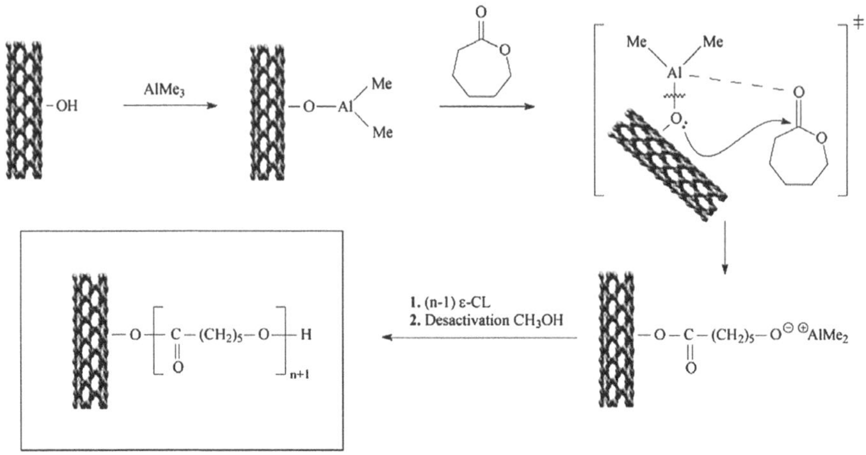

Fig. 6.20 Schematic representation of different steps leading to PCL-g-CNT CPC. (Reprinted with permission, Castro et al. 2009)

with PCL-g-CNT, leading to good dispersion and finally stabilizing the conducting network of nanocomposite. In terms of signal sensitivity, selectivity, reproducibility, and stability, the electro-chemical properties of Conductive Polymer Composite (CPC) sensors were analyzed towards sensing behaviour of water, methanol, toluene, tetrahydrofuran and chloroform vapours which resulted in the improvement in the sensor response using PCL grafted CNT (PCL-g-CNT). It was observed that for methanol vapour, CPC responses showed low amplitude below 1, whereas, it could reach up to 6 or even more than 100 for toluene and chloroform vapours, respectively. The ranking of the resistance relative amplitude (A_R) for all CPC followed the order: water $<$ methanol $<$ toluene $<$ tetrahydrofuran $\ll$ chloroform. Thus, CNT based PCL-g-CNT can be employed as a smart sensing material towards the sensitivity of water, methanol, and toluene.

Polyaniline-tungsten based nanocomposite materials with honeycomb like morphology was synthesized using in situ one pot chemical oxidative method (Chaudhary and Kaur 2015). A cost-effective spin coating technique was employed to fabricate sensing device for the monitoring of SO_2 at levels as low as 5 ppm at room temperature. FTIR and UV-Vis. established the formation of heterojunctions which led the enhancement in sensing response of hybrid nanocomposite. The sensing response of the nanocomposite based sensing device was found to be higher (~10.6%) as compared to its parent devices (~4% for polyaniline nanofibres and is negligible for tungsten oxide nanostructures) towards SO_2 (10 ppm). Enhancement in sensitivity may be due to the optimum porosity, branched structure and formation of heterojunctions of the nanocomposite. Gas sensing behaviour of PANI was found to be entirely different in comparison to the inorganic semiconductor. In PANI, the mechanism of SO_2 sensing may be explained regarding charge compensation between the p-PANI and the electron acceptor (SO_2). The electrons were transferred from the p-PAN to the electron accepting SO_2 by the interaction of PAN with SO_2 which decline the resistance of PANI. Alternatively, metal oxides (WO_3), grains are always covered with adsorbed oxygen (O_2) molecules which captured electrons from the conduction band of the metal oxide, resulting formation of a chemisorbed oxygen species (O_2^-, O^- and O^{2-}) depending upon the sensor's operating temperature. Below 100 °C, the oxygen ions exist in O_2^-, however, other forms (O^- and O^{2-}) was also achieved above 100 °C. By exposing nanocomposite material to sulphur dioxide gas, the molecules of SO_2 extract electrons from the conduction band of WO_3, and react with O_2 which increases the width of the depletion layer and a decreased the conductivity of sensor. It was the first attempt to use PANI-WO_3 nanocomposite as a sensor for monitoring SO_2 at room temperature. Thus, PANI based cost effective nano-hybrid sensor has the potential to consume low power can be easy to fabricate and possesses all the essential characteristics of a reliable sensing device.

Polypyrrole base; PPy/a-Fe_2O_3 hybrid nanocomposite was synthesized by varying weight percentages (10–50%) of CSA (Navale et al. 2014). Nanomaterial was deposited on glass substrates using a spin coating technique to employ it as a sensing material. Morphological characterization of material was performed using XRD, FESEM, and TEM analyses (Fig. 6.21). Thin films of CSA doped PPy/aFe_2O_3 nanocomposite were examined towards the sensing behaviour of NO_2, Cl_2, NH_3, H_2S, CH_3OH and C_2H_5OH gases. Among all gases, nanocomposite material was found to be highly sensitive and selective for NO_2 gas at room temperature (i.e., with a chemiresistive response of 64% at 100 ppm with a reasonably fast response time of 148 s). FESEM micrograph of PPy/a-Fe_2O_3 hybrid nanocomposite (30% CSA doped) showed spherical granular morphology. The high porosity of PPy/a-Fe_2O_3 provides a high surface area to volume ratio and gas diffusion. Therefore, increasing the reaction between gas molecules and the surface of the films excepted to be a higher sensing responses (Tai et al. 2007; Wang et al. 2012). TEM image of PPy/a-Fe_2O_3 nanocomposite thin film demonstrated approximately spherical shaped nanoparticles in aggregated form. The dark shaded a-Fe_2O_3 nanoparticles were embedded into a light shaded matrix of PPy with an average diameter of 32 nm.

Gas sensing response of PPy/aFe_2O_3 hybrid nanocomposite thin films (30% CSA doped) for the detection of NO_2 gas with different concentrations (5–100 ppm) at 25 ± 2 °C indicated that the gas response increases linearly by increasing level of NO_2 gas (from 5 ppm to 100 ppm) and the reaction of CSA doped PPy/a-Fe_2O_3 sensor was found to 7%, 12%, 14%, 26%, 40%, 55% and 64%. Over 100 ppm NO_2 gas concentration, the response of the sensor was remained constant, which confirmed the equilibrium state of the nanocomposite sensor (Hsu et al. 2008). On the basis of the above finding, 100 ppm is considered as an optimum or saturated concentration for the adsorption of a NO_2 gas. The higher response of NO_2 gas can be explained on the basis of different surface interactions between active layer of the film and adsorbed gas (NO_2). The lower concentration of NO_2 result in the lower

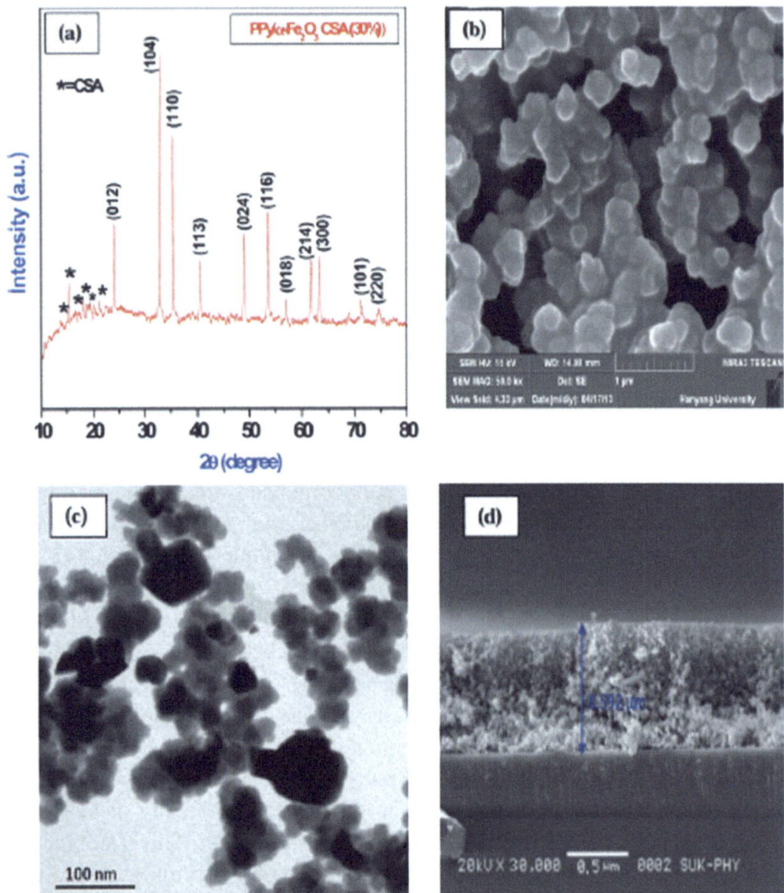

Fig. 6.21 (**a**) XRD pattern of CPF (30%) hybrid nanocomposite, (**b**) FESEM image of CPF (30%) nanocomposite thin film, (**c**) TEM image of CPF (30%) nanocomposite thin film and (**d**) cross-section SEM image of CPF (30%) nanocomposite thin film. (Reprinted with permission, Navale et al. 2014)

surface covered by gas molecules which reduced surface interactions between NO_2 gas molecules and coating of nanocomposite film. Thus, the maximum gas response (64%) was obtained at 25 ± 2 °C by exposing 100 ppm NO_2 gas. The relationship between response and NO_2 gas concentration of 30% CSA doped PPy/a-Fe2O3 hybrid nanocomposite films is shown in Fig. 6.22. The CSA doped PPy/a-Fe2O3 nanocomposites demonstrated a better response, stability and shorter recovery times as compared to PPy and PPy/a-Fe_2O_3 nanocomposites alone. Therefore, PPy/a-Fe_2O_3 is considered as an excellent gas sensor at room temperature which could be used for selective detection of NO_2 sensors. PANI based nanocomposite ion-exchange materials and their sensing applications is shown in Table 6.1.

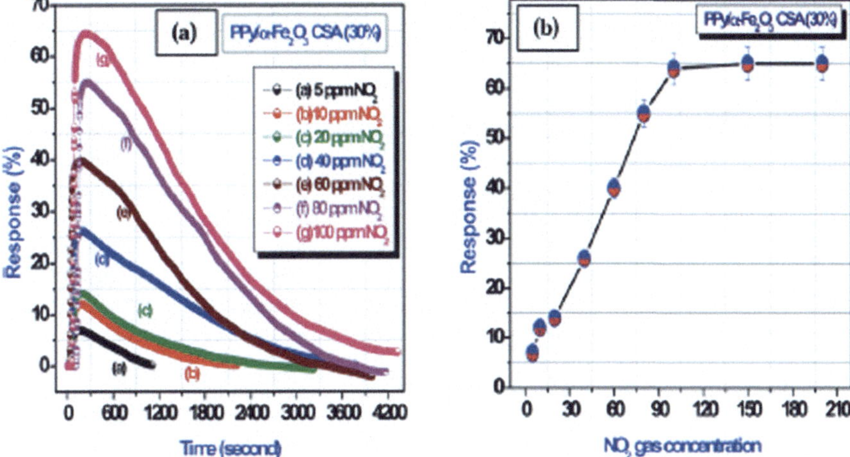

Fig. 6.22 (**a**) Response to 30% CSA doped PPy/a-Fe_2O_3 thin film for various concentrations of NO_2 gas and (**b**) relationship between response and NO_2 gas concentration to 30% CSA doped PPy/a-Fe_2O_3 thin film. (Reprinted with permission, Navale et al. 2014)

6.3 PANI Based Nanocomposites and Their Sensing Applications in Diverse Fields

6.3.1 Detection of Volatile Organic Compounds (VOC)

A number of volatile organic compounds (VOC) have a wide range of applications which are generally used in the form of solvents, preservatives and antiseptics and fuels. However, their inhalations (in the form of vapour) have serious health effects. In this regards, detection of alcohol vapours using PANI-metal oxide based inclusions PANI-Pd nanocomposite has been synthesized (Athawale et al. 2006). The material was used for the sensing behaviour of methanol, and the results revealed that nanocomposite material covered a very high sensing response (~10^4 magnitudes) towards the equilibrium level of methanol. FTIR analysis confirmed the inclusions of Pd in the matrix of PANI which act as a catalyst for reduction of imine nitrogen by methanol. However, in a mixture of VOCs, the PANI-Pd nanocomposite selectively detected methanol with an identical magnitude of response which covered longer response time.

Choudhury et al., fabricated Ag nanoparticles embedded in the matrix of PANI for the detection of ethanol (Choudhury 2009). Nanocomposite materials demonstrated more rapidly protonation–deprotonation responses as compared to pure PANI. In addition, PANI-Ag based nanoparticle have also been used for the sensing behaviour of toluene and triethylamine (Li et al. 2013a). Sensing responses of the material was investigated regarding chemisorption and diffusion mode.

Table 6.1 Some reported PANI based nanocomposite ion-exchange materials

S. no	Nanocomposite	Analysis	Application	References
1.	Polyaniline–titanium(IV)phosphate	FTIR, SEM, TEM, TGA	Sensing of ammonia	Khan et al. (2011)
2.	PANI–TiP	FTIR, XRD, SEM, TGA	Sensing of methanol	Khan et al. (2010b)
3.	V_2O_5@polyaniline nanofiber	SEM, TEM, XRD, TGA-DTA-DSC, XPS, UV–vis.	Sensing of ammonia	Hasan et al. (2015)
4.	P3MTh–TMP	FTIR, SEM, TEM, XRD, TGA	Sensing of ammonia	Khan and Baig (2013a)
5.	PANI/WO_3/GR/GCE	CVs, SWVs, XRD, AFM, Raman spectroscopy, EIS, HRTEM, FTIR	Phenanthrene	Tovide et al. (2014)
6.	Ppy/ZSI	FTIR, XRD, SEM, TEM, EDX,TGA	Formaldehyde	Khan et al. (2013a)
7.	POT-TiP	TEM, XRD, SEM, FTIR	Humidity	Khan and Baig (2013b)
8.	POA-SAP	SEM, TEM, XRD, TGA-DTA	Humidity	Khan et al. (2013b)
9.	PANI-TMP	FTIR, SEM, TEM, XRD, TGA, UV-Vis.	Methanol, ethanol, 1-propanol	Khan et al. (2013a)
10.	PANI-ZSM-5	FTIR, SEM, TEM, XRD	epinephrine, paracetamol, and folic acid	Kaur and Srivastava (2015)
11.	Pani@MWCNT	FTIR, XRD, Raman spectroscopy, SEM, and TEM	Ammonia	Ansari et al. (2014)
12.	PANI/MWCNTs	UV–vis., FT-IR, Raman Spectroscopy, XRD, XPS, HRTEM	Ammonia	Abdulla et al. (2015)
13.	LDH/PANI	UV–vis., SEM, AFM	Ammonia	Xu et al. (2013a)
14.	PANI/TiO_2	SEM,	Ammonia, CO	Tai et al. (2007)
15.	PmAP/c-MWCNT	FTIR, XRD, SEM, TEM	Alcohol	Verma et al. (2015)
16.	PANI/In_2O_3	SEM, TEM, XRD	H_2, CO, NO_2	Sadek et al. (2006)
17.	PHTh, PEDT, PHTh–PEDT/SnO_2	UV, FTIR	NO_2	Ram et al. (2005a)
18.	PANI/Mn_3O_4	–	Humidity	Singla et al. (2007)
19.	Ti–PTFE	SEM, AFM, XPS	Volatile organic compound	Rujisamphan et al. (2016)
20.	PCL-g-CNT	NMR, SEM, AFM	Water, methanol, toluene	Castro et al. (2009)
21.	PANI-WO_3	FTIR, SEM, TEM	CO_2	Chaudhary and Kaur (2015)
22.	PPy/a-Fe_2O_3	XRD, FESEM, TEM	NO_2	Navale et al. (2014)

Nanomaterial showed sensing responses towards triethylamine and toluene, however, deprotonation of PANI and swelling of the polymer results drop in its sensitivity.

Chloroform is an anaesthetic widely used solvent and its inhalation cause depression of the central nervous system. The monitoring of chloroform was reported by the synthesis of PANI-Cu nanocomposite (Sharma et al. 2002). An adsorption–desorption mechanism was adequately described at the surface of metal clusters. The FTIR analysis established interaction of chloroform with metal clusters of the composite. It was observed that PANI supported PANI-Cu demonstrated excellent performance in the detection of chloroform.

Besides the detection of methanol, ethanol, and chloroform, PANI based nanomaterials have also been used as sensors for the detection of reducing gases, namely NH_3 and H_2S which generally depend on deprotonation of PANI. Deprotonation of PANI was achieved by exposing reducing gases through electron donation which result in the enhancement of PANI resistance. However, the addition of metal impurity improves the sensitivity of the sensor. Detection of gases using PANI supported metal oxide nanocomposite have been extensively studied. It was found that addition of metal with PANI forms a p-n heterojunction with a depletion layer. Owing to the formation of the depletion region, nanocomposite materials demonstrated adsorption of gases in the form of alternation in electrical responses. The change in electrical conductivity of nanocomposites altered its sensing behaviour of nanocomposite towards the gases. Thus, PANI-TiO_2 nanocomposite was fabricated by spin coating technique for detection of NH_3. Nanocomposite sensor was examined by using different concentrations of NH_3 (Pawar et al. 2011). The thin film of nanocomposite sensor exhibited good sensing response towards NH_3 below 20 ppm. Sensing characteristic of the sensor was occurred due to the formation of a positively charged depletion layer at the heterojunction of PANI and TiO_2. On increasing concentration of NH_3, the gas sensing response and recovery time increases simultaneously while response time shows opposite behavior. XRD pattern of PANI changes the morphology of nanocomposite as amorphous nature after addition of PANI. XRD diffraction pattern shows sharp and well defined peaks which confirmed the crystalline nature of nanocomposite as shown in Fig. 6.23.

The SEM image of the PANI film demonstrated fibrous morphology with many pores and gaps among the fibers that covered an average grain size of about 60 nm. The image of the nanocomposite (Fig. 6.24c) exhibits no agglomeration of particle and a uniform distribution of the TiO_2 particles in the matrix of PANI which established that the particle of TiO_2 is embedded within the netlike structure built by PANI chains (Fig. 6.24).

A similar finding for the detection of NH_3 reported by Dhingra et al. (2013). However, a different observation was reported by Deshpande and coworkers who studied the response of PANI-SnO_2 nanocomposite towards NH_3 (Deshpande et al. 2009). They observed that pure PANI was found to be reduced in the absence of NH_3 while the nanocomposite showed an oxidative in the presence of NH_3. The I–V characteristic of the nanocomposite, shown in Fig. 6.25c, revealed a diode-like character associated with electrical conductance through the hopping mechanism. Due to the existence of n-type SnO_2 in the matrix of PANI, the formation of

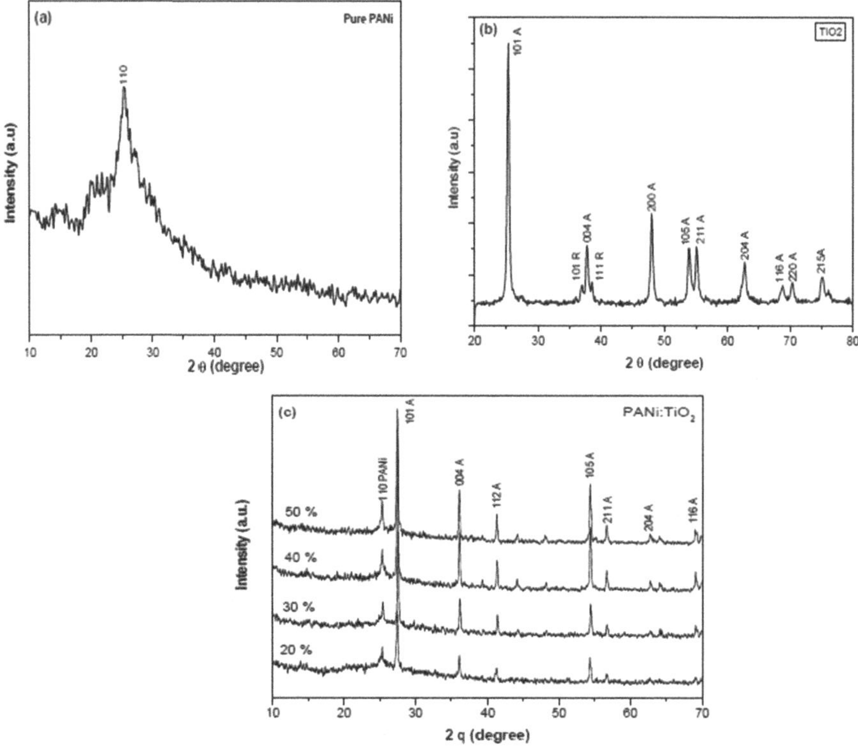

Fig. 6.23 X-ray diffraction patterns of Pure PANi (**a**), TiO_2 (**b**) and PANi/TiO_2 (0–50 wt) (**c**). (Reprinted with permission, Pawar et al. 2011)

localized p–n heterojunction carried out which make PANI-SnO_2 nanocomposite electrically more insulator. Exposure to NH_3 gas resulted in polarization of NH_3 molecules by the depletion region, which provided positive charge on the surface of PANI and the mobility of these charges along the PANI chain made the nanocomposite relatively more conducting. Thus, the formation of p–n heterojunction between PANI and nano-metal oxide plays an essential role as reported by other researchers (Patil et al. 2011; Khuspe et al. 2013).

Aligned poly(methyl methacrylate) (PMMA) fibers were synthesized via electrospinning, and the chain of PANI was grown on the surface of the PMMA fibers to develop PANI/PMMA nanocomposite (Zhang et al. 2014). PANI/PMMA nanocomposite was employed for trace level detection of NH_3 in the concentration as low as 1 ppm. Morphology of PANI/PMMA composite was examined by SEM and Raman spectrometry. Sensing behaviour study revealed that arranged fibers of PANI/PMMA exhibited high sensitivity and fast response upon exposure to ammonia vapor of 1–30 ppm. The high sensitivity response of nanocomposite was accredited due to the highly aligned PMMA microfibers coated with PANI which facilitated faster diffusion of gas molecules, thereby, accelerating electron transfer.

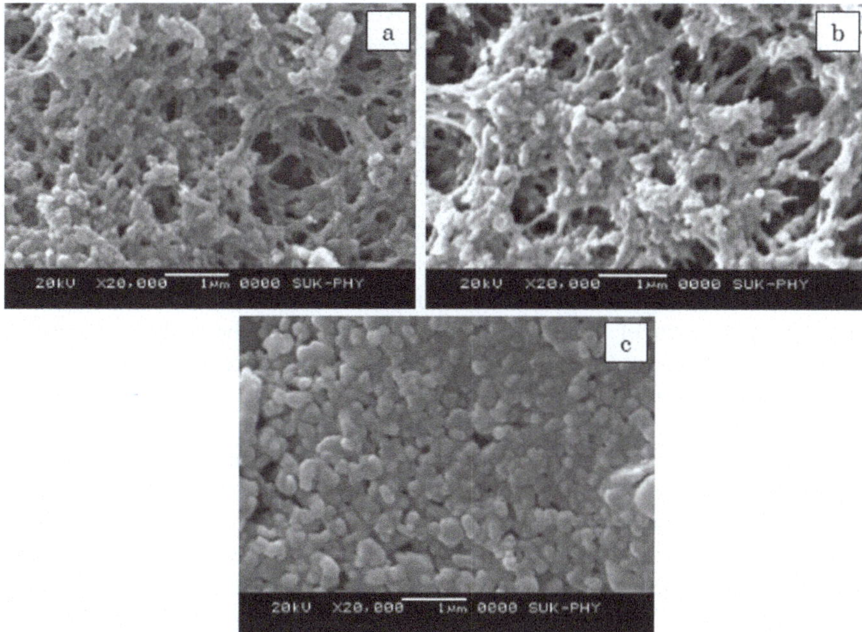

Fig. 6.24 Scanning electron micrographs of (**a**) PANi, (**b**) TiO_2 and (**c**) PANi/TiO_2 (50 wt) films. (Reprinted with permission, Pawar et al. 2011)

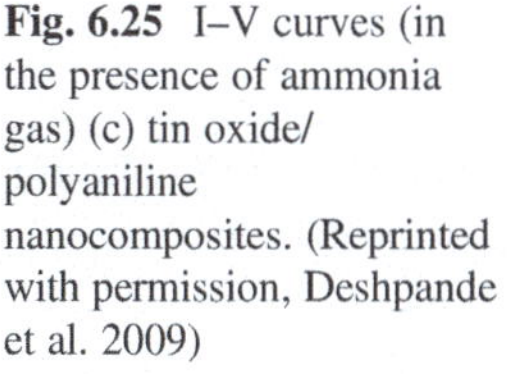

Fig. 6.25 I–V curves (in the presence of ammonia gas) (c) tin oxide/polyaniline nanocomposites. (Reprinted with permission, Deshpande et al. 2009)

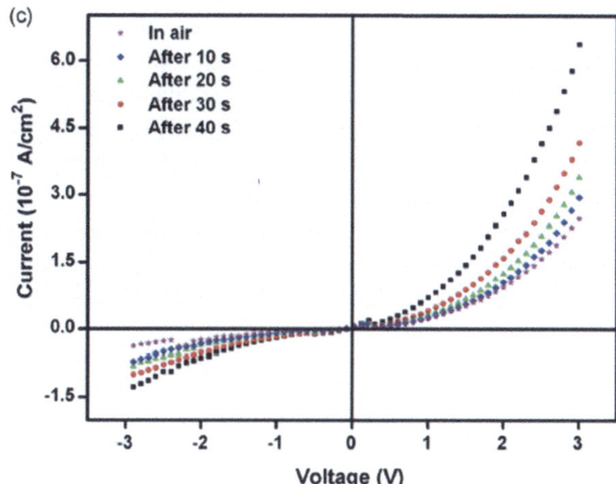

Another porous thin film nanocomposite of PANI/sulphonated Ni phthalocyanine (PANI-NiTSPc) was fabricated for the detection of NH_3 gas (Zhihua et al. 2016). Ammonia gas-sensing behaviour of nanocomposite films was examined under optimum conditions at room temperature. The observed response value (S) of the composite film up to 100 ppm NH_3 was found as 2.75 in short response time (10 s).

Over a concentration range of 5–2500 ppm, PANI/NiTSPc films demonstrated a fast recovery rate, good reproducibility, and acceptable long-term stability. The outstanding sensing performance was achieved because of porous, ultra-thin film structure and the "NH_3-capture" effect of the "flickering" NiTSPc molecules. Thus, the proposed PANI-NiTSPc nanocomposite thin film sensors were found to be excellent potential candidates for the detection of NH_3.

Crowley and co-workers synthesized PANI-$CuCl_2$ nanocomposites sensor printed on screen printed interdigitated electrodes for the detection of H_2S gas at trace level (Crowley et al. 2010). The sensing material was composed of screen printed silver interdigitated electrode (IDE) on a flexible PET substrate along with inkjet printed layers of PANI and copper (II) chloride. The finding of sensing behaviour explored a different phenomenon in which protonation of PANI nanocomposite occurred during the exposure of H_2S gas. It was found that PANI gets deprotonated by complexation of $CuCl_2$ due to binding of $CuCl_2$ with S^{2-} ion caused the evolution of HCl which improved the electrical conductivity of PANI. On exposure to hydrogen sulphide; 2.5 ppmv (ppm by volume), nanocompoite sensor showed a linear relationship between measured current and concentration over 10–100 ppmv region. Thus, PANI-$CuCl_2$ based sensor act as chemiresistor concerning measured current and concentration. Sarfraz et al. also reported a similar study based on the synthesis of PANI-$CuCl_2$ nanocomposite printed on interdigitated electrodes for H_2S detection (Sarfraz et al. 2013). Raut and co-workers described the synthesis of PANI based PANI–CdS nanocomposite film by a simple spin coating technique and its application for detection of H_2S gas. Nanocompoite sensor achieved a maximum response (~48%) for 100 ppm H_2S concentration. It was also mentioned that the sensing mechanism of the nanocomposite was achieved by the modifications in nanocomposite at the depletion region (Raut et al. 2012).

PANI based PANI-Au nanoparticles nanowires with a diameter of 250–320 nm had synthesized via electrochemical technique (a two step galvanostatic technique). PANI nanowires were then electrochemically functionalized with gold nanoparticles using cyclic voltammetry technique. Prepared nanocomposite sensor was examined for detection of H_2S gas and fast selective, gas response was observed at room temperature. Proposed mechanism for the formation of AuS and subsequent protonation of PANI for H_2S detection by PANI-Au nanocomposites is given by Eq. 6.1).

$$H_2S + Au/AuS + 2H^+ \tag{6.1}$$

It was suggested that donation of electrons to the protons is followed by the reaction Eq. 6.1) which led to the alternation in the resistance of the PANI-Au nanocomposite. The sensing response of PANI-Au nanocomposite and PANI as a function of time towards different concentrations of H_2S gas is shown in Fig. 6.26. These chemiresistive sensors show an excellent limit of detection (0.1 ppb) along with excellent selectivity and reproducibility over a wide dynamic range (0.1–100 ppb).

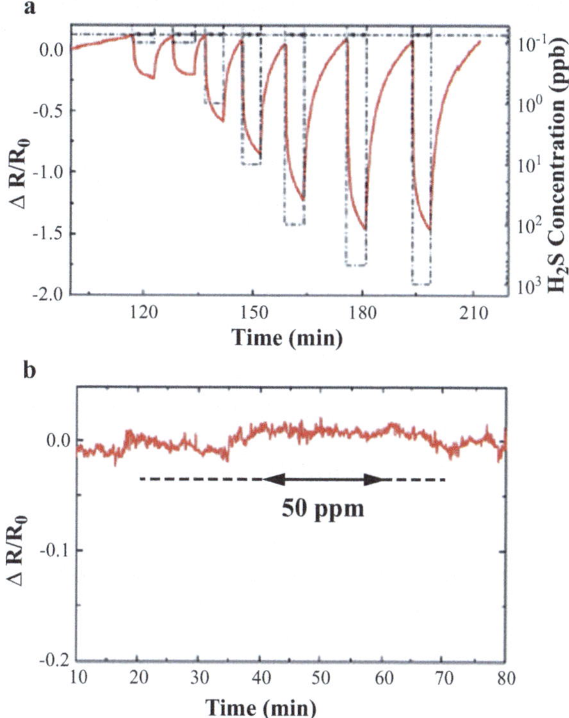

Fig. 6.26 (**a**) Response and recovery transients (solid line) of gold nanoparticles functionalized PANI nanowire network based chemiresistive sensor toward 0.1 ppb, 1 ppb, 10 ppb, 100 ppb, 500 ppb, and 1 ppm (dashed line) concentrations of H_2S gas, (**b**) response of unfunctionalized PANI nanowire network toward 50 ppm of H_2S gas (-- indicates carrier gas and 4 indicates gas analyte) (reprinted with permission) (Shirsat et al. 2009)

6.3.2 LPG Sensing Application

PANI based nanocomposite materials play vital role for safety purposes in the detection of LPG sensing (Joshi et al. 2007). The use of electroactive PANI as a physical transducer at room temperature sensing provides a safer option as opposed to metal oxide based high temperature detectors. In this regard, synthesis of n-CdSe/p-PANI nanocomposite was carried out by a simple electrodeposition technique to examine its sensing response towards LPG. The junction diode was tested to detect the sensitivity of liquefied petroleum gas (LPG) at 25 ± 2 °C. Biased current–voltage characteristics of the junction diode demonstrated a considerable shift by exposing different concentrations of LPG. Nanocomposite sensor showed maximum response up to 70% for 0.08% (by volume) LPG. On the basis of LPG concentration, the response time was found in between the range of 50 and 100 s, while optimal recovery was achieved in 200 s.

A similar finding was reported by the fabrication of PANI/g-Fe_2O_3 nanocomposite to detect sensing of LPG at room temperature (Sen et al. 2014). The nanocomposite films were examined for their response to LPG in the concentration range 50–200 ppm. Morphological characterization was performed by FTIR, UV-vis. XRD, FESEM analyses. The nanoscale morphology of the composites provided a large surface area which enhance the gas sensitivity resulting adsorption of gas molecules,. The sensing mechanism alters the depletion region of the p–n junction between PANI and γ-Fe_2O_3 which result in an electronic charge transfer between the gas molecules and nanocomposite sensor and optimum responses were obtained for 200 ppm LPG together with response times as low as 60 s. Thus, on the basis of above mentioned observations detection of LPG increase in the depletion depth owing to the adsorption of gas molecules at the depletion region of the p–n heterojunction.

A series of PANI based nanocomposite sensors were fabricated by Dhawale and co-workers (Dhawale et al. 2008, 2010a, b). The materials were used for the sensing responses of LPG at room temperature which demonstrated significant selectivity towards LPG as compared to N_2 and CO_2. The improvement in gas sensing response was found due to a change in the barrier potential of nanocomposite heterojunction.

6.3.3 Sensing Behaviour Towards Humidity

Water vapour plays essential role in the electrical conductance of PANI. Therefore a number of PANI supported nanocomposite were fabricated for the sensing behaviour of humidity (Shukla et al. 2012). PANI-ZnO based nanocomposite electrochemical humidity sensor was developed by the wet-chemical method at ambient condition. The nanocomposite sensor (~200 nm thickness) was examined for the sensing of humidity. The results of sensing study revealed that the particle crystalline homogenous ZnO existed in the matrix of PANI which enhanced electrical conductance in the range of 10^{-2} scm^{-1} and covered thermal stability up to 280 °C. The response and recovery time of the sensor was found between 32 and 45 s, respectively which demonstrated better sensing characteristics than pure PANI and other reported humidity sensors. The adsorption of water molecules on the sensor results in efficient directional charge conduction at the heterojunction formed between PANI and ZnO. Another PANI based humidity sensor of nanostructure Co dispersed in polyaniline deposited as a clad; PANI-Co was reported to determine humidity (Vijayan et al. 2008). The sensor exhibited the quick response of 8 s (20–95% RH) and recovery time in 1 min (95–20% RH) on fibre optical waveguide (in the range of 20–100% RH). In another study, PANI-Ag nanocomposite sensor was developed on an optical fiber clad to examine humidity (Vijayan et al. 2008). The sensing behaviour of PANI-Ag was observed in the range of 5–95% relative humidity (RH). It demonstrated sensing response of humidity with complete reversible nature having almost 1% of standard deviation. Response time of the sensor was found in very short period (30 s) and recovery was achieved in 90 s. By decreasing size of nanoparticles, the improvement in the sensor responses was occurred. Such high responses towards humidity have also been observed with other PANI based nanocomposites as well (Parvatikar et al. 2006; Sajjan et al. 2013).

A fast, responsive humidity sensor was designed by the synthesis of silver vanadium oxide nanospheres dispersed in different ultrathin layers of polyaniline (PANI–β-$AgVO_3$) using in-situ oxidative polymerization (Diggikar et al. 2013). The XRD analysis showed a monoclinic crystallographic form of silver vanadium oxide

(SVO) which was dispersed in ultrathin layers of PANI. Morphological characterization was carried out by FESEM, HRTEM, and AFM analyses. FTIR and EDAX spectroscopic investigations confirmed the existence of SVO in the PANI layer. The SVO dispersed in a layer of PANI-NC revealed excellent humidity sensing characteristics. The response and recovery times are found to be 4 and 8 s, respectively. Cavallo and coworkers demonstrated a high level of relative humidity (65–90%) swelling of the polymer due to continuous absorption of water. The sensing property of PANI increased due to its inter chain distance thus, hindering the charge hopping process and decreasing electrical conductance. It is, therefore, imperative that the influence of humidity on the electrical behaviour of PANI and its pathways be investigated (Cavallo et al. 2015).

6.3.4 Analysis of Miscellaneous Gases

PANI also shows low sensitivity towards certain gasses; methane (CH_4) and carbon monoxide (CO), because these gases do not show redox properties at room temperature (Yan et al. 2009). These analytes may not undergo chemical reactions with PANI, however, have weak physical interaction with the matrix of the polymer. The existence of metal/metal oxide nanoparticles in the matrix of PANI improved its sensing response towards these gases. Polyaniline/Indium(III) Oxide (PANi/In_2O_3) nanocomposite thin films was synthesized and fabricated on AT-Cut quartz crystal microbalance (QCM) of Ag electrodes by the combining two techniques (electrostatic self-assembly and in-situ oxidation polymerization) at 10 °C. The material was used for sensing the study of CH_4 and CO gases. The observed results revealed that PANi/In_2O_3 thin film gas sensor demonstrated good linear sensitivity towards CH_4 and CO, while higher sensing behaviour was achieved for CH_4, (i.e., the sensitivity is 0.386 Hz per ppm when the sensor is exposed to 500 ppm CH_4 while only 0.16 Hz per ppm).

Ram et al. have also proposed theory regarding sensing behaviour of gases detection via catalytic pathway for the detection of CO by PANI-SnO_2 nanocomposite (Ram et al. 2005b). Other PANI-SWCNT nanocomposite based sensor was also synthesized which exhibited good sensing response towards CO (Kim et al. 2010). The mobility of these positive charges generated on the amine nitrogen of PANI increases its conductivity. As compared to NH_3 gas which was also included in the study, the sensor showed a propensity towards CO absorption. In most cases, detection of CO by PANI nanocomposite was described using the particle electron transfer model – the stable resonance structure of $^{+}C{=}O^{-}$ withdraws the lone pair of an electron from the amine nitrogen in PANI thus, creating a positive charge on it. The mobility of these positive charges generated on the amine nitrogen of PANI increases its conductivity.

The highly sensitive and rapid NO gas sensor polyaniline/TiO_2/carbon nanotube composites (PANI-TiO_2-MWCNT) was synthesized (Yun et al. 2013). It was used for NO detection through the photocatalytic behaviour of TiO_2. The gas sensing property of nanocomposite was measured by the changing of electrical resistance

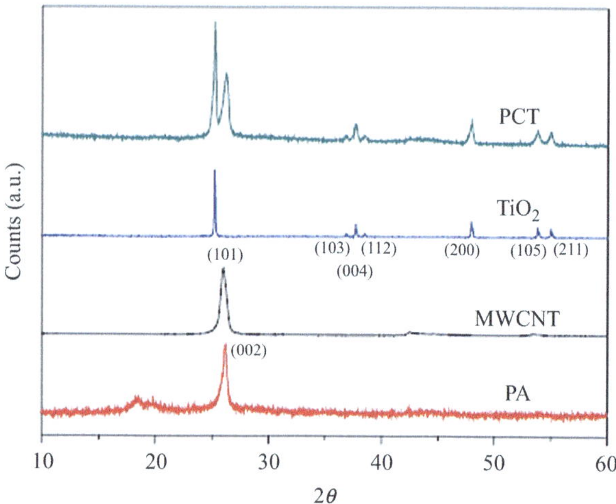

Fig. 6.27 XRD patterns of the prepared samples. (Reprinted with permission, Yun et al. 2013)

without or with UV irradiation for the photo-degradation of NO. The photo-degraded products such as HNO_2, NO_2, and HNO_3, were adsorbed on the PANI-coated carbon nanotubes, resulted decline in electrical resistance due to the p-type semiconductors of carbon nanotube and polyaniline. The XRD pattern of TiO_2 exhibited the characteristics of anatase TiO_2 phase with a tetragonal structure which possessed the photocatalytic properties as shown in Fig. 6.27. Morphological characterization was carried out by FE-SEM analysis. The FE-SEM images of nanocomposite show aggregated PANI phases of the uniform sized spherical type. The average size of PANI spherical phases was measured as 193 ± 27 nm (average value measured five times). The uniformly coated PANI on MWCNTs has observed with the average thickness of the PANI coating layer was around 23 ± 8 nm considering the average diameter of MWCNTs (130 ± 15 nm). PANI was also coated on TiO_2 particles; it was aggregated to some extent as shown in Fig. 6.28c. PCT sample showed the TiO_2 particles dispersed on the PANI-coated MWCNTs as shown in Fig. 6.28d. It was examined that under UV irradiation, the NO gas gets decomposed by the photocatalytic effect of TiO_2 resulted HNO_2, NO_2, and HNO_3. These decomposition products adsorbed onto the surface of PANI/MWCNT owing to its high specific surface area and hydrophilicity of PANI. Such combining effects alter the electrical resistance of the nanocomposite, thus, facilitating detection. Figure 6.29 shows the sensitivities (S) of the nanocomposites towards NO under UV irradiation. PANI/TiO_2/MWCNT nanocomposite (PCT) exhibited a high response of NO, while PANI (PA) showed the least. PANI-MWCNT (PC) and PANI-TiO_2 (PT) composites showed better response as compared to its components.

PANI has also been used for the detection of a number of gases including NO_2 and H_2 supported nanocomposites (Xu et al. 2013b; Srivastava et al. 2010). Introduction of CNT or metal oxides into PANI have shown response towards H_2 gas. Interdigitated electrodes based on PANI-TiO_2 and PANI/MWCNTs were used for the detection of H_2 gas with both of the nanocomposite materials demonstrated eliciting a high sensing response.

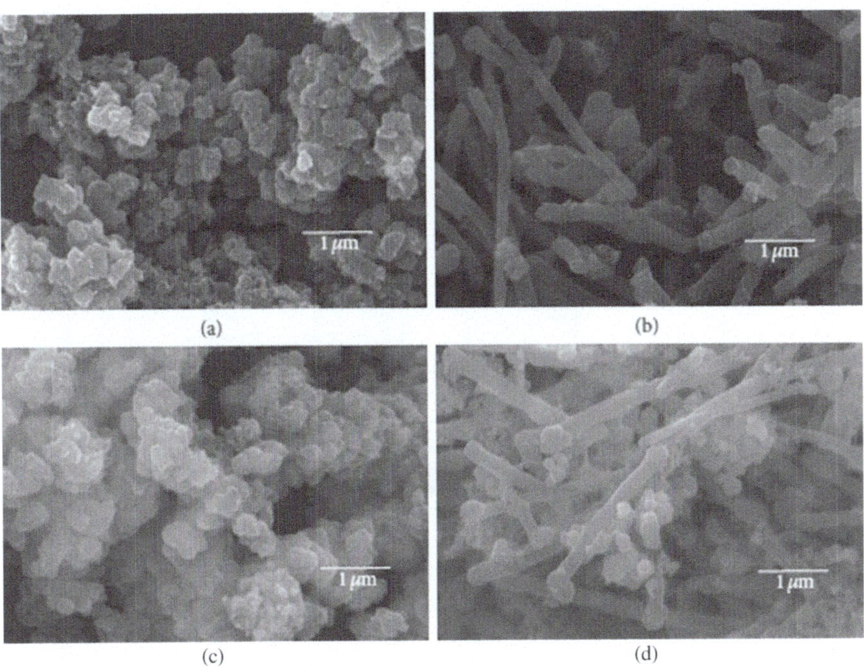

Fig. 6.28 FE-SEM images of (**a**) PA, (**b**) PC, (**c**) PT, and (**d**) PCT samples. (Reprinted with permission, Yun et al. 2013)

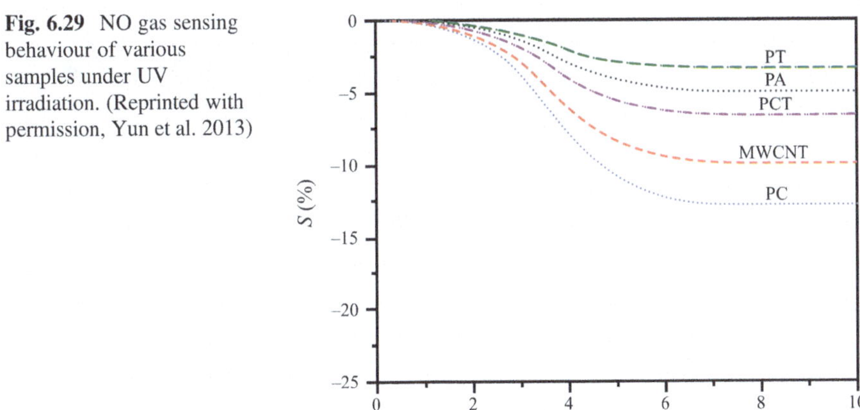

Fig. 6.29 NO gas sensing behaviour of various samples under UV irradiation. (Reprinted with permission, Yun et al. 2013)

The structural characterization and gas sensing behaviour of PANI-MWCNTs based surface acoustic wave (SAW) sensor deposited on lithium tantalite SAW transducers for H_2 gas was also reported (Arsat et al. 2011). The SEM analysis explored the dense growth of nanocomposite which was much denser than that of PANI nanofibers. Another SAW based PANI-WO_3 sensor deposited on layered

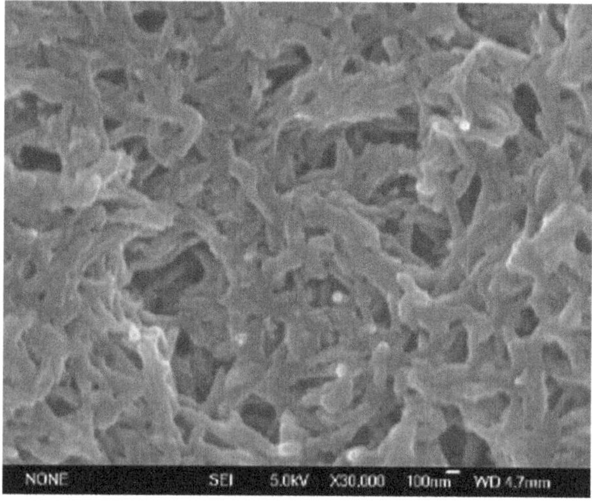

Fig. 6.30 SEM image of polyaniline/WO3 nanofibers. (Reprinted with permission, Sadek et al. 2008)

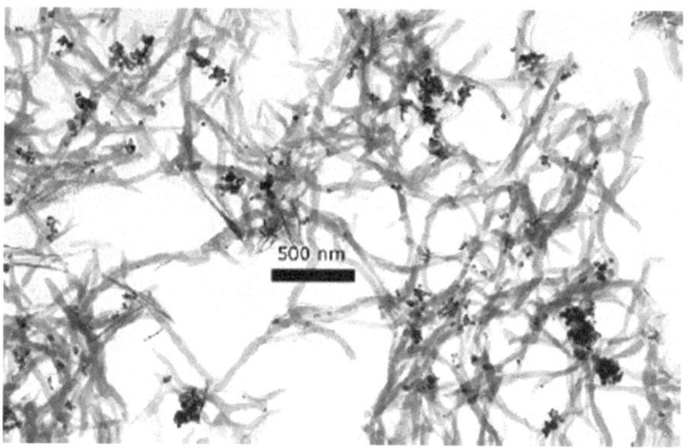

Fig. 6.31 TEM image of polyaniline/WO3 nanofiber composite. (Reprinted with permission, Sadek et al. 2008)

ZnO/64° YX $LiNbO_3$ SAW transducer using oxidative chemical polymerization which showed fast sensing response together with good repeatability (Sadek et al. 2008). XRD, SEM and TEM analyses characterized it. The average of the PANI-WO_3 nanofiber layer on the substrate measured by profilemeter was found as 0.4 nm. The average diameter of the nanofibers was 90 nm and shows a connecting network with each other (Fig. 6.30). Nanoparticles of WO_3 embedded in the backbone of PANI (Fig. 6.31). TEM image shows the embedded WO_3 nanoparticles are integrated into the polyaniline nanofibers backbone. From SEM and TEM images, it can be concluded that the deposited thin film is porous. SEM and TEM analyses confirmed the porous nature of PANI-WO_3 nanocomposite. The XRD shows sharp line-peaks in the bottom which confirmed the existence of WO_3 (from JCPDS database) in the matrix of PANI and the peaks in the above are due to the PANI-WO_3 nanocomposite (Fig. 6.32).

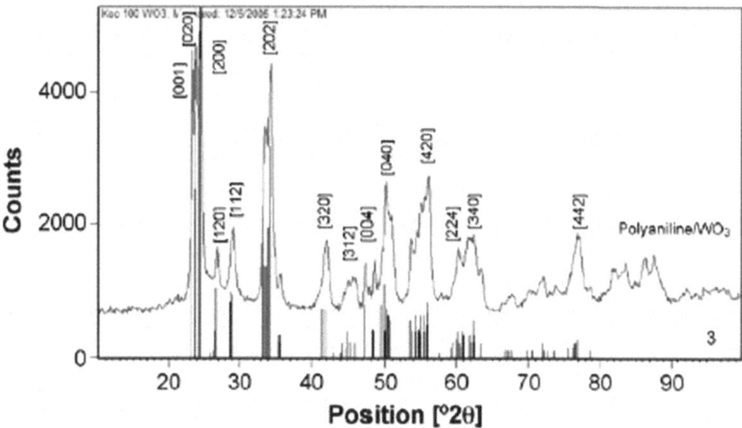

Fig. 6.32 XRD image of polyaniline/WO3 nanofiber composite. (Reprinted with permission, Sadek et al. 2008)

A PANI-PtO_2 based thin film sensor has also been reported for H_2 sensing (Conn et al. 1998). The authors stated that during the exposure of H_2 gas under the sensing condition, the PtO_2 present in the matrix of PANI was reduced to PtO. It catalytically oxidizes to water which decreases the resistance of PANI.

An interdigitated electrode of PANI–chloroaluminium phthalocyanine (PANI-ClAlPc) for CO_2 detection has been prepared by spin coating method from PANI as the base of composites and ClAlPc (with different concentrations) as the second component (Azim-Araghi and Jafari 2010). To determine the sensitivity, reversibility, response and recovery time, these nanocomposite thin films were exposure to different concentrations (0–2000 ppm) of CO_2 gas. For the practical application, the composites sensors were investigated at different temperatures. The sensitivity factor of composites has covered a range between 0.05–7.20. An optimum concentration of 10% ClAlPc was chosen in the matrix of PANI for maximum sensitivity at 300 K. Nanocomposite sensor demonstrated unexpected behaviour and the conductivity of thin films that was increased on exposure lower RH % and decreased on higher RH %. The results of sensing behaviour confirmed that the CO_2 mixtures reduced the sensitivity of thin films in compression with pure CO_2 (1000 ppm).

The sensing of gas can perform using different pathways – from nanometal catalyzed reaction to modification of barrier height. The response extracted from a nanocomposite sensor on exposure to an analyte (gas) generally governed: (i) charge transfer phenomenon between the constituents of the nanocomposite, and (ii) reaction between the gas and the nanocomposite. For example, H_2S can either have a reducing or an oxidizing effect on PANI depending on the interaction of PANI with the secondary component in the nanocomposite. Moreover, taking into consideration, the humidity factor is found necessary because it significantly decreases the sensitivity of PANI sensors. Therefore, for future PANI based sensing materials, a deeper understanding of the sensing behaviour is required to develop high performance sensors.

6.3.5 PANI Based Biosensors Nanocomposites

Biosensor system which functioned through immobilized sensor activity was first time reported by Lowe (Lowe 1984). It was the beginning of modern biosensors that has been widely used in nowadays. Their system comprised of an enzyme (glucose oxidase) immobilized on a gel which measured the concentration of glucose in biological solutions. A number of biosensors have come up since, which find finding applications in various fields including; industry, clinical diagnostics and environment monitoring (Baraton 2008).

A biosensor typically consists of a biorecognition element (or bioreceptor), a transducer element, and electronic components for signal processing. The schematic of a biosensor operation is shown in Fig. 6.33. It operates in three stages – (i) recognition of a specific analyte (gas) by the bioreceptor, (ii) transformation of biochemical reaction into the transducer-type reaction, and (iii) processing of transducer signal (Baraton 2008).

The enzyme based electrode is a miniature chemical transducer which functions by combining an electrochemical procedure with immobilized enzyme activity. This particular model employed glucose oxidase immobilized on a gel to measure the concentration of glucose in biological solutions as well as in the tissues in vitro.

PANI is particularly attractive as biosensors because it gives a conducting matrix for immobilization of bioreceptors (i.e., confined movement of bioreceptors in a defined space) onto it. Its electroactivity allows it to act as a mediator for electron transfer through a redox or enzymatic reaction. Such direct communication with the bound bioreceptors leads to a range of analytical signals which provides a measurement of the sensor activity (Dhand et al. 2011). This conducting polymer showed

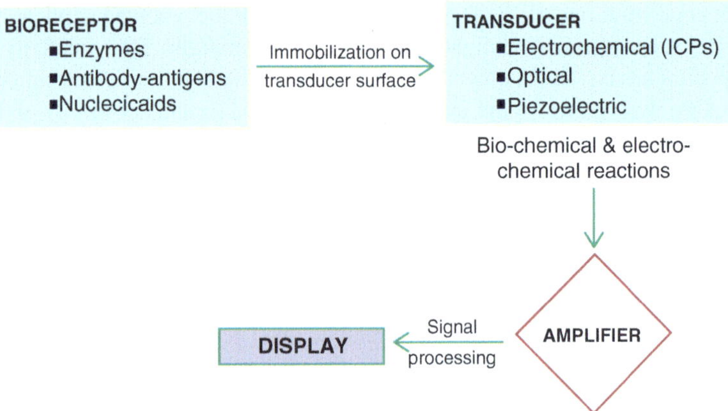

Fig. 6.33 Schematic representation of biosensor operation. (Reprinted with permission, Baraton 2008)

significant binding capacity with biomolecules along with fast electron transfer to produce new analytical applications. Apart from its electronic properties, PANI has shown excellent stability and strong biomolecular interactions (Di and Ivaska 2006; Imisides et al. 1996) that is found necessary for biosensor applications. On the basis of electrical conductivity, electrochromism, and pH sensitivity PANI has been successfully used for detection of different biological compounds (Muhammad-Tahir and Alocilja 2003; Malinauskas et al. 2004; Hoa et al. 1992). PANI based nanocomposite materials provide a scope to further assess their potential as biosensors. A secondary component in the matrix of PANI is often seen to increase bioreceptor binding onto PANI surface. Based on the response of bioreceptors, biosensors are broadly classified as enzymatic biosensors, immunosensors, and DNA/nucleic acid biosensors:

6.3.5.1 Enzymatic Biosensors

Clark and Lyons has introduced the concept of enzymatic biosensors by using the enzyme glucose oxidase (GOx) (Clark and Lyons 1962). An amperometric biosensor was developed in which the enzyme catalyzed the oxidation of glucose on the surface of the Pt electrode. Because of the enzymatic reaction, the oxygen flux on electrode surface varied as a function of glucose concentration, thus enabling its detection. However, the enzyme activity carried out in solution as opposed to the recent biosensors which have immobilized enzymes.

6.3.5.2 Glucose Biosensor

The biosensor of glucose was first introduced by significant modifications with reference to its detection technique and development of newer materials for enzyme immobilization. A novel glucose biosensor based on a composite of Au nanoparticles (NPs)–conductive polyaniline (PANI) nanofibers (PANI-Au) has been synthesized by Xian et al. (Xian et al. 2006). Nanocomposite material was immobilized with glucose oxidase (GOx) and Nafion on the surface by electrochemical oxidation of H_2O_2. It was found that the PANI-Au nanocomposite observed a much higher anodic current as compared to pure PANI. This resulted a better response, possibly due to fast electron transfer between electrode and H_2O_2 facilitated by Au in PANI matrix. It demonstrated a linear calibration curve over the concentration range from 1.0×10^{-6} to 8.0×10^{-4} mol/L, with a slope and detection limit (S/N = 3) of 2.3 mA/M and 5.0×10^{-7} M, respectively. In addition, the glucose biosensor system indicates excellent reproducibility (less than 5% R.S. D.) and good operational stability (over 2 weeks).

Other PANI based PANI/MWCNTs nanocomposites have also been reported, and these were used as biosensors for the detection of glucose (Gopalan et al. 2009; Zhong et al. 2011; Fu and Yu 2014). A novel glucose biosensor was developed based on the direct electrochemistry of glucose oxidase (GOD) adsorbed in graphene/polyaniline/gold nanoparticles (AuNPs) nanocomposite modified glassy carbon electrode (GCE) (Fu and Yu 2014). In comparison to graphene, polyanilline (PANI) or graphene/PANI, the graphene/PANI/AuNPs nanocomposite, graphene/PANI/AuNPs was found was more biocompatible and it offered a favourable microenvironment for facilitating the direct electron transfer between GOD and electrode. The adsorbed GOD displayed a pair of well-defined quasi-reversible redox peaks with a formal potential of −0.477 V (vs. SCE) and an apparent electron transfer rate constant (ks) of 4.8 s^{-1} in 0.1 M pH 7.0 PBS solution. The amperometric response of GOD graphene/PANI/AuNPs modified electrode showed improvement in amperometric response with respect the concentration of glucose (in the range of 4.0 M–1.12 mM) at lower detection limit; 0.6 M (at signal-to-noise of 3). Addition of GOD to prepare graphene/PANI/AuNPs nanocomposite showed its selective behaviour for detection of glucose. Le et al., developed a PANI/MWCNTs biosensor deposited on interdigitated planar Pt-film electrode over which the GOx was immobilized via glutaraldehyde (Le et al. 2013). The rapid amperometric response of PANI/MWCNTs concerning with the changing in glucose concentration is shown in Fig. 6.28. The porous nature of the nanocomposite facilitated stronger binding to GOx, and the nanocomposite proved to be an efficient transducer with a response time of 5 s. A GOx immobilized PANI–TiO_2 nanotube composite as an electrochemical biosensor reported by Zhu and co-workers (Zhu et al. 2015). Nanocomposite materials demonstrated excellent biocompatibility, better electrical conductivity, low electrochemical interferences and high signal-to-noise ratio for the development of electrochemical biosensors.

TiO_2 nanoparticles were initially transformed into TNTs, on which aniline was polymerized by oxidative polymerization to form an intimate and uniform PANI–TNT composite using a hydrothermal method. Characterization of the nanocomposite was carried out by different spectroscopic techniques. The PANI–TNT nanocomposite was used to immobilize glucose oxidase (GOD) for the construction of an electrochemical biosensor. Cyclic voltammetry studied the direct electrochemistry, and electro-catalytic performance of the biosensors based on PANI-TNTs and TNTs. The performance studies revealed that the nanoscaled tub-like morphology facilitated direct electron transfer of GOx, giving a sensitivity of 11.4 mA mM1 at a low detection range of 0.5 mM at a high signal-to-noise ratio (Fig. 6.34).

Although most of the biosensors utilize immobilized GOx for the detection of glucose, other enzymes can also be used for this purpose. In this regards, Ozdemir et al., reported a biosensor based on pyranose oxidase immobilized Au/PANI/AgCl/gelatin nanocomposite for glucose detection, wherein sensing was facilitated by amperometric detection of consumed O_2 during the enzymatic reaction (Ozdemir et al. 2010). The importance of glucose biosensors lies in the simpleistic and accurate monitoring of blood glucose levels, which might help in controlling the growing issue of diabetes around the world.

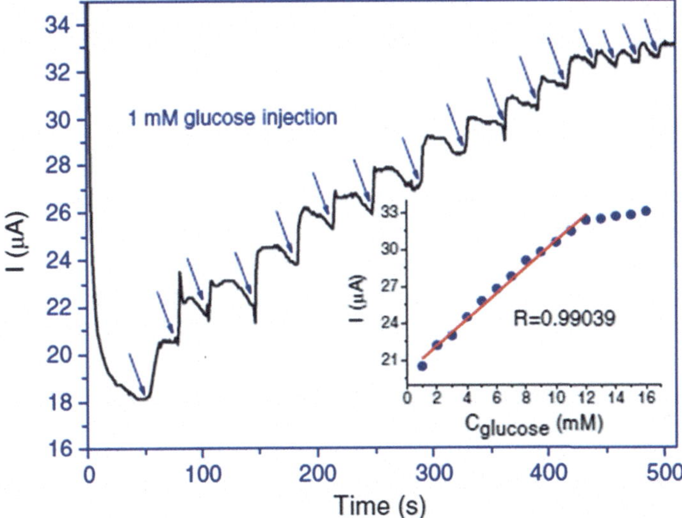

Fig. 6.34 Amperometric response of GOx/PANI-MWCNT/IDmE upon increasing the glucose concentration in steps of 1 mM at +0.6 V (versus SCE) in PBS. The inset shows the calibration plot156 (Reproduced with permission from IOPscience). (Reprinted with permission, Le et al. 2013)

6.3.5.3 PANI Based Immunosensors

Immunosensor is made of the combination of antigen–antibody specificity and transducer forms. For the detection of salbutamol (SAL), an ultrasensitive electrochemical immunosensor based on nanogold particles (nano-Au), Prussian Blue (PB), polyaniline/poly (acrylic acid) (PANI (PAA)) and Au-hybrid graphene nanocomposite (AuGN) was developed by Huang et al., (Huang et al. 2011). Nano-Au, PB, and PANI (PAA)-incorporated film was used to enhance the electroactivity, stability and catalytic activity for hydrogen reduction. AuGN was employed to immobilise chitosan, nano-Au and horseradish peroxidase–anti-SAL antibody (HRP–AAb). The resulting nanostructure (AuGN–HRP–AAb) was used as the label for the immunosensor. A label of chitosan coated graphene with the nano-Au shell was used to immobilize HRP-anti-SAL antibody on this multicomponent nanocomposite. The immunosensor showed excellent catalytic activity for hydrogen reduction on the electrode.

In another study, Liu and co-workers prepared a novel graphene oxide nanosheets/polyaniline (GO/PANI/CdSe) nanowires for detection of interleukin-6 (Liu et al. 2013). The GO/PANi/CdSe nanocomposites demonstrated excellent biocompatibility, dispersity, and solubility. Electrochemiluminescence (ECL) of CdSe quantum dots (QDs) was greatly enhanced by combining with GO/PANi nanocomposites. The nanocomposite of GO/PANi/CdSe was employed to develop an ultrasensitive ECL immunosensor for the detection of IL-6. The nanocomposite showed high specificity, long term stability and reproducibility with a detection limit as low as 0.17 pg mL^{-1}.

Polyaniline nanowires (PANI) and gold nanoparticle (AuNPs) based hybrid nanocomposite film has been fabricated onto an Au substrate. The prepared nanocomposite was used to develop a mediator free immunosensor for the detection of PSA (Dey et al. 2012). Microscopic studies revealed uniform distribution of the AuNPs in the PANI backbone provides desired microenvironment for antibody immobilization. The results of electrochemical studies proposed that AuNPs improve the electro-active surface area of PANI and resulted in high electron transport in a mediator free electrolyte. The sensor performance was measured using differential pulse voltammetry (DPV) which studied the electrochemical changes resulting from the biochemical reactions on the surface. Microscopic studies established a uniform distribution of the AuNPs on the PANI backbone resulting in a nanoporous morphology that provides the desired microenvironment for antibody immobilization. The nanocomposite provided a high surface area for immobilization of anti-PSA, which improved electron transfer and results high performance of the immunosensor. The electrochemical response studies of the immunosensor to different concentrations of PSA are shown in Fig. 6.35. The sensor exhibited good linearity, high sensitivity and excellent response at very low concentrations. The immunosensor showed excellent stability up to 5 weeks after that stability decreases and current response became low.

Synthesized PANI-Au nanoparticles were studied using UV-vis. spectroscopy. The broadening of the observed absorption peak is accredited to the presence of different size particles. It is generally found that Au nanoparticles are known to exhibit a strong surface plasmon absorbance band at about 520 nm which depends on size, shape, material properties, surrounding media, and proximity to other nanoparticles. The particle size of Au nanoparticles was measured at 50 nm using the UV absorbance peak at 535 nm. The particle size, shape, and distribution of the Au nanoparticles were studied using TEM analysis (inset, Fig. 6.36a). TEM images exposed crystalline nature of Au nanoparticles with minimal agglomeration among the particles, and they appear in icosahedra shape with an estimated particle size of ~50 nm. SEM studies (Fig. 6.30) of synthesized PANI (image a) exhibited very smooth surface with a homogenous diameter estimated as 100–200 nm with length varying from 1 to 2 microns. After the electrophoretic deposition of PANI nanowires onto the Au substrate, the SEM study (image b) exhibited the uniform deposition of PANI nanowires over a large surface area without a significant change in length and diameter of the nanowires. The SEM image of the Au-NPs-PANI–Au nanocomposite (image c) depicted that AuNPs are uniformly distributed onto the PANI surface via self-assembly due to differences in surface charges resulting in strong electrostatic interactions. Moreover, AuNPs are dispersed over PANI with minimal agglomeration, which is expected to improve the electrochemical properties of PANI.

A multilayer nanocomposite of Au-PANI-cMWCNTs-chitosan was used to design a highly sensitive immunosensor for detecting an organophosphate insecticide (i.e., chlorpyrifos) (Sun et al. 2013). PANI-coated MWCNTs were fabricated via in-situ oxidative polymerization. Carboxylated MWCNTs played a significant role to achieve the thin and uniform coating of PANI, thus, resulting in the improved

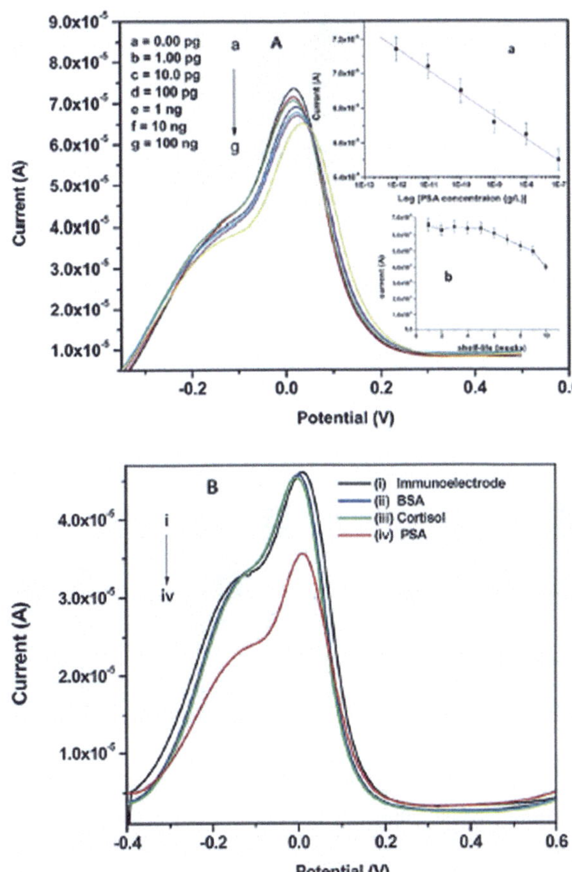

Fig. 6.35 (**A**) Electrochemical response studies of the BSA/anti-PSA/AuNP-PSA/Au immunoelectrode as a function of PSA (1 pg mL1–100 ng mL^{-1}) in PBS (10 mM, pH 7 containing 0.9% NaCl) using DPV technique. Inset (**a**) calibration curve between the magnitude of electrochemical response current vs. logarithm of PSA concentration and inset (**b**) shelf-life studies of BSA/ anti-PSA/AuNP-PSA/Au immunoelectrode. (**B**) Interference studies of BSA/ anti-PSA/AuNP-PSA/Au immunoelectrode using BSA (10 ng mL^{-1}) and cortisol (10 ng mL^{-1}) concerning with PSA (10 ng mL^{-1}). Reproduced from178 with permission from The Royal Society of Chemistry. (Reprinted with permission, Dey et al. 2012)

immunosensor response. The immunosensor showed higher sensitivity and specific immunoreactions for analysis of real samples as well. Besides, storage at low temperatures affords the sensor a longer shelf life as in case of other biosensors. Experimental parameters affecting the preparation process and analytical performance of the immunosensor were optimized. Under optimal conditions (antibody concentration: 5 μg/mL, working buffer pH: 6.5, incubation time: 40 min, incubation temperature: 25 °C), the immunosensor showed a wide linear range from 0.1 to 40×10^{-6} mg/mL and from 40×10^{-6} mg/mL to 500×10^{-6} mg/mL. The detection limit was found to be 0.06×10^{-6} mg/mL which provided a valuable tool for the chlorpyrifos detection in real samples.

Graphene/PANI nanocomposite (GR–PANI) based immunosensor has been utilized for the detection of estradiol (Li et al. 2013b). GR–PANI composites was used to enhance the electroactivity and stability of the electrode. During the synthesis, Horseradish peroxidase–graphene oxide–antibody (HRP–GO–Ab) conjugates was constructed by using the carboxylated GO (as the carrier of the antibody), resulting

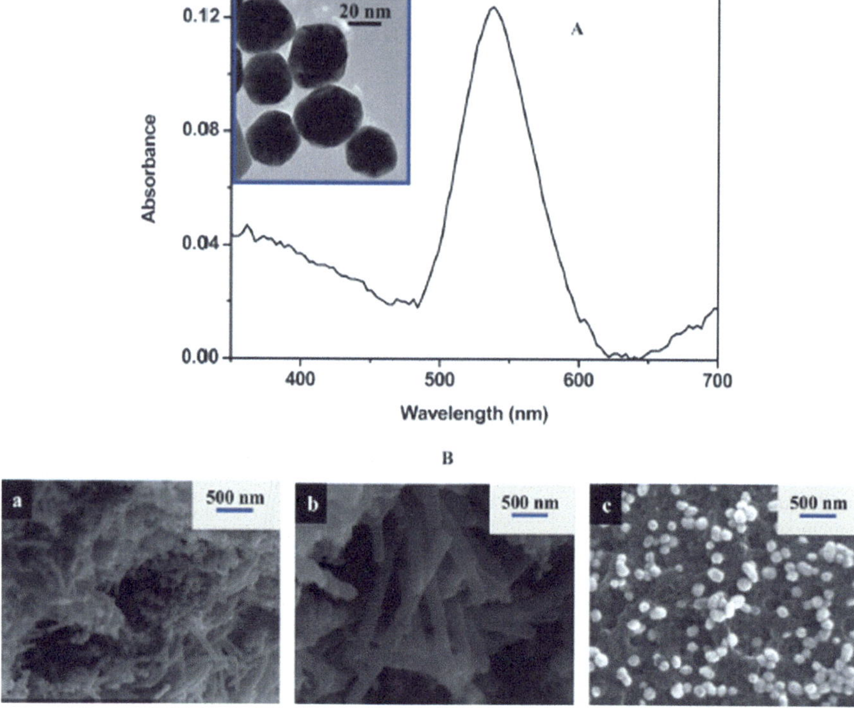

Fig. 6.36 (**a**) UV-visible absorption spectra of AuNPs, (inset) TEM image of AuNPs synthesized by co-precipitation method, (**b**) SEM images of PANI nanowires (image a) prepared by chemical oxidative method, (image b) electrophoretically deposited PANI nanowires onto the Au substrate, (image c) self assembled AuNPs on the nanostructured surface charged PANI/Au electrode. (Reprinted with permission, Dey et al. 2012)

in the improved catalytic activity of the hydrogen reduction of the electrode. The current response of the immunosensor was remarkably enhanced due to the synergistic effects of GR and PANI. The immunosensor revealed a wide range of linearity with a low detection limit of 0.02 ng mL^{-1}. Therefore, a competitive immunoassay was established between 17β-estradiol and the HRP–GO–Ab conjugates which showed a wide linear response to estradiol in the range 0.04–7.00 ng/mL and a limit of detection of 0.02 ng/mL (S/N = 3). The developed immunosensor was successfully applied for the detection of estradiol in real samples.

The detection of benzo[a] pyrene (BaP) was performed using a multi-enzyme antibody of RP-HCS secondary antibody immobilized on Fe_3O_4/PANI/Nafion sensor (Lin et al. 2012). The nanocomposite provided an active electron transfer pathway for reducing H_2O_2. The stable film of Fe_3O_4/PANI/Nafion not only immobilized biomolecules but also catalyzed the reduction of hydrogen peroxide which confirmed an accelerated electron transfer pathway of the platform. On the basis of a competitive immunoassay, the current charge was found to be proportional to the logarithm of BaP concentration encompassing the range of 8 pM and 2 nM with the detection limit of 4 pM. Thus, the proposed immunosensor exhibited acceptable reproducibility and stability.

A label-free immunosensor has been designed for detecting low density lipoproteins (LDL) (Lin et al. 2012). The AuNPs–AgCl@PANI hybrid material was synthesized by reacting silver chloride@polyaniline (PANI) core–shell nanocomposites (AgCl@PANI) combined with Au nanoparticles (AuNPs). Apolipoprotein B-100 antibody was immobilized on the surface of Au–AgCl/PANI nanocomposite and the sensor performance was monitored via electrochemical impedance spectroscopy (EIS). The nanocomposite material provided a cover for high antibody loading due to the high surface-to-volume ratio. The poor conductivity of LDL and the negative charges performed brought about a shift in electron transfer resistance which facilitated its detection. The immunosensor had a very low detection limit (0.34 pg mL^{-1}) at an optimized condition, i.e., incubation time and incubation temperature (50 min, 37 °C). PANI based hybrid materials provide a useful transducer to which the immunochemical reactions can be coupled. The advantages of fabricated immunosensor include simplicity of design and tunability of PANI properties by a combination of nanomaterials, thus, making facile label-free detection of antigens possible.

6.3.5.4 DNA Based Biosensors

PANI based biosensors were also employed for the detection of DNA to provide high sensitivity as well as accurate and rapid detection. These biosensors have been used in the fields of forensics, biological warfare detection, gene analysis, and DNA diagnostics (Dhand et al. 2011). A single stranded DNA (ssDNA) probe immobilized on the surface of transducer recognizes its complementary DNA target by hybridization. A PANI-graphite oxide based nanocomposite was fabricated on a carbon paste electrode (CPE) for monitoring DNA hybridization (Wu et al. 2005). Both ssDNA and dsDNA (double-strand DNA) change the redox properties of the nanocomposite electrode. To monitor hybridization and detect the complementary ssDNA of the biosensor, Square wave voltammetry (SWV) was utilized. The responses were influenced by pH and incubation time. DNA sensing behaviour of graphene and PANI nanocomposites have also been reported by other researchers (Bo et al. 2011; Du et al. 2012). Graphene sheets on GCE followed by electropolymerization of PANI and electrodeposition of Au nanoparticles were employed as a layered biosensor in order to detect BCR/ABL fusion gene in chronic myelogenous leukemia (CML) (Wang et al. 2014). After hybridization, the 30 -biotin site face moved away from the electrode surface, binding the streptavidin–alkaline phosphatase. The hydrolysis of 1-naphthyl phosphate to 1-naphthol was monitored using DPV. The biosensor showed high selectivity having a detection limit of 2.11 pM.

Another label-free electrochemical DNA sensor using PPY/PANI/ Au nanocomposite was employed as ssDNA labeled with 6-mercapto-1- hexane for hybridization (Wilson et al. 2012). Compared to the lone Au nanoparticle based transducer, the biopolymer improved the hybridization efficiency of the ssDNA, highly sensitive towards the complementary ssDNA with a low detection limit of 1013 M. Forster and co-workers used A PANI nanocomposite with surface deposited Au nanoparticles as DNA biosensor (Spain et al. 2011, 2013). In one of their reports, thiolated ssDNA, which act as complementary to a sequence of Staphylococcus aureus, was immobilized onto the surface of the nanocomposite. HRP labeled probe strand was hybridized to an unbound section of the capture strand. The target concentration was determined by measuring current during reduction of benzoquinone through HRP.

PANI/MWCNT nanocomposite was prepared for detecting the phosphinothricin acetyltransferase gene (PAT), one of the screening detection genes for the transgenic plants (Yang et al. 2009; Prathap et al. 2013). A ssDNA probe was immobilized onto the nanocomposite and hybridization was detected via EIS. It was successfully applied to the real sample of genetically modified soybean.

PANI/ZrO_2 nanocomposites were used for detecting polytyrosine Phosphinothricin acetyl transferase (PAT) (Yang et al. 2012). The nanocomposite aided in ssDNA immobilization and hybridization was investigated by CV and EIS techniques. The electron transfer resistance was found to increase with the increase in the concentration of target DNA. It was also found that nanocomposites, i.e., PANI- MWCNT and PANI-ZrO_2, exhibited the same detection limit.

In another study, the detection of a DNA base (i.e., guanine, adenine, thymine, and cytosine) was performed using PANI/MnO_2 nanocomposite (Prathap et al. 2013). Radhakrishnan et al., reported the fabrication of PPY/PANI nanocomposite with ssDNA immobilized onto it via glutaraldehyde, using methyl blue dye as an electrochemical indicator (Radhakrishnan et al. 2013). The biosensor showed remarkable sensitivity with a detection limit of 50 fM. PANI based DNA biosensors were also utilized for detection of pesticide.

A nucleic acid based biosensor was reported by Prabhakar et al., for pesticide detection (Prabhakar et al. 2008). The calf thymus dsDNA immobilized onto PANI-polyvinyl sulphonate (PVS) nanocomposite showed superior stability. Hybridization with the target DNA changes the doping level of PANI, and as a result alters its conductivity. PANI based nanocomposites and their sensing applications presented in Table 6.2.

6.4 Conclusion

PANI based nanocomposite ion-exchange materials and nanocomposites have demonstrated excellent behaviour towards detection and sensing of gases. Also, these nano-materials have facilitated the immobilization of range of bioreceptors (such as enzymes, antigen–antibodies, and nucleic acids) onto their surfaces for exposure of

Table 6.2 PANI based nanocomposite ion-exchange materials and their sensing applications

S. no	Nanocomposite	Analysis	Application	Reference
1.	PANI-Pd	UV-vis., TEM	Methanol	Athawale et al. (2006)
2.	PANI-Ag	UV-vis., TEM	Ethanol	Choudhury (2009)
3.	PANI-Ag	SEM, TEM	Triethylamine, toluene	Li et al. (2013a)
4.	PANI-Cu	TEM, FTIR	Chloroform	Sharma et al. (2002)
5.	PANI-TiO_2	XRD, SEM	NH_3	Pawar et al. (2011)
6.	PANI-SnO_2	FTIR, XRD, SEM, AFM	NH_3	Deshpande et al. (2009)
7.	PANI-PMMA	SEM, RAMAN	NH_3	Zhang et al. (2014)
8.	PANI-NiTSPc	SEM, AFM, RAMAN, FTIR	NH_3	Zhihua et al. (2016)
9.	PANI-$CuCl_2$	UV-vis., SERS	H_2S	Crowley et al. (2010)
10.	PANI–CdS	XRD, SEM	H_2S	Raut et al. (2012)
11.	PANI–Au	SEM	H_2S	Shirsat et al. (2009)
12.	n-CdSe-p-PANI	SEM, AFM	LPG	Joshi et al. (2007)
13.	PANI/g-Fe_2O_3	FTIR, UV-vis. XRD, FESEM	LPG	Sen et al. (2014)
14.	PANI-TiO2	XRD, SEM	LPG	Dhawale et al. (2008)
15.	PANI-CdSe	XRD, SEM	LPG	Dhawale et al. (2010a)
16.	PANI-ZnO	XRD, SEM, TGA	LPG	Dhawale et al. (2010b)
17.	PANI-ZnO	by XRD, SEM, TGA and UV-Vis	Humidity	Shukla et al. (2012)
18.	PANI-Co	XRD, SEM, FTIR	Humidity	Vijayan et al. (2008)
19.	PANI-Ag	XRD, UV-vis. FTIR, SEM	Humidity	Vijayan et al. (2008)
20.	PANI–β-$AgVO_3$	FESEM, HRTEM, AFM analyses. FTIR and EDAX	Humidity	Diggikar et al. (2013)
21.	PANI-In_2O_3	FTIR, SEM	CH_4, CO	Yan et al. (2009)
22.	PANI-SnO_2	UV-vis., SEM	CO	Ram et al. (2005b)
23.	PANI-SWCNT	FTIR, SEM	CO	Kim et al. (2010)

(continued)

Table 6.2 (continued)

S. no	Nanocomposite	Analysis	Application	Reference
24.	PANI-TiO_2-MWCNT	XRD, FESEM	NO	Yun et al. (2013)
25.	PANI- SnO_2–ZnO	FTIR, XRD, FESEM	NO_2	Xu et al. (2013b)
26.	MWNT-PANI and TiO_2-PANI	XRD, SEM	H_2	Srivastava et al. (2010)
27.	PANI-WO_3	SEM, TEM, XRD	H_2	Sadek et al. (2008)
28.	PANI-PtO_2	–	H_2	Conn et al. (1998)
29.	PANI-ClAlPc	–	CO	Azim-Araghi and Jafari (2010)
30.	PANI-Au	TEM	Glucose	Xian et al. (2006)
31.	Graphene/PANI/Au	FTIR, SEM	Glucose	Fu and Yu (2014)
32.	PANI–TiO_2	FTIR, XRD, TEM	Glucose	Zhu et al. (2015)
33.	PANI (PAA)-Au	–	Salbutamol	Huang et al. (2011)
34.	GO-PANI-CdSe	TEM	Interleukin-6	Liu et al. (2013)
35.	PANI/Au	UV-vis., SEM, TEM, AFM	Prostate-specific antigen	Dey et al. (2012)
36.	Au-PANI-cMWCNTs-chitosan	XRD, SEM	Chlorpyrifos	Sun et al. (2013)
37.	GR–PANI	SEM, FTIR, UV-vis.	Estradiol	Li et al. (2013b)
38.	Fe_3O_4-PANI-Nafion	UV-vis., TEM	Benzo[a] pyrene	Lin et al. (2012)
39.	AuNPs–AgCl@PANI	TEM, XRD, TGA, XPS	Lipoproteins	Lin et al. (2012)
40.	PANI-GO	–	DNA hybridization	Wu et al. (2005)
41.	Au-PANI on GCE	SEM	BCR/ABL fusion gene (in chronic myelogenous leukemia)	Wang et al. (2014)
42.	PPY-PANI-Au	SEM	DNA hybridization	Wilson et al. (2012)
43.	PANI-MWCNT	SEM	Phosphinothricin acetyltransferase gene	Yang et al. (2009)
44.	PANI-cMWCNTs	XRD, SEM, TEM, FTIR	DNA bases	(Prathap et al. 2013)

an array of biological agents via a combination of biochemical and electrochemical reactions. However, their commercialization remains a severe obstacle to scientific community because PANI itself suffers from some drawbacks, including their synthesis. Controlled synthesis of PANI at optimized pH and temperature is necessary to avoid secondary growth while retaining its nanostructure. In addition, dopant and its electronic properties play a crucial role in the synthesis of PANI. PANI can be highly conducting, self-doped, and/or soluble in aqueous and non-aqueous organic solvents based on dopant nature. PANI in doped form can have a relatively short shelf life. Besides, merging the secondary component in PANI showed a synergistic effect in improving its shelf life in doped form, and various other properties, and further applications. Therefore, fabrication of PANI nanocomposites with extended experience in doped structure is essential as far as its durability is concerned. Compared to pure PANI, its nanocomposites have shown improved characteristics regarding thermal and chemical stability. Moreover, PANI based nanocomposites have been used for safer, selective and sensitive detection of number combustible and toxic gases at room temperature. These materials halt the response variation resulting from structural changes in the sensing material at elevated temperatures as well as prevent sensor instability.

One of the disadvantages of a large number of nanocomposites based sensors is that their response is influenced by humidity which may result in false responses in the material. Thus, development of PANI supported nanocomposites resistant to humidity is one of the critical challenges. The enhancement in electron transfer capability of PANI nanocomposites is found attractive for detecting biological agents. Immobilization of biological agents onto PANI transducers becomes possible after incorporation of a secondary component with PANI. A better interaction between the components of the nanocomposite will improve the effectiveness and performance of the sensor. In this respect, there is a definite need for new approaches in order to understand the mechanism involved in analyte sensing by PANI and its various exciting nanocomposites for fabricating and commercializing highly selective sensors for specific agents.

Acknowledgement The authors would like to express their appreciations to Science and Engineering Research Board (DST) fast tract young scientist scheme (SB/FT/CS-122/2014) for providing Postdoctoral Fellowship to Mohammad Shahadat.

References

Abdulla S, Mathew TL, Pullithadathil B (2015) Highly sensitive, room temperature gas sensor based on polyaniline-multiwalled carbon nanotubes (PANI/MWCNTs) nanocomposite for trace-level ammonia detection. Sensors Actuators B Chem 221:1523–1534

Adams P et al (1996) Low temperature synthesis of high molecular weight polyaniline. Polymer 37 (15):3411–3417

Ampuero S, Bosset J (2003) The electronic nose applied to dairy products: a review. Sensors Actuators B Chem 94(1):1–12

Ansari MO, Mohammad F (2011) Thermal stability, electrical conductivity and ammonia sensing studies on p-toluenesulfonic acid doped polyaniline: titanium dioxide (pTSA/Pani: TiO2) nanocomposites. Sensors Actuators B Chem 157(1):122–129

Ansari MO et al (2013) Thermal stability in terms of DC electrical conductivity retention and the efficacy of mixing technique in the preparation of nanocomposites of graphene/polyaniline over the carbon nanotubes/polyaniline. Compos Part B 47:155–161

Ansari MO et al (2014) Ammonia vapor sensing and electrical properties of fibrous multi-walled carbon nanotube/polyaniline nanocomposites prepared in presence of cetyl-trimethylammonium bromide. J Ind Eng Chem 20(4):2010–2017

Arrad O, Sasson Y (1989) Commercial ion exchange resins as catalysts in solid-solid-liquid reactions. J Org Chem 54(21):4993–4998

Arsat R et al (2011) Hydrogen gas sensor based on highly ordered polyaniline/multiwall carbon nanotubes composite. Sens Lett 9(2):940–943

Athawale AA, Kulkarni MV (2000) Polyaniline and its substituted derivatives as sensor for aliphatic alcohols. Sensors Actuators B Chem 67(1):173–177

Athawale AA, Bhagwat S, Katre PP (2006) Nanocomposite of Pd–polyaniline as a selective methanol sensor. Sensors Actuators B Chem 114(1):263–267

Ayad MM, El-Hefnawey G, Torad NL (2009) A sensor of alcohol vapours based on thin polyaniline base film and quartz crystal microbalance. J Hazard Mater 168(1):85–88

Azim-Araghi M, Jafari M (2010) Electrical and gas sensing properties of polyaniline-chloroaluminium phthalocyanine composite thin films. Eur Phys J Appl Phys 52(01):10402

Baraton M-I (2008) Sensors for environment, health and security: advanced materials and technologies. Springer, Dordrecht

Bo Y et al (2011) A novel electrochemical DNA biosensor based on graphene and polyaniline nanowires. Electrochim Acta 56(6):2676–2681

Bushra R et al (2014) Synthesis, characterization, antimicrobial activity and applications of polyanilineTi (IV) arsenophosphate adsorbent for the analysis of organic and inorganic pollutants. J Hazard Mater 264:481–489

Castro M et al (2009) Carbon nanotubes/poly (ε-caprolactone) composite vapour sensors. Carbon 47(8):1930–1942

Cavallo P et al (2015) Understanding the sensing mechanism of polyaniline resistive sensors. Effect of humidity on sensing of organic volatiles. Sensors Actuators B Chem 210:574–580

Chaudhary V, Kaur A (2015) Enhanced room temperature sulfur dioxide sensing behaviour of in situ polymerized polyaniline–tungsten oxide nanocomposite possessing honeycomb morphology. RSC Adv 5(90):73535–73544

Choudhury A (2009) Polyaniline/silver nanocomposites: dielectric properties and ethanol vapour sensitivity. Sensors Actuators B Chem 138(1):318–325

Clark LC, Lyons C (1962) Electrode systems for continuous monitoring in cardiovascular surgery. Ann N Y Acad Sci 102(1):29–45

Conn C et al (1998) A polyaniline-based selective hydrogen sensor. Electroanalysis 10(16):1137–1141

Crowley K et al (2010) Fabrication of polyaniline-based gas sensors using piezoelectric inkjet and screen printing for the detection of hydrogen sulfide. Sens J IEEE 10(9):1419–1426

Deshpande N et al (2009) Studies on tin oxide-intercalated polyaniline nanocomposite for ammonia gas sensing applications. Sensors Actuators B Chem 138(1):76–84

Dey A et al (2012) Mediator free highly sensitive polyaniline–gold hybrid nanocomposite based immunosensor for prostate-specific antigen (PSA) detection. J Mater Chem 22(29):14763–14772

Dhand C et al (2011) Recent advances in polyaniline based biosensors. Biosens Bioelectron 26 (6):2811–2821

Dhawale D et al (2008) Room temperature liquefied petroleum gas (LPG) sensor based on p-polyaniline/n-TiO2 heterojunction. Sensors Actuators B Chem 134(2):988–992

Dhawale D et al (2010a) Room temperature LPG sensor based on n-CdS/p-polyaniline heterojunction. Sensors Actuators B Chem 145(1):205–210

Dhawale D et al (2010b) Room temperature liquefied petroleum gas (LPG) sensor. Sensors Actuators B Chem 147(2):488–494

Dhingra M et al (2013) Impact of interfacial interactions on optical and ammonia sensing in zinc oxide/polyaniline structures. Bull Mater Sci 36(4):647–652

Di W, Ivaska A (2006) Electrochemical biosensors based on polyaniline. Chem Anal 51(6):839–852

Diggikar RS et al (2013) Formation of multifunctional nanocomposites with ultrathin layers of polyaniline (PANI) on silver vanadium oxide (SVO) nanospheres by in situ polymerization. J Mater Chem A 1(12):3992–4001

Dimitriev O (2003) Interaction of polyaniline and transition metal salts: formation of macromolecular complexes. Polym Bull 50(1–2):83–90

Docquier N, Candel S (2002) Combustion control and sensors: a review. Prog Energy Combust Sci 28(2):107–150

Du M et al (2012) Fabrication of DNA/graphene/polyaniline nanocomplex for label-free voltammetric detection of DNA hybridization. Talanta 88:439–444

Dubbe A (2003) Fundamentals of solid state ionic micro gas sensors. Sensors Actuators B Chem 88 (2):138–148

Feng J, MacDiarmid A (1999) Sensors using octaaniline for volatile organic compounds. Synth Met 102(1):1304–1305

Fu L, Yu A (2014) Carbon nanotubes based thin films: fabrication, characterization and applications. Rev Adv Mater Sci 36:40–61

Fuke MV et al (2008) Evaluation of co-polyaniline nanocomposite thin films as humidity sensor. Talanta 76(5):1035–1040

Fuke MV et al (2009) Ag-polyaniline nanocomposite cladded planar optical waveguide based humidity sensor. J Mater Sci Mater Electron 20(8):695–703

Gangopadhyay R, De A (2000) Conducting polymer nanocomposites: a brief overview. Chem Mater 12(3):608–622

Genies E et al (1990) Polyaniline: a historical survey. Synth Met 36(2):139–182

Gopalan AI et al (2009) An electrochemical glucose biosensor exploiting a polyaniline grafted multiwalled carbon nanotube/perfluorosulfonate ionomer–silica nanocomposite. Biomaterials 30(30):5999–6005

Hasan M et al (2015) Ammonia sensing and DC electrical conductivity studies of p-toluene sulfonic acid doped cetyltrimethylammonium bromide assisted V2O5@ polyaniline composite nanofibers. J Ind Eng Chem 22:147–152

Hoa D et al (1992) A biosensor based on conducting polymers. Anal Chem 64(21):2645–2646

Hsu YF et al (2008) Undoped p-type ZnO Nanorods synthesized by a hydrothermal method. Adv Funct Mater 18(7):1020–1030

Hu H et al (2002) Adsorption kinetics of optochemical NH 3 gas sensing with semiconductor polyaniline films. Sensors Actuators B Chem 82(1):14–23

Huang W-S, Humphrey BD, MacDiarmid AG (1986) Polyaniline, a novel conducting polymer. Morphology and chemistry of its oxidation and reduction in aqueous electrolytes. J Chem Soc Faraday Trans 82(8):2385–2400

Huang J et al (2011) Electrochemical immunosensor based on polyaniline/poly (acrylic acid) and Au-hybrid graphene nanocomposite for sensitivity enhanced detection of salbutamol. Food Res Int 44(1):92–97

Imisides M, John R, Wallace G (1996) Microsensors based on conducting polymers. ChemTech 26 (5):19–25

Jain S et al (2003) Humidity sensing with weak acid-doped polyaniline and its composites. Sensors Actuators B Chem 96(1):124–129

Joshi S, Lokhande C, Han S-H (2007) A room temperature liquefied petroleum gas sensor based on all-electrodeposited n-CdSe/p-polyaniline junction. Sensors Actuators B Chem 123(1):240–245

Kaur B, Srivastava R (2015) Simultaneous determination of epinephrine, paracetamol, and folic acid using transition metal ion-exchanged polyaniline–zeolite organic–inorganic hybrid materials. Sensors Actuators B Chem 211:476–488

Khan AA (2006) Applications of Hg (II) sensitive polyaniline Sn (IV) phosphate composite cation-exchange material in determination of Hg^{2+} from aqueous solutions and in making ion-selective membrane electrode. Sensors Actuators B Chem 120(1):10–18

Khan AA, Baig U (2013a) Electrical conductivity and ammonia sensing studies on in situ polymerized poly (3-methythiophene)–titanium (IV) molybdophosphate cation exchange nanocomposite. Sensors Actuators B Chem 177:1089–1097

Khan AA, Baig U (2013b) Electrical conductivity and humidity sensing studies on synthetic organic–inorganic poly-o-toluidine–titanium (IV) phosphate cation exchange nanocomposite. Solid State Sci 15:47–52

Khan AA, Khalid M, Niwas R (2010a) Humidity and ammonia vapor sensing applications of polyaniline–polyacrylonitrile composite films. Sci Adv Mater 2(4):474–480

Khan AA, Khalid M, Baig U (2010b) Synthesis and characterization of polyaniline–titanium (IV) phosphate cation exchange composite: methanol sensor and isothermal stability in terms of DC electrical conductivity. React Funct Polym 70(10):849–855

Khan AA, Baig U, Khalid M (2011) Ammonia vapor sensing properties of polyaniline–titanium (IV) phosphate cation exchange nanocomposite. J Hazard Mater 186(2):2037–2042

Khan AA, Baig U, Khalid M (2013a) Electrically conductive polyaniline-titanium (IV) molybdophosphate cation exchange nanocomposite: synthesis, characterization and alcohol vapour sensing properties. J Ind Eng Chem 19(4):1226–1233

Khan AA et al (2013b) Ion-exchange and humidity sensing properties of poly-o-anisidine sn (IV) arsenophosphate nano-composite cation-exchanger. J Environ Chem Eng 1(3):310–319

Khuspe G et al (2013) Ammonia gas sensing properties of CSA doped PANi-SnO2 nanohybrid thin films. Synth Met 185:1–8

Kim I et al (2010) Gas sensor for CO and NH 3 using polyaniline/CNTs composite at room temperature. In: Nanotechnology (IEEE-NANO), 10th IEEE conference on 2010, IEEE

Koul S, Chandra R (2005) Mixed dopant conducting polyaniline reusable blend for the detection of aqueous ammonia. Sensors Actuators B Chem 104(1):57–67

Le TH et al (2013) Electrosynthesis of polyaniline–mutilwalled carbon nanotube nanocomposite films in the presence of sodium dodecyl sulfate for glucose biosensing. Adv Nat Sci Nanosci Nanotechnol 4(2):025014

Li Z-F et al (2013a) Understanding the response of nanostructured polyaniline gas sensors. Sensors Actuators B Chem 183:419–427

Li J et al (2013b) Electrochemical immunosensor based on graphene–polyaniline composites and carboxylated graphene oxide for estradiol detection. Sensors Actuators B Chem 188:99–105

Lin M et al (2012) Electrochemical immunoassay of benzo [a] pyrene based on dual amplification strategy of electron-accelerated Fe3O4/polyaniline platform and multi-enzyme-functionalized carbon sphere label. Anal Chim Acta 722:100–106

Liu P-Z et al (2013) Electrochemiluminescence immunosensor based on graphene oxide nanosheets/polyaniline nanowires/CdSe quantum dots nanocomposites for ultrasensitive determination of human interleukin-6. Electrochim Acta 113:176–180

Lowe CR (1984) Biosensors. Trends Biotechnol 2(3):59–65

Malinauskas A et al (2004) Electrochemical response of ascorbic acid at conducting and electrogenerated polymer modified electrodes for electroanalytical applications: a review. Talanta 64(1):121–129

Matsuguchi M et al (2002) Effect of NH 3 gas on the electrical conductivity of polyaniline blend films. Synth Met 128(1):15–19

Muhammad-Tahir Z, Alocilja EC (2003) A conductometric biosensor for biosecurity. Biosens Bioelectron 18(5):813–819

Nabi S et al (2010) Development of composite ion-exchange adsorbent for pollutants removal from environmental wastes. Chem Eng J 165(2):405–412

Nabi S et al (2011a) Synthesis and characterization of nano-composite ion-exchanger; its adsorption behavior. Colloids Surf B: Biointerfaces 87(1):122–128

Nabi S et al (2011b) Heavy-metals separation from industrial effluent, natural water as well as from synthetic mixture using synthesized novel composite adsorbent. Chem Eng J 175:8–16

Nabi S et al (2011c) Synthesis and characterization of polyanilineZr (IV) sulphosalicylate composite and its applications (1) electrical conductivity, and (2) antimicrobial activity studies. Chem Eng J 173(3):706–714

Navale S et al (2014) Camphor sulfonic acid doped PPy/α-Fe2O3 hybrid nanocomposites as NO2 sensors. RSC Adv 4(53):27998–28004

Novák P et al (1997) Electrochemically active polymers for rechargeable batteries. Chem Rev 97 (1):207–282

Ozdemir C et al (2010) Electrochemical glucose biosensing by pyranose oxidase immobilized in gold nanoparticle-polyaniline/AgCl/gelatin nanocomposite matrix. Food Chem 119(1):380–385

Parvatikar N et al (2006) Electrical and humidity sensing properties of polyaniline/WO3 composites. Sensors Actuators B Chem 114(2):599–603

Patil S et al (2011) Fabrication of polyaniline-ZnO nanocomposite gas sensor. Sens Transducer 134 (11):120

Pawar S et al (2011) Fabrication of polyaniline/TiO2 nanocomposite ammonia vapor sensor. J Nano Electron Phys 3(1):1056

Prabhakar N et al (2008) Improved electrochemical nucleic acid biosensor based on polyaniline-polyvinyl sulphonate. Electrochim Acta 53(12):4344–4350

Prathap MA, Srivastava R, Satpati B (2013) Simultaneous detection of guanine, adenine, thymine, and cytosine at polyaniline/MnO2 modified electrode. Electrochim Acta 114:285–295

Radhakrishnan S et al (2013) Polypyrrole nanotubes–polyaniline composite for DNA detection using methylene blue as intercalator. Anal Methods 5(4):1010–1015

Raj AD et al (2010) Self assembled V2O5 nanorods for gas sensors. Curr Appl Phys 10(2):531–537

Ram MK, Yavuz O, Aldissi M (2005a) NO2 gas sensing based on ordered ultrathin films of conducting polymer and its nanocomposite. Synth Met 151(1):77–84

Ram MK et al (2005b) CO gas sensing from ultrathin nano-composite conducting polymer film. Sensors Actuators B Chem 106(2):750–757

Raman NK, Anderson MT, Brinker CJ (1996) Template-based approaches to the preparation of amorphous, nanoporous silicas. Chem Mater 8(8):1682–1701

Raut B et al (2012) Novel method for fabrication of polyaniline–CdS sensor for H2S gas detection. Measurement 45(1):94–100

Riegel J, Neumann H, Wiedenmann H-M (2002) Exhaust gas sensors for automotive emission control. Solid State Ionics 152:783–800

Rujisamphan N et al (2016) Co-sputtered metal and polymer nanocomposite films and their electrical responses for gas sensing application. Appl Surf Sci 368:114–121

Sadek A et al (2006) A layered surface acoustic wave gas sensor based on a polyaniline/In2O3 nanofibre composite. Nanotechnology 17(17):4488

Sadek A et al (2008) A polyaniline/WO3 nanofiber composite-based ZnO/64 YX LiNbO3 SAW hydrogen gas sensor. Synth Met 158(1):29–32

Sajjan K et al (2013) Humidity sensing property of polyaniline-cromium oxide nanocomposites. In: Proceeding of international conference on recent trends in applied physics and material science: RAM 2013, AIP Publishing

Santhanam K, Gupta N (1993) Conducting-polymer electrodes in batteries. TRIP 1:284–289

Sarfraz J et al (2013) Printed hydrogen sulfide gas sensor on paper substrate based on polyaniline composite. Thin Solid Films 534:621–628

Sen T et al (2014) Polyaniline/γ-Fe2O3 nanocomposite for room temperature LPG sensing. Sensors Actuators B Chem 190:120–126

Shahadat M et al (2012) Synthesis, characterization, photolytic degradation, electrical conductivity and applications of a nanocomposite adsorbent for the treatment of pollutants. RSC Adv 2 (18):7207–7220

Shahadat M et al (2015) Titanium-based nanocomposite materials: a review of recent advances and perspectives. Colloids Surf B: Biointerfaces 126:121–137
Shahadat M et al (2017) A critical review on the prospect of polyaniline-grafted biodegradable nanocomposite. Adv Colloid Interf Sci 249:2–16
Sharma S et al (2002) Chloroform vapour sensor based on copper/polyaniline nanocomposite. Sensors Actuators B Chem 85(1):131–136
Shirsat MD et al (2009) Polyaniline nanowires-gold nanoparticles hybrid network based chemiresistive hydrogen sulfide sensor. Appl Phys Lett 94(8):083502
Shukla S et al (2012) Fabrication of electro-chemical humidity sensor based on zinc oxide/ polyaniline nanocomposites. Adv Mater Lett 3(5):421–425
Singh V et al (2008) Synthesis and characterization of polyaniline–carboxylated PVC composites: application in development of ammonia sensor. Sensors Actuators B Chem 132(1):99–106
Singla M, Awasthi S, Srivastava A (2007) Humidity sensing; using polyaniline/Mn3O4 composite doped with organic/inorganic acids. Sensors Actuators B Chem 127(2):580–585
Spain E et al (2011) High sensitivity DNA detection using gold nanoparticle functionalised polyaniline nanofibres. Biosens Bioelectron 26(5):2613–2618
Spain E, Keyes TE, Forster RJ (2013) Vapour phase polymerised polyaniline–gold nanoparticle composites for DNA detection. J Electroanal Chem 711:38–44
Srivastava S et al (2010) TiO_2/PANI And MWNT/PANI composites thin films For hydrogen gas sensing. In: AIP conference proceedings
Sun X, Qiao L, Wang X (2013) A novel immunosensor based on Au nanoparticles and polyaniline/ multiwall carbon nanotubes/chitosan nanocomposite film functionalized interface. Nano-Micro Lett 5(3):191–201
Sutar D et al (2007) Preparation of nanofibrous polyaniline films and their application as ammonia gas sensor. Sensors Actuators B Chem 128(1):286–292
Syed AA, Dinesan MK (1991) Review: polyaniline – a novel polymeric material. Talanta 38 (8):815–837
Tai H et al (2007) Fabrication and gas sensitivity of polyaniline–titanium dioxide nanocomposite thin film. Sensors Actuators B Chem 125(2):644–650
Tovide O et al (2014) Graphenated polyaniline-doped tungsten oxide nanocomposite sensor for real time determination of phenanthrene. Electrochim Acta 128:138–148
Vatutsina O et al (2007) A new hybrid (polymer/inorganic) fibrous sorbent for arsenic removal from drinking water. React Funct Polym 67(3):184–201
Verma SK et al (2015) Poly (m-aminophenol)/functionalized multi-walled carbon nanotube nanocomposite based alcohol sensors. Sensors Actuators B Chem 219:199–208
Vijayan A et al (2008) Optical fibre based humidity sensor using co-polyaniline clad. Sensors Actuators B Chem 129(1):106–112
Wang X et al (2012) Synthesis of nestlike ZnO hierarchically porous structures and analysis of their gas sensing properties. ACS Appl Mater Interfaces 4(2):817–825
Wang L et al (2014) Graphene sheets, polyaniline and AuNPs based DNA sensor for electrochemical determination of BCR/ABL fusion gene with functional hairpin probe. Biosens Bioelectron 51:201–207
Wilson J et al (2012) Polypyrrole–polyaniline–Au (PPy–PANi–Au) nano composite films for label-free electrochemical DNA sensing. Sensors Actuators B Chem 171:216–222
Wu J et al (2005) A biosensor monitoring DNA hybridization based on polyaniline intercalated graphite oxide nanocomposite. Sensors Actuators B Chem 104(1):43–49
Xian Y et al (2006) Glucose biosensor based on Au nanoparticles–conductive polyaniline nanocomposite. Biosens Bioelectron 21(10):1996–2000
Xu D-M et al (2013a) Multilayer films of layered double hydroxide/polyaniline and their ammonia sensing behavior. J Hazard Mater 262:64–70
Xu H et al (2013b) NO 2 gas sensing with SnO 2–ZnO/PANI composite thick film fabricated from porous nanosolid. Sensors Actuators B Chem 176:166–173

Yan X et al (2009) Preparation and characterization of polyaniline/indium (III) oxide (PANi/In2O3) nanocomposite thin film. In: 4th international symposium on advanced optical manufacturing and testing technologies: advanced optical manufacturing technologies. International Society for Optics and Photonics

Yang T et al (2009) Synergistically improved sensitivity for the detection of specific DNA sequences using polyaniline nanofibers and multi-walled carbon nanotubes composites. Biosens Bioelectron 24(7):2165–2170

Yang J, Wang X, Shi H (2012) An electrochemical DNA biosensor for highly sensitive detection of phosphinothricin acetyltransferase gene sequence based on polyaniline-(mesoporous nanozirconia)/poly-tyrosine film. Sensors Actuators B Chem 162(1):178–183

Yun J, Jeon S, Kim H-I (2013) Improvement of NO gas sensing properties of polyaniline/MWCNT composite by photocatalytic effect of TiO2. J Nanomater 2013:3

Zhang H-D et al (2014) High-sensitivity gas sensors based on arranged polyaniline/PMMA composite fibers. Sensors Actuators A Phys 219:123–127

Zhihua L et al (2016) Fast response ammonia sensor based on porous thin film of polyaniline/ sulfonated nickel phthalocyanine composites. Sensors Actuators B Chem 226:553–562

Zhong H et al (2011) In situ chemo-synthesized multi-wall carbon nanotube-conductive polyaniline nanocomposites: characterization and application for a glucose amperometric biosensor. Talanta 85(1):104–111

Zhu J et al (2015) Preparation of polyaniline–TiO2 nanotube composite for the development of electrochemical biosensors. Sensors Actuators B Chem 221:450–457

Chapter 7
A New Polyoxovanadate Based Hybrid Materials: A Promising Sensor for Picric Acid and Pd^{2+} Found in the Aqueous Environment

Mukul Raizada, M. Shahid, and Farasha Sama

Abstract Apart from traditional metal-organic frameworks as sensitive materials, discrete cages or clusters to sense hazardous species are uncommon. Keeping this view in mind, a new hybrid discrete material of decavanadate anion and copper complex cations is designed for the purpose. A novel polyoxovanadate (POV)-based inorganic-organic hybrid compound showing the unique combination of first anagostic(V···H) interaction was synthesized. Single crystal X-ray data ascertained the bonding modes and geometry of the complex along with novel anagostic weak intermolecular interactions in the complex material. X-ray crystallography confirmed the composition of the cluster to be $\{Cu(Pyno)_4\}\{NEt_3H\}_2[H_2V_{10}O_{28}]$ (**1**), containing decavanadate as an anion with square planar copper(II) complex and triethylammonium as cations. The compound was further characterized by FTIR, time decay and magnetic studies. Magnetic studies confirmed the presence of the Cu^{2+} state in the complex at RT as well as low temperature. The cluster displayed rare intermolecular V···H, lp···π, V–O···H, π···π and C–H···H interactions, which generate a supramolecular framework. Hirshfeld surface analyses have verified these interactions. The hybrid material is disclosed as the first aqueous phase sensor for picric acid (PA) as well as Pd^{2+}. The complex shows highly sensitive, discriminative and selective sensing behavior for the said species and is the first example of its type in discrete molecule category. The sensing pathways are investigated by spectral titrations, time decay, and DFT (B3LYP/def2–SVP) studies. The lowest detection limit has been discovered for the present POV towards the sensing of both PA and Pd^{2+} ions with ~0.18 and ~0.80 ppb, respectively.

Keywords Polyoxovanadate · Anagostic · Magnetic studies · Aqueous phase sensor · DFT

M. Raizada (✉) · M. Shahid · F. Sama
Department of Chemistry, Aligarh Muslim University, Aligarh, India

M. Oves et al. (eds.), *Modern Age Waste Water Problems*,
https://doi.org/10.1007/978-3-030-08283-3_7

7.1 Introduction

Polyoxovanadates (POVs), as a significant furcation of polyoxometalates (POMs), have fascinated the increasing interest of chemists owing to variable oxidation states of vanadium (+3, +4, and + 5) and various coordination spheres of vanadium oxide (VO) polyhedra including VO_4 tetrahedron, VO_5 square pyramid or trigonal bipyramid, and VO_6 octahedron (Miras et al. 2012; Long et al. 2007; Basler et al. 2002; McGlone et al. 2012; Li et al. 2014; Zhou et al. 2014; Kögerler et al. 2010). POMs are a category of anionic metal oxo clusters with large topological diversity and inherent multifunctional nature (Pope 1983; Pope and Müller 1994, 2001; Cronin and Long 2012; Cronin and Müller 2012). Generally, vanadium oxide polyhedra can be self-condensed to form a variety of vanadium oxide units, including chains, layers, clusters, three-dimensional (3D) frameworks and high-dimensional open structures (Miras et al. 2014; Du et al. 2014; Venegas-Yazigi et al. 2011; Müller et al. 1998; Cameron et al. 2013; Xi and Mak 2012; Chen et al. 2013; Breen and Schmitt (2008); Yin et al. 2011; Truflandier et al. 2010; Li et al. 2008). As a result, POVs are unremarkably utilized as excellent building blocks for fabricating POVs-based hybrid compounds (Zhang and Schmitt 2011; Li et al. 2015a, b; Fernández de Luis et al. 2013; Wutkowski et al. 2013; Kanoo et al. 2011). Nowadays, POM-based crystalline solids with permanent porosity are of great interest because essential features of POMs such as reversible redox properties or high catalytic site density can be combined with the characteristics resulting from open-framework structures (Zhang and Schmitt 2011; Li et al. 2015a, b; Fernández de Luis et al. 2013; Wutkowski et al. 2013; Kanoo et al. 2011; Uchida and Mizuno 2007; Eguchi et al. 2012a, b; Kawahara et al. 2014). Among POMOFs, four diverse approaches for assembling clusters into extended lattices can be identified according to Wang et al.: (2010a) (a) organically derivatized POM units linked by metal ions (Han and Hill 2007); (b) POM clusters directly connected by metal ion linkers in fully inorganic open frameworks (Liu et al. 2013a; Zhang et al. 2014, 2015; An et al. 2015); (c) metal-substituted POMs connected through organic bridging ligands (Nohra et al. 2011; Qin et al. 2015; Zheng et al. 2008); (d) POMs connected through metal-organic ligand-metal linking fragments (Fu et al. 2012; Wang et al. 2010b; An et al. 2006). A fifth subclass closely related to type (b) above could also be proposed: that in which the metal ion linkers belong to distinct coordination complexes with peripheral organic ligands (Zhang and Schmitt 2011; Li et al. 2015a, b; Fernández de Luis et al. 2013; Wutkowski et al. 2013; Kanoo et al. 2011; Dolbecq et al. 2010). Thus, ornamenting POVs by incorporating rational organic ligands is an efficient synthetic approach to prepare POVs-based hybrid materials with engrossing structure motifs (Mahimaidoss et al. 2014; Lan et al. 2008; Liu et al. 2015; Li et al. 2015a, b; Aronica et al. 2008). Polyoxovanadates form an important POM subfamily in which $[H_nV_{10}O_{28}]^{(6-n)-}$ decavanadates are considered among the most appropriate members, as they are the predominant species in acidic aqueous solution (Hayashi 2011). As known, the terminal and bridging oxygen atoms on the surface of POMs can not only act as versatile proton acceptors and donors, but also can coordinate with other metal ions. Therefore, POMs can form either supramolecular assemblies with organic or metal-organic moieties through weak interactions such as van der Waals and H-bonding interactions (Gatteshi et al. 1994), or coordination complexes with metal ions. Crystal engineering of the metal-mediated self-assembly of coordination compounds afforded by S···H, O···H, N···H, S···S, and C–H···π (chelate, CS_2M) secondary interactions plays key roles in the organization of supramolecular networks (Desiraju 2005; Kaiser et al. 2009; Zhong et al. 2008; Takahashi et al. 2006; Sokolov et al. 2006; Steed and Atwood 2000; Armstrong et al. 1995; Tiekink and Schpector 2011; Newman and White 1972a, b; Martin et al. 1972; Brookhart et al. 2007; Huynh et al. 2008; Yao et al. 1997; Saßmannshausen 2012; Zhang et al. 2006; Siddiqui and Tiekink 2013; Schöler et al. 2014; Holaday et al. 2014). The presence of such important C–H···M bonding interactions between the hydrogen atom of C–H on the ligand fragment and metal centres facilitating agostic, anagostic or preagostic and hydrogen bonding interactions is less frequently observed in the organometallic complexes (Brookhart et al. 2007; Huynh et al. 2008; Yao et al. 1997; Saßmannshausen 2012; Zhang et al. 2006; Siddiqui and Tiekink 2013; Schöler et al. 2014; Holaday et al. 2014). In comparison to agostic bonding, the nature of anagostic and hydrogen bonding interactions have been less explored in the literature. In fact, the agostic interactions constituted of 3c-2e bonds are the strongest and supported by a noticeable upfield chemical shift of the participating protons. While, somewhat weaker anagostic interactions are largely electrostatic and show a downfield shift of the uncoordinated C–H protons (Brookhart et al. 2007; Huynh et al. 2008; Yao et al. 1997; Saßmannshausen 2012; Zhang et al. 2006; Siddiqui and Tiekink 2013; Schöler et al. 2014; Holaday et al. 2014). Electron-deficient early transition metals mostly stabilise the agostic interactions whereas the anagostic M···H–C interactions are typically associated with d^8 transition metals centers that are square planar prior to the interaction, as illustrated by a variety of rhodium(I) phosphinate complexes synthesized by Bergman and coworkers (Mukhopadhyay and Pal 2006). Prompted from the conclusion that the M–H–C interaction is not unambiguously designated as a hydrogen bond, the nature of the anagostic M···H–C interaction in a variety of square planar d^8 compounds was recently addressed theoretically by Bergman, Ellman, Oldfield, and coworkers (Zhang et al. 2006). The agostic M···H–C interactions are characterized by comparatively short M–H distances (~1.8–2.3 Å) and small M–H–C bond angles (~90–140°), whereas anagostic interactions are characterized by relatively long M···H distances (~2.3–2.9 Å) and large M···H–C bond angles (~110–170°) (Fig. 7.1) (Zhang et al. 2006; Braga et al. 1996).

The critical point is that M···H–C interactions come in more than one variety, and the nature of the interaction depends critically on the metal centre. Till date, most of the anagostic interactions involve d^8 metal, e.g., Ni (Yadav et al. 2015). However, no report belongs to interpret V···H interactions. Moreover, apart from the structure features, the POMs clusters usually show similar semiconductor photochemical behaviors because of analogous electronic features between band gap transition of semiconductors and HOMO-LUMO transition of POMs (Li et al. 2012; Chen et al. 2014; Barea et al. 2014; Ahmed et al. 2013; Tanaka et al. 2012; Lv et al. 2014; Parrot

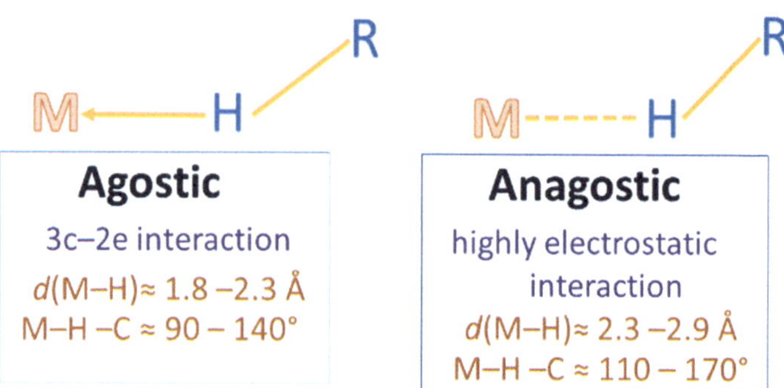

Fig. 7.1 Structural differences between agostic and anagostic interactions

et al. 2013). As a result, POMs are good candidates as inexpensive and green photo catalysts for eliminating organic pollutants from water, in optical, electronic, and magnetic materials. Currently, organic dyes and heavy metal cations (like Pd^{2+}) have been given marvelous attention for their adverse effects on human health and the environment (Tahmasebi et al. 2015; Heng et al. 2014; Shiraishi et al. 2013; He et al. 2015; Liu et al. 2013b; Tan et al. 2014). Moreover, their fluorescence probe for heavy metal ions (Saha et al. 2012; Wang et al. 2014; Cho et al. 2014; Stavila et al. 2014) and electrochemical behaviors were also explored. The rapid and precise detection of explosives could also be achieved through their fluorescent probe properties as it becomes a significant concern in the view of the hurriedly growing use of explosives in international terrorist attacks (Germain and Knapp 2009; Salinas et al. 2012; Shanmugaraju and Mukherjee 2015; Chowdhury and Mukherjee 2015), for security operations, environmental protection, mine field analysis and forensic research (Hu et al. 2014; Cui et al. 2012). The constituents used in most of the industrial explosives are based on nitroaromatics, nitroaliphatics, and organic peroxide compounds, such as 2,4,6-trinitrotoluene (TNT), 3-Nitrophenol (MNP), 2,4-dinitrotoluene (2,4-DNT), 2,4,6-trinitrophenol (TNP) or Picric Acid (PA), 1,3,5-trinitro-1,3,5-triazacyclohexane (RDX), 4-Nitrophenol (PNP), 2,3-dimethyl-2,3-dinitrobutane (DMNB), Nitrobenzene and nitromethane (NM) (Germain and Knapp 2009; Salinas et al. 2012; Shanmugaraju and Mukherjee 2015; Chowdhury and Mukherjee 2015; Toal and Trogler 2006; Meaney and McGuffin 2008; Gole et al. 2014a, b). Amongst these nitro complexes, TNT or picric acid, with remarkable explosive power, is generally used in matches, fireworks, glass, dyes, and leather industries (Hu et al. 2015). TNT disengages into the environment during commercial production and use, leading to the contamination of soil and aquatic systems (He et al. 2009). Thus, the selective and sensitive detection of TNT present in soil and ground water is very significant for tracing interred explosives and environmental monitoring surrounding industrial areas. However, selective and sensitive detection of TNT in the presence of other nitro compounds in the aqueous phase is a challenge due to their strong electron affinities, leading to false responses (Xu et al.

2011). This present hybrid material proves to be an efficient sensor not only for picric acid but also heavy metal Pd^{2+} in the aqueous phase.

7.2 Experimental

Most of the reagents in this study were analytical grade and obtained from commercial sources and used without further purification. $VOSO_4 \cdot 5H_2O$, 4-picoline N-oxide (Pyno), Copper sulphate, Benzene-1,2,4-tricarboxylic acid, Triethylamine (NEt_3), all the nitro aromatic compounds and metal salts were obtained from the Sigma-Aldrich Chemical Co. India.

Caution! *Picric acid is highly explosive and should be handled carefully and in small amounts.*

7.2.1 Physical Methods

The FT-IR spectra of the compounds were recorded within the range 4000–400 cm^{-1} using KBr pellets on a Perkin Elmer Model spectrum GX spectrophotometer. Melting points were determined by the open capillary method and uncorrected. The elemental C, H, and N analysis were obtained from Micro-Analytical Laboratory of Central Drug Research Institute (CDRI), Lucknow, India. The electronic spectrum of 10^{-3} M solution in H_2O was recorded using a Perkin Elmer λ-45 UV visible spectrophotometer with cuvettes of 1 cm path length. Fluorescence measurements were made on a Hitachi F-2700 spectrophotometer. Solid state fluorescence spectra were recorded on a Jobin Yvon Horiba Fluorolog-3 spectrofluorimeter at room temperature. PXRD patterns have been recorded by "Miniflexll X-ray diffractometer" with Cu-Kα radiation. Time-resolved fluorescence spectra were executed in a time-resolved single photon counting setup on FL920 spectrometer (Edinburgh Instruments, UK). The instrument response function (IRF) was obtained using LudoxTM suspension. The emission decay data was analyzed using FAST software available freely online. Thermal gravimetric analysis (TGA) data were recorded from room temperature up to 600 °C at a heating rate of 20 °C min^{-1}. The data were obtained using a Shimadzu TGA-50H instrument. Cyclic voltammetry (CV) was performed on EG&G PAR 273 Potentiostat/Galvanostat an IBM PS2 computer with EG&G M270 software for data acquisition. The three-electrode cell configuration comprised of, a platinum sphere, a platinum plate and Ag(s)/AgCl were used as working, auxiliary and reference electrodes, respectively. The supporting electrolyte used was $[^nBu_4N]ClO_4$. Platinum sphere electrode was sonicated for 2 min in dilute nitric acid, dilute hydrazine hydrate and then in double distilled water to remove the impurities. The solutions were deoxygenated by bubbling research grade nitrogen gas, and an atmosphere of nitrogen was maintained over the solution during measurements.

7.2.2 X-Ray Crystal Structure Determination and Refinements

Single crystal X-ray diffraction of the **1** was performed at 296 K on a Bruker SMART APEX CCD diffractometer; X-ray data were collected using graphite monochromated Mo-$K\alpha$ radiation (l = 0.71073 Å). The linear absorption coefficients, scattering factors for the atoms, and the anomalous dispersion corrections were taken from the International Tables for X-ray Crystallography (Ibers and Hamilton 1974). The data integration and reduction were processed using SAINT Software (SMART & SAINT 2003). Empirical absorption correction was applied to the collected reflections using SADABS (Sheldrick 2002) and the space group was determined using XPREP (XPREP 1995). The structure was solved by direct methods using SIR-97 (Sheldrick 2008) and refined on F^2 by full matrix least squares using the SHELXL-2016/6 program package (Sheldrick 2015). All non-hydrogen atoms were refined with anisotropic displacement parameters. A summary of the crystallographic data and the structure refinement for the complexes is given in Table 7.1. CCDC reference numbers for **1** is 1555755.

7.2.3 Magnetic Measurements

All the magnetic measurements have been performed in Quantum Design made a vibrating sample magnetometer (VSM), (PPMS, Quantum Design). All the magnetic susceptibility data were collected at 0.1 Tesla field in the temperature range of 1.8–300 K on a crystalline solid sample of **1**. To avoid torquing, samples were fixed in the vaseline. Pascal's constant was used for the diamagnetic data correction. For the magnetic relaxation measurements, the sample has been cooled in zero fields to the desired temperatures. After proper thermal stabilization, a magnetic field (H) is applied and magnetization has been measured as a function of time (t).

7.2.4 Fluorescence Experiments

In typical experimental setup, 1 mM solution of **1** was formed in H_2O. In a 1 cm quartz cuvette, 3 mL solution of **1** in water was placed and the fluorescence response upon excitation at 320(**1**) nm was measured in-situ after incremental addition of

Table 7.1 Crystal structure refinement data for **1**

Compound	**1**
Empirical formula	$C_{36}H_{60}CuN_6O_{32}V_{10}$
Formula weight	1661.84
Crystal system	Triclinic
Space group	P-1
Color	Brown
a, Å	10.8029(6)
b, Å	13.5597(7)
c, Å	21.3263(11)
α	96.201(3)
β	90.085(3)
γ	113.148(3)
h	$-12 \leq h \leq 12$
k	$-16 \leq k \leq 16$
l	$-24 \leq l \leq 24$
V, $Å^3$	2852.4(3)
Z	2
ρ calcd (g/cm^3)	1.935
λ, Å	0.71073
T, K	296(2)
μ, mm^{-1}	2.019
F(000)	1666
Reflections collected	9576
Independent reflections	6718
$R^{a,b}$ indices [I > 2σ(I)]	R1 = 0.0818,
	wR2 = 0.2281
R indices (all data)	R1 = 0.1122,
	wR2 = 0.2515
Goodness-of-fit on F^2	1.012

$^{a}R_1 = \Sigma\, ||F_o| - |F_c|| / \Sigma |F_o|$ with $F_o^2 > 2\sigma(F_o^2)$
$^{b}wR_2 = [\Sigma w(|F_o^2|-|F_c^2|)^2 / \Sigma |F_o^2|^2]^{1/2}$

freshly prepared nitro analyte solutions in the range of 330–575 nm, while keeping 2 nm slit width for both source and detector.

7.2.5 DFT Studies

The density functional theory (DFT) calculations were done with the ORCA 3.0.3 computational package (Neese 2009, 2012). The geometry optimization was carried out by hybrid B3LYP functional (Lee et al. 1988) using Aldrich's def2-SVP basis set for C, H, O, N atoms (Weigend and Ahlrichs 2005; Schaefer et al. 1992, 1994). The optimized structure was further re-calculated using a def2-TZVP basis set for all atoms to calculate the HOMO and LUMO energies. To accelerate the calculations, we utilized the resolution of identity (RI) approximation with the decontracted auxiliary def2-SVP/J coulomb fitting basis sets and the chain-of-spheres (RIJCOSX) approximation to exact exchange as implemented in ORCA (Grimme et al. 2010; Steffen et al. 2010). Frontier molecular orbitals (HOMO and LUMO) play an important role to exemplify the chemical reactivity, kinetic stability and electrical transport properties of the molecule. Systems having high *E*HOMO are good electron donors while those having low *E*LUMO are good electron acceptors.

7.2.6 Hirshfeld Surface Analysis

Hirshfeld surfaces (Spackman and Jayatilaka 2009; Seth et al. 2011a, b) mapped with various properties and 2D fingerprint plots (Seth 2014; Seth et al. 2015; Dey et al. 2016) were generated using Crystal Explorer 3.0 (Wolff et al. 2012). This has been proven to be a useful visualization tool for the analysis of intermolecular interactions in the crystal packing (Seth 2014; Seth et al. 2015; Dey et al. 2016). The Hirshfeld surface can be defined in a crystal as the region around a molecule where the ratio of the electron distribution of a sum of spherical atoms for the molecule (the promolecule) to the corresponding amount over the crystal (the procrystal) equals to 0.5. The shape and nature of the surface are directly influenced by the surrounding environment of the molecule in the crystal. For each point on the Hirshfeld isosurface, two distances d_e, the distance from the point to the nearest nucleus external to the surface, and d_i, the distance to the nearest nucleus internal to the surface, are defined. The normalized contact distance (d_{norm}) based on d_e and d_i is given by

$$d_{norm} = \frac{d_i - r_i^{vdW}}{r_i^{vdW}} + \frac{d_e - r_e^{vdW}}{r_e^{vdW}} \tag{7.1}$$

Where r_i^{vdW} and r_e^{vdW} being the van der Waals radii of the atoms. The value of d_{norm} may be negative or positive depending on the intermolecular contacts being shorter or longer than the van der Waals separations (Seth 2014; Seth et al. 2015; Dey et al. 2016). The parameter d_{norm} is indicative of a surface with a red-white-blue coloring scheme (bright-red spots highlight shorter contacts, white colour areas represent contacts around the van der Waals separation, and blue colour regions are devoid of close contacts). The electrostatic potential (ESP) were mapped on the Hirshfeld surface (Zhang and Schmitt 2011; Li et al. 2015a, b; Fernández de Luis et al. 2013; Wutkowski et al. 2013; Kanoo et al. 2011) over the range − 0.136 au (red), through 0 (white), to 1.185 au (blue) for **1**. For this purpose, ab initio wave functions were obtained at HF/STO-3G** using Crystal Explorer 3.0 software. Molecular geometries were taken directly from the relevant crystal structure with H atoms at their neutron distances. 3D-deformation density (DD) maps were also plotted using Crystal Explorer 3.0 for the molecule at the crystal geometry over the electron density iso-surface (the value is 0.008 e/au^{-3}) using Crystal Explorer 3.0. For this purpose, ab initio wave functions were also obtained at HF/STO-3G**.

7.2.7 *Synthesis of {Cu(Pyno)$_4$}{NEt$_3$H}$_2$[H$_2$V$_{10}$O$_{28}$] (1)*

A mixture containing Cu(OAc)$_2$ (2 mmol) and 4-picoline N-oxide (4 mmol) was dissolved in hot H_2O (20 ml) and stirred for 1 h, then $VOSO_4{\cdot}5H_2O$ (4 mmol) was added and refluxed for 4 h at 80 °C. After cooling down at room temperature 0.5 ml of NEt$_3$ was added and further stirred for 20 min, after filtration it is allowed to slow evaporation for 25 days, brownish colored X-ray quality crystals of **1** were obtained. The synthesized complex is soluble in methanol, ethanol, acetone, and water. The complex decomposes at 220 °C. Yield 48%, anal. calcd for $C_{36}H_{60}CuN_6O_{32}V_{10}$: C = 26.02, H = 3.64, N = 14.00; found: C = 26.21, H = 3.28, N = 14.41. IR spectra (KBr pellets, cm^{-1}): ν(C=C): 1620 cm^{-1}, ν(C=N): 1454 cm^{-1}, ν(C–H): 3100, 3026, 2974 cm^{-1}, ν(Cu–O–Cu): 826 cm^{-1}.

7.3 Results and Discussion

7.3.1 *Synthetic Strategy*

Our initial synthesis strategy was to follow the methodology proposed by Shivaiah et al. (Shivaiah and Das 2005) to synthesize a hybrid material with vanadium based Anderson type POM having Cu at the centre in spite of Al. Thus, copper acetate was employed in this synthesis, and the reaction yielded block shaped brown colored crystals with a decavanadate POM with two triethylammonium ions and a Cu-pyno complex cations having the molecular formula {Cu(Pyno)$_4$}{NEt$_3$H}$_2$[H$_2$V$_{10}$O$_{28}$] denoted in the rest of the paper as V$_{10}$O$_{28}$-Cu-pyno-NEt. As, it has already been reported that decavanadate was formed in solution within the pH range of 2–6 (Evans 1966), but the change in pH from acidic to slight basic medium in the last phase of synthesis doesn't affect the formation of decavanadate unit inspite of this it helps in entrapment of metal/organic hybrid. V$_{10}$O$_{28}$-Cu-pyno-NEt crystals are stable in the reaction solution as well in open atmosphere as yellowish-brown colour. X-ray diffraction data have been collected on the same crystal.

7.3.2 *Crystal Structure Investigations of {Cu(Pyno)$_4$} {NEt$_3$H}$_2$[H$_2$V$_{10}$O$_{28}$] (1)*

The crystal structure of {Cu(Pyno)$_4$}{NEt$_3$H}$_2$[H$_2$V$_{10}$O$_{28}$] is shown in Fig. 7.2 along with its packing diagram. The asymmetric unit of the structure {Cu(Pyno)$_4$}

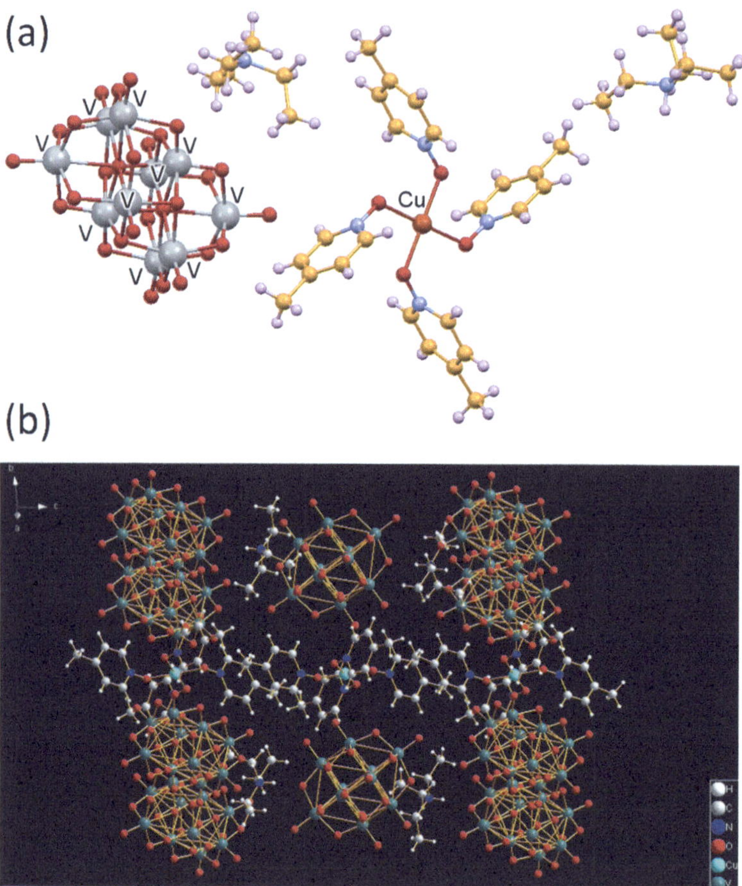

Fig. 7.2 (**a**) Asymmetric structural unit of $V_{10}O_{28}$-Cu-pyno-NEt. (**b**) Packing diagram of the complex

$\{NEt_3H\}_2[H_2V_{10}O_{28}]$ is made up of two units of half decavanadate unit, two units of half Cu-pyno complex and two NEt_3H molecules. The NEt_3H units are arranged diagonally opposite to each other extending along the [1 0 0] direction.

A decavanadate polyanion is made up of highly condensed VO_6 octahedra containing a total of 10 vanadium atoms and 28 oxygen atoms. It is observed as a group of six VO_6 octahedra arranged in a 3 × 2 rectangular arrays in which three VO_6 octahedra are connected by sharing edges, whereas two VO_6 octahedra are joined from above and below by sharing sloping edges with the central six octahedra. There are three kinds of VO_6 octahedra: two octahedra are in the centre of the rectangular group and share seven edges each with adjacent octahedra; four octahedra are at the corners of the rectangular group sharing four edges with neighbours, and four octahedra are at the top and bottom of the group sharing five edges each

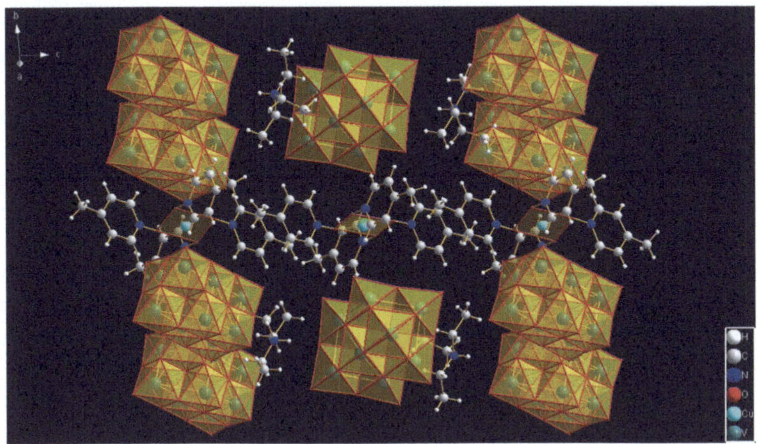

Fig. 7.3 Polyhedral representation of the complex

with neighbors. Two oxygen atoms (Oc) lie inside the $V_{10}O_{28}{}^{6-}$ group and are synchronized to six vanadium atoms each and four oxygen atoms (Ob1) lie on the surface of the group and coordinate to three vanadium atoms each. Further, 14 oxygen atoms (Ob2) lying on the surface coordinate with two vanadium atoms each; and eight oxygen atoms (Ot) lying on the outer corners synchronized only one vanadium atom each (Evans 1966). The Cu-pyno complex has, Cu in a square planar geometry wherein the bonding of Cu is with the four oxygen atoms of the 4-picoline N-oxide in equatorial fashion. Two molecules of triethylamine also entrapped in the crystal lattice in the form of triethylammonium ions (Fig. 7.2). The polyhedral representation of the above complex is shown in Fig. 7.3.

To determine the various protonation sites, the power function $s = (R/1.791)^{-5.1}$ was used which relates the V–O distance and the bond number as R and s, respectively (Brown 1981). The Σs values calculated were 1.18 for the terminal O2 atom connecting V6 and V8 atoms and 1.166 for the terminal O10 atom connecting V1 and V2 atoms in the $V_{10}O_{28}$-Cu-pyno-NEt complex suggesting these sites as protonated. The oxidation state of the vanadium is also confirmed through BVS calculation (Table 7.2) (Brown and Altermatt 1985; Liu and Thorp 1993). The orientation of the pyno rings and the triethylammonium ions in the crystal lattice is such that they provide extra stability to the crystal structure through different kinds of interactions like π···π, lp···lp, O···H, N···H and V···H which give rise to the formation of the supramolecular network (Fig. 7.4a, b). Interestingly, C–C distances between two [Cu-pyno] complexes are in the range of 3.262 Å suggesting that the structure is more stabilized by a possible π· · ·π stacking which may alone lead to the formation of 1D chain (Fig. 7.4c), while O···H and N···H along with π· · ·π interactions lead to the formation of 2D sheet (Fig. 7.4d). The O···H and N···O interactions between NEt_3H and $H_2V_{10}O_{28}{}^{6-}$ lead to the creation of 1D chain (Fig. 7.4b).

Table 7.2 Bond valence sum calculation of Vanadium (V) oxidation state in the crystal structure of **1**[a]

Atom	V^{V}	V^{IV}	V^{III}
V0	**4.985**	4.735	4.239
V1	**4.948**	4.701	4.208
V2	**5.041**	4.789	4.287
V3	**4.920**	4.674	4.184
V4	**4.964**	4.716	4.221
V6	**4.985**	4.736	4.239
V7	**4.953**	4.706	4.212
V8	**4.991**	4.741	4.244
V9	**4.948**	4.701	4.208
V10	**5.023**	4.776	4.228

[a]The Values in **bold** are the closest to the charge for which it was calculated; the nearest whole number can be taken as the oxidation state of that atom

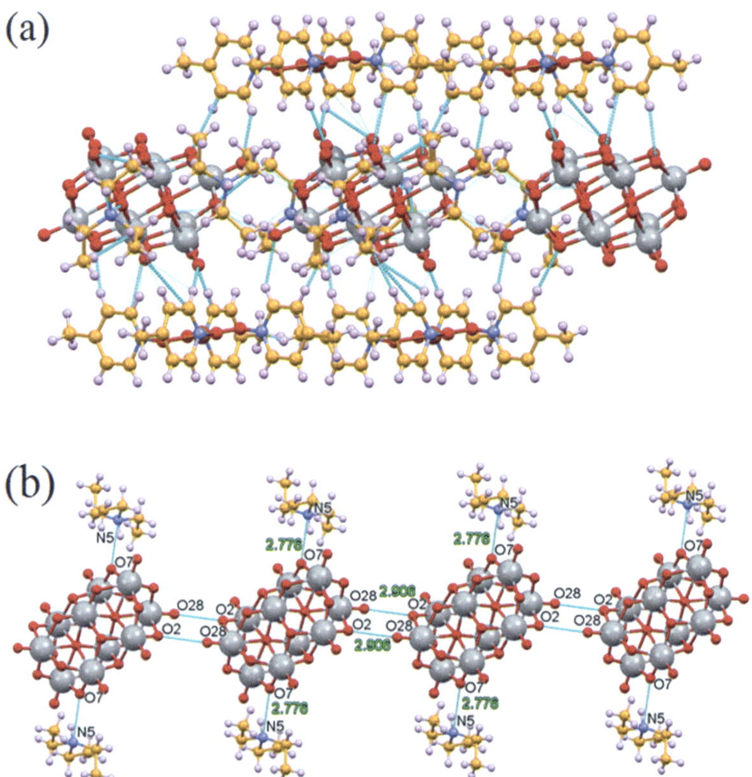

Fig. 7.4 (**a**) 3D Supramolecular network due to different kinds of interactions, (**b**) 1D Chain formed by O···H and N···O interactions

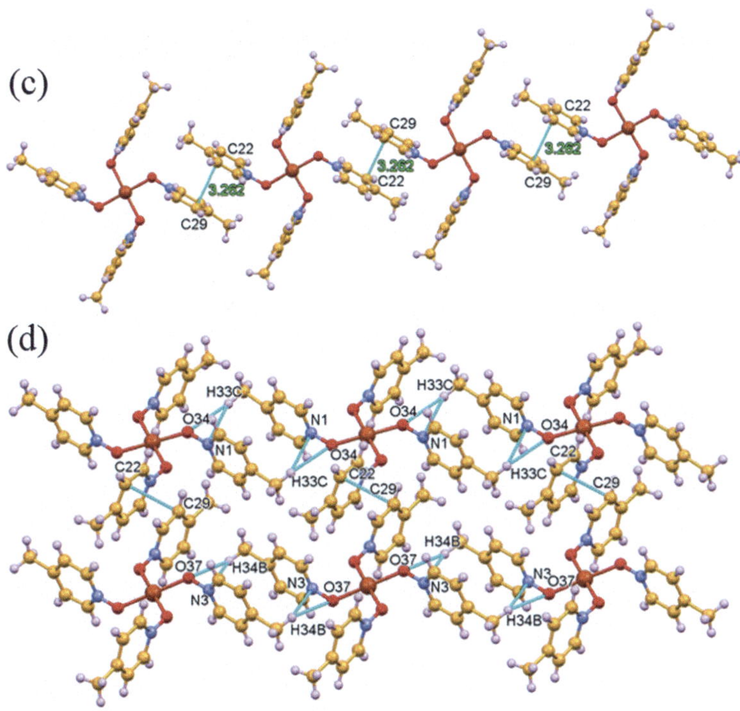

Fig. 7.4 (**c**) 1D chain formed due to π· · ·π stacking, (**d**) 2D sheet formed by O···H, N···H, and π· · ·π interactions

The C–H···O hydrogen bonds between the oxygen of decavanadate anion and the carbon of the cation {$(NEt_3H^+)_2$ or $[Cu(pyno)_4]^{2+}$} occur in the range 3.092–3.213 Å in C···O distance. The mean value (3.168 Å) indicate that C···O distance is not longer than the sum of van der Waals radii of the oxygen atom of decavanadate anion and the carbon of the counter cations. This is in contrast with the other reports (Nakamura and Ozeki 2008; Ito et al. 2012; Ito and Yamase 2010; Nyman et al. 2005, 2009) where longer C···O distances were observed as little contact with the mean value ~3.5 A°. The most critical unprecedented interaction found is V···H interaction whose bond angle is in the range (141.87–149.63°) while the bond length is in the range (3.086–3.180 A°), little more than the required for anagostic interactions (Fig. 7.5).

This is the first report of M···H interactions involving vanadium atom and slight variations as longer distances of V···H anagostic interactions are observed. As per Bergman et al. (Brookhart et al. 2007), anagostic interaction is typically associated with d^8 species (e.g., Ni^{2+}) while in the present case, it is d^0 species (V^{5+}) which shows nearly anagostic interactions. The list of important covalent and noncovalent interactions is given in Table 7.3.

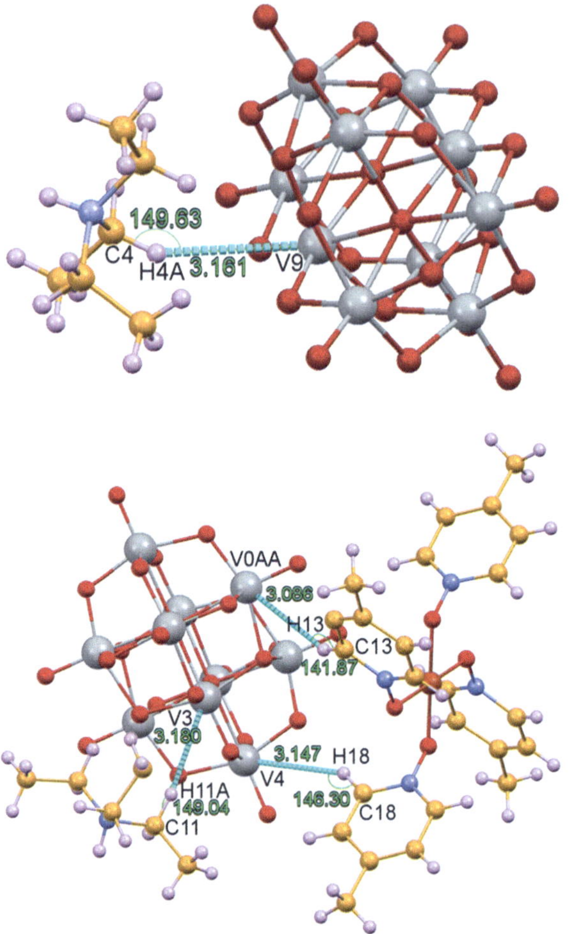

Fig. 7.5 Intermolecular C–H···V interactions on the verge of anagostic interactions

Hydrogen bonding and π···π stacking are playing a significant role in the formation of an extended 3D supramolecular network in the hybrid materials (Fig. 7.4a). In contrast to the structures reported (Dolbecq et al. 2010; Rakovsky et al. 2005; Murphy et al. 1997) containing a Cu complex, $[Cu(pyno)_4]^{2+}$ exists as a square planar arrangement (Fig. 7.4c). These have been further verified through Hirshfeld surface analysis (vide infra). The compound $V_{10}O_{28}$-Cu-pyno-NEt contains V in hexa-coordination having [1 + 4 + 1] bonding with one vanadyl bond, four equatorial bonds and one trans bond.

The vanadyl bond is shorter in length due to the presence of V=O bonds with bond lengths in the range of 1.70–1.95 Å; the equatorial bond lengths are in the range 1.75–2.10 Å, and trans bonds are in the range of 2.30 to 2.35 Å. Baur et al. explained that a substantial difference in bond lengths is an indication of mixed

Table 7.3 Distances and angles of non-covalent interactions in **1**

D–H···A	d(D–H)	d(H···A)	d(D···A)	∟(DHA)
C13–H13···V0AA	0.930	3.086	3.861	141.87
C18–H18···V4	0.930	3.147	3.955	146.30
C4–H4A···V9	0.969	3.161	4.028	149.63
C11– H11A···V3	0.970	3.180	4.043	149.04
N5–C9···O11	1.513	3.193	4.033	113.02
C23–H23···O18	0.930	2.377	3.177	143.98
C24–H24···O28	0.930	2.465	3.272	145.12
C28–C29···C22	1.386	3.262	3.508	88.40
C9–H9A···O11	0.971	2.654	3.193	115.46
C18–H18···O16	0.930	2.171	3.092	170.72
C23–H23···O18	0.930	2.377	3.177	143.98
C5–H5B···O3	0.970	2.610	3.213	120.55

valent vanadium in the compound (Schindler et al. 2000). However, the detailed magnetic susceptibility measurements showed the paramagnetic nature of $V_{10}O_{28}$-Cu-pyno-NEt indicating that vanadium atoms are in the +5-oxidation state and the paramagnetic nature of the sample can be speculated as a contribution from magnetic copper in the +2 state, which correlates with the calculated magnetic moment and brown colour of the crystals (Fig. 7.10a). The selected bond lengths for the compounds $V_{10}O_{28}$-Cu-pyno-NEt are shown in Table 7.4.

7.3.3 FT-IR Spectra, PXRD Analyses, Cyclic Voltammetry and TGA

Vibrational spectra of $V_{10}O_{28}$-Cu-pyno-NEt show the number of characteristics vibrations owing to the presence of a secondary amine, metal oxygen bonding and aromatic carboxylic acid ligand. It provides ample information about the nature of ligand and their coordination modes (Fig. 7.6a). The peaks in the region 400–1000 cm^{-1} correspond to the fingerprint region of the decavanadate ($V_{10}O_{28}{}^{6-}$) anion and the peaks in the range 1100–1650 cm^{-1} are characteristic of the (Cu-pyno) complex. The shift of the C=N absorption band from 1422 cm^{-1} in pyno to 1390 cm^{-1} for the hybrid materials is due to the coordination of nitrogen associated oxygen with copper as confirmed from the crystal structure. The absorption band at 956 cm^{-1} is a significant peak attributed to the stretching vibration of the vanadyl bond (V=O), which correlates with the number of hydrogen bonds in which terminal O atoms are involved. One strong band is observed at 950 cm^{-1} in $V_{10}O_{28}$-Cu-pyno-NEt. This gives some insight into the protonation sites of this bridging O atom. The protonation is much lower as indicated by the predominantly strong peak (Wery et al. 1996; Griffith and Lesniak 1969; Corigliano and di Pasquale 1975; Fuchs et al. 1976). The hydrogen bonds to the terminal O atoms diminish the force

Table 7.4 Selected bond lengths and angles for **1**

Bond length (1)			
O0AA–V6	1.766(6)	O15–V2	1.930(6)
O0AA–V10	1.923(6)	O15–V3	1.946(6)
O2–V6	1.949(6)	O15–V0AA	2.016(6)
O2–V8	2.026(6)	O16–V2	1.762(6)
O3–V9	1.919(6)	O16–V4	1.908(6)
O3–V7	1.925(6)	O17–V7	1.608(7)
O3–V6	2.072(6)	O18–V8	1.770(6)
O4–V9	2.112(6)	O18–V7	1.910(6)
O4–V6	2.268(6)	O19–V10	1.602(6)
O4–V7	2.283(6)	O20–V6	1.600(7)
O4–V8	2.309(6)	O21–V6	1.934(6)
O4–V10	2.314(5)	O21–V9	1.957(6)
O5–V9	1.681(6)	O21–V7	2.022(6)
O5–V10	2.074(6)	O22–V1	1.612(7)
O6–V9	1.702(6)	O23–V2	1.597(7)
O6–V8	2.022(6)	O24–V0AA	1.601(6)
O7–V7	1.816(6)	O25–V1	1.783(7)
O7–V10	1.873(7)	O25–V0AA	1.916(6)
O8–V10	1.795(6)	O26–V0AA	1.832(6)
O8–V8	1.853(6)	O26–V4	1.887(6)
O9–V3	1.706(6)	O27–V4	1.619(6)
O9–V1	2.018(6)	O28–V8	1.613(6)
O10–V2	1.959(6)	O34–Cu1	1.921(9)
O10–V1	2.026(6)	O35–Cu1	1.898(9)
O11–V0AA	1.931(6)	O37–Cu2	1.947(10)
O11–V3	1.933(6)	O38–Cu2	1.927(9)
O11–V2	2.058(6)	N1–O34	1.291(12)
O12–V3	2.117(6)	N2–O35	1.314(12)
O12–V2	2.261(6)	N3–O37	1.289(12)
O12–V0AA	2.297(6)	C32–N4	1.352(16)
O12–V1	2.301(6)	C27–N3	1.307(16)
O12–V4	2.315(5)	C28–N4	1.363(15)
O13–V3	1.680(6)	C24–N3	1.392(16)
O13–V4	2.075(6)	C1–N6	1.521(13)
O14–V4	1.791(6)	C5–N6	1.490(14)
O14–V1	1.863(6)	C17–N1	1.333(15)
V6–O0AA–V10	114.8(3)	V3–O11–V2	108.9(3)
V6–O2–V8	112.2(3)	V3–O12–V3	100.9(2)
V9–O3–V7	108.3(3)	V2–O12–OA	170.6(3)
V9–O3–V6	108.2(3)	V3–O12–V1	170.7(3)
V7–O3–V6	98.0(3)	V2–O12–V1	92.8(2)
V9–O4–V6	91.0(2)	V3–O12–V4	170.5(3)

(continued)

Table 7.4 (continued)

Bond length (1)			
V9–O4–V7	90.3(2)	V2–O12–V4	85.42(19)
V6–O4–V7	170.5(3)	V3–O13–V4	110.6(3)
V9–O4–V8	170.9(3)	V3–O14–V1	114.1(3)
V6–O4–V8	92.3(2)	V2–O15–V3	107.1(3)
V7–O4–V8	85.17(19)	V2–O16–V4	115.5(3)
V9–O4–V8	88.6(2)	N1–O34–Cu	121.2(7)
V9–O5–V10	110.6(3)	N4–O38–Cu	121.4(8)
V9–O6–V8	110.0(3)	O35–Cu1–O34	90.1(4)
V7–O7–V10	115.0(3)	O38–Cu2–O37	90.3(4)
V10–O8–V8	114.5(3)	N3–O37–Cu2	119.8(7)
V3–O9–V1	110.1(3)	N4–O38–Cu2	121.4(8)
V2–O10–V1	112.0(3)	O34–Cu1–O34	180.000(2)
V0AA–O11–V3	108.0(3)	O38–Cu2–O38	180.0(7)

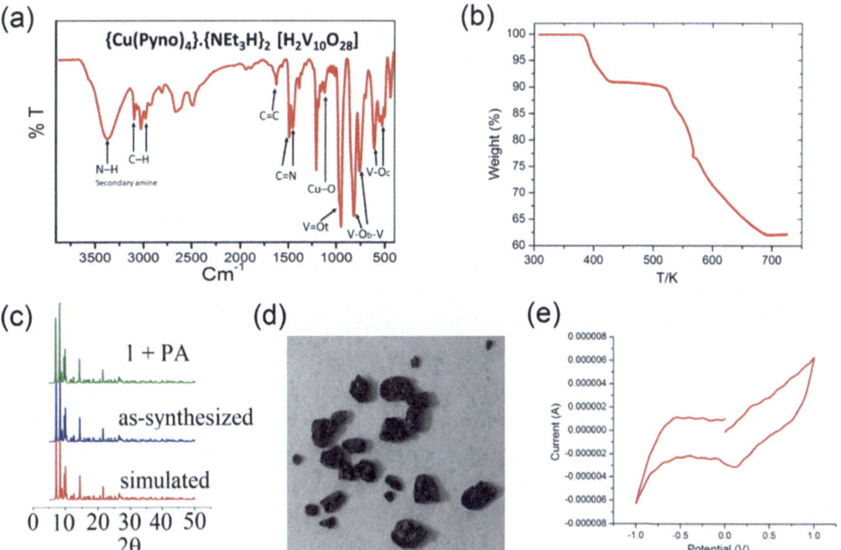

Fig. 7.6 (**a**) FT-IR spectrum, (**b**) thermogram, (**c**) PXRD patterns, (**d**) dark brown block-shaped crystals and (**e**) cyclic voltammogram of the hybrid material $V_{10}O_{28}$-Cu-pyno-NEt

constants of the V=O bonds, and hence the intensity reduction occurs. Therefore, these regions in the IR spectra of the decavanadate system can be regarded as fingerprint regions for characterizing protonated $V_{10}O_{28}{}^{6-}$ polyanions (Griffith and Lesniak 1969).

The shift of the V–Ob–V stretching vibration band to a higher energy region (826–442 cm^{-1} region) is formally due to the increase in the bond strength during the formation of decavanadate polyhedra from VO_3^{-1}. Also, one additional weak band was observed for the hybrid $V_{10}O_{28}$-Cu-pyno-NEt below 500 cm^{-1} due to the vibration of the V^{5+}-O central bond which has a large bond length in this polyhedron (Onodera and Ikegami 1980). The weak C=C vibrational band (1624 cm^{-1}) generally observed for pyno was shifted considerably towards the lower energy region in the case of hybrid and formed as a short band at 1620 cm^{-1} indicating a substantial influence of conjugation on the three aromatic rings of pyno.

A single broad peak at 3360 cm^{-1} clearly indicates the presence of secondary amine formed due to the protonation of triethylamine.

Furthermore, the powder X-ray diffraction (PXRD) measurements were performed on crystalline samples of **1** for the confirmation of the phase purity of the samples. The PXRD patterns of **1** before and after immersing in PA for 48 h are identical and in good agreement with the simulated ones confirming the structural integrity and phase purity of sample **1**. Further, the identical PXRD patterns of the POM and the mixture POM + PA ascertains that there is no degradation of POM after immersing in PA (Fig. 7.6c). The thermal stability of the compound $V_{10}O_{28}$-Cu-pyno-NEt was investigated using TGA analysis as shown in Fig. 7.6b. The TGA curves of the complex exhibit one weight loss step between 380 K and 430 K. The weight loss measured for the compound $V_{10}O_{28}$-Cu-pyno-NEt was 9.8% (calc. 12.3%). These can be attributed to the loss of unbound lattice NEt_3H molecules. The number of triethylammonium ions in the molecular formula of the compound $V_{10}O_{28}$-Cu-pyno-NEt is two. The decomposition of 4-picoline N-oxide and the POM starts from 510 K for the compound $V_{10}O_{28}$-Cu-pyno-NEt. $V_{10}O_{28}$-Cu-pyno-NEt was found to be completely decomposed at 690 K after which a flat plateau followed.

The electrochemical properties of the cation anion combination of $V_{10}O_{28}$-Cu-pyno-NEt have been studied at 298 K. Although the ligands involved are redox-inactive in the applied potential range, the electrochemical processes may arise from the copper metal-based redox reactions. The electrochemical redox properties of **1** are given as (Fig. 7.6e) at 200 V/s scan rate. The voltammogram showed a cathodic peak which is coupled with an anodic peak forming a quasi-reversible redox couple at $E^0_{1/2} = +0.22$ V (Samara et al. 1992; Siddiqi and Mathew 1994; Dimitrou et al. 2001; Sheldon and Kochi 1981; El-Maali et al. 2005; Neelakantana et al. 2008).

$$Cu^{II} \xrightleftharpoons{E^0{}_{1/2}=+0.22} Cu^{I}$$

7.3.4 *Study of Non-covalent Interactions*

The Hirshfeld surfaces (Spackman and Jayatilaka 2009; Seth et al. 2011a, b) of **1** are illustrated in Figs. 7.7 and 7.8, revealing surfaces mapped over a d_{norm} range of −1.083 to 1.465 Å and − 0.455 to 1.391 Å.

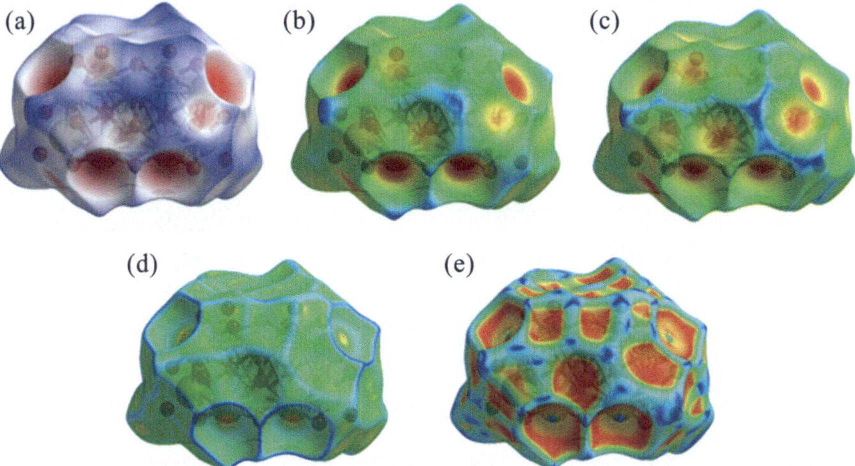

Fig. 7.7 (**a**) Hirshfeld surface of **1** mapped with d_{norm} (**a**), d_i (**b**), d_e (**c**), shape index (**d**) and curvedness (**e**) for decavanadate unit

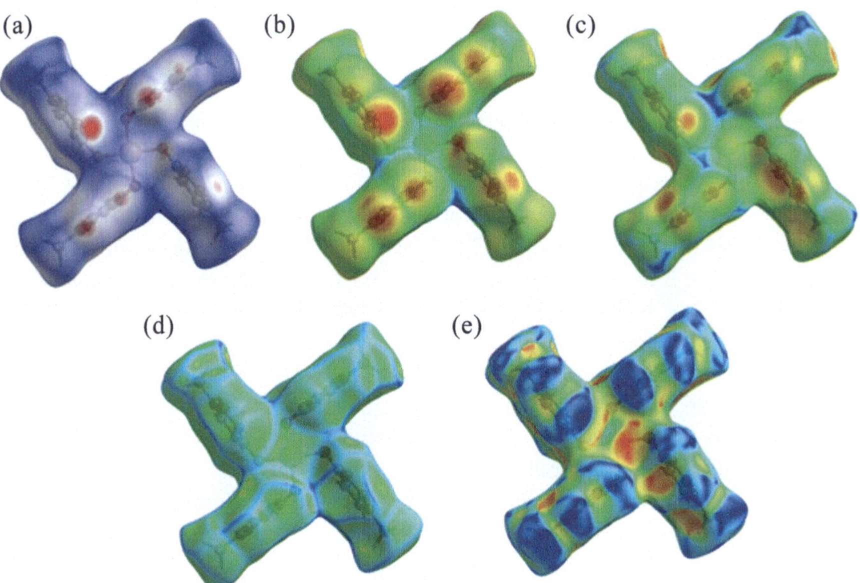

Fig. 7.8 (**a**) Hirshfeld surface of **1** mapped with d_{norm} (**a**), d_i (**b**), d_e (**c**), shape index (**d**) and curvedness (**e**) for the copper unit

These surfaces are transparently shown to visualize the moieties around which these were calculated. The deep red depressions (Sama et al. 2017a; Ansari et al. 2016) visible on the d_{norm} surfaces are indicative of hydrogen bonding contacts. The dominant interactions viz. C–H···O, N–H···O, C–N···O, π···π, C–N···H are

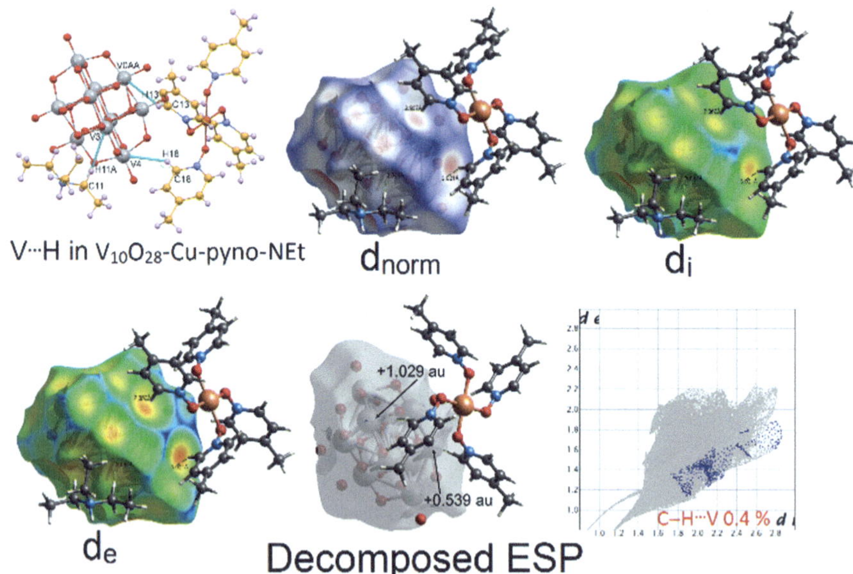

Fig. 7.9 C–H···V interactions on the verge of anagostic interactions through light red spots shown by Hirshfeld surface mapping over dnorm, d_i and d_e, and a decomposed ESP map

shown in the Hirshfeld surface plots as the red-shaded area in Figs. 7.7 and 7.8, and the most unprecedented C–H· · ·V is also confirmed (Fig. 7.9). The 2D-fingerprint plots provide quantitative information about the decomposition of the Hirshfeld surfaces into contributions from the various intermolecular interactions present in the crystal structures (Sama et al. 2017a; Ansari et al. 2016). The fingerprint plots are presented in Fig. 7.10. The intermolecular C–H· · ·O interactions appear as one long sharp spike, π ···π interactions appear as two broad middle side peaks while C–H· · ·V interactions appear as patches in the lower side of d_e 2D fingerprint plots in the region 1. 0 Å < (d_e + d_i) < 5.20 Å shown as a light sky-blue pattern.

The intermolecular C–H· · ·O interactions mapped over d_{norm}, d_i, d_e, shape index, curvedness are shown in Fig. 7.11. Complementary regions are visible in the fingerprint plots where one molecule acts as a donor ($d_e > d_i$) and the other as an acceptor ($d_e < d_i$). The fingerprint plots can be used to highlight a particular atom-pair close contacts. This enables the separation of contributions from different interaction types, which overlap in the full fingerprint.

The electrostatic potential (ESP) for the complex was mapped on the Hirshfeld surfaces displaying the presence of a positive and negative ESP on the V atom of $V_{10}O_{28}{}^{6-}$ of magnitude −0.136 au to +1.185 au for the $V_{10}O_{28}$-Cu-pyno-NEt unit (Fig. 7.12).

This electronegative region on the V atoms lead to the formation of unusual interaction with the electropositive region over the H atoms of pyno and NEt_3H. This results in the creation of the short C–H···V interaction (Fig. 7.13a). The electronegative region over the oxygen atoms of $V_{10}O_{28}{}^{6-}$ gives rise to the formation other

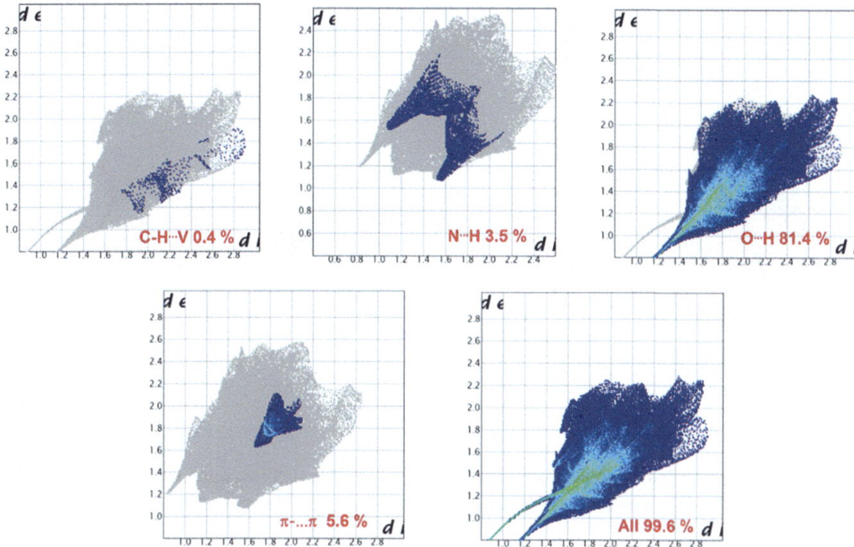

Fig. 7.10 2D Fingerprint plots for various interactions present in the $V_{10}O_{28}$-Cu-pyno-NEt unit

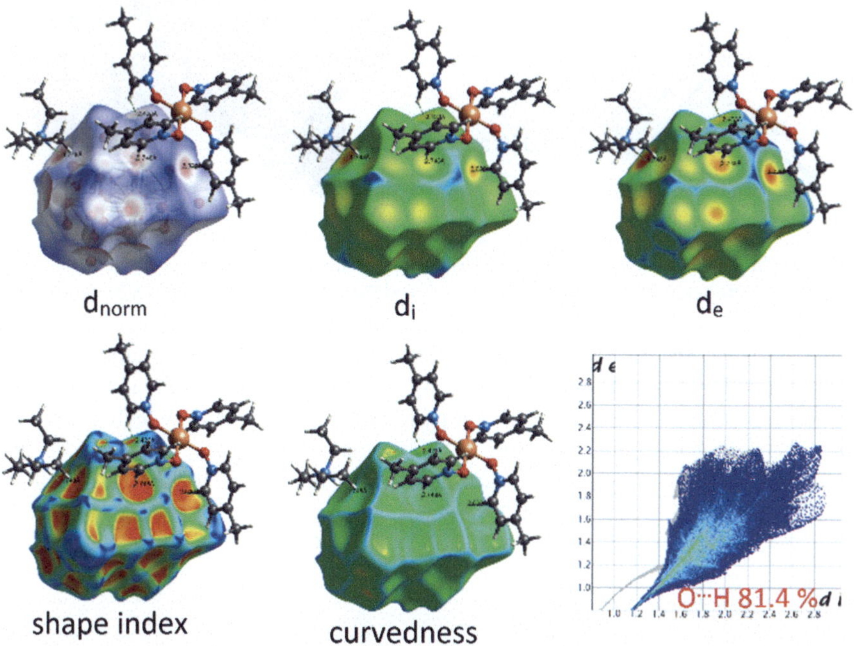

Fig. 7.11 C–H···O interactions through red spots shown by Hirshfeld surface mapping over d_{norm}, d_i and d_e, and a decomposed ESP map

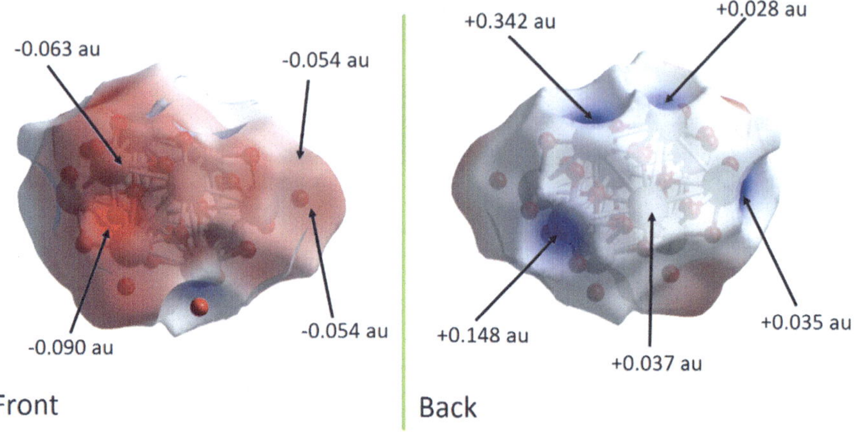

Fig. 7.12 Front and back views of the electrostatic potential (ESP) mapped over the Hirshfeld surface for $V_{10}O_{28}$-Cu-pyno-NEt over the range − 0.136 au (red) through 0.000 (white) to +1.185 au (blue)

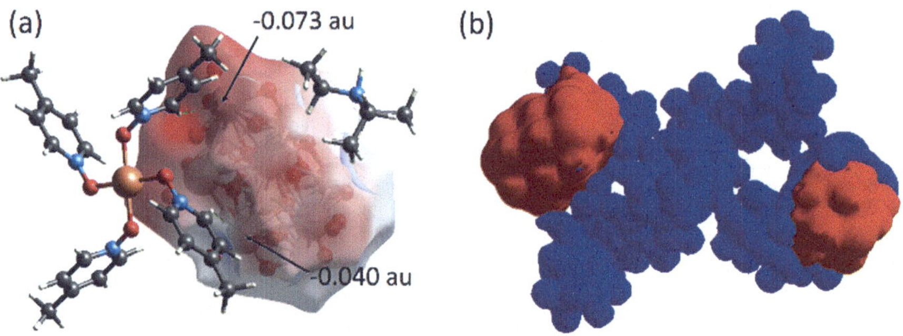

Fig. 7.13 (**a**) Hirshfeld surface mapped with electrostatic potential (ESP) for $V_{10}O_{28}$-Cu-pyno-NEt, showing the σ-hole on the aryl and 2°. (**b**) 3D-deformation density plot for the hybrid showing the presence of charge depletion (CD) region (red) and the charge concentration (CC) region (blue)

contacts like C–H···O, N–H···O in the crystal packing. The deformation density calculations for **1** were also performed (Fig. 7.13b) revealing that the presence of a charge depletion (CD) region on the Cu-pyno and NEt_3H moieties, which is directed towards the charge concentration (CC) region over $V_{10}O_{28}{}^{6-}$ atoms of the decavanadate facilitating formation of the C–H···V contacts in the crystal structure.

7.3.5 Magnetic Studies

The temperature dependence of molar magnetic susceptibility and molar inverse magnetic susceptibility for the compound $V_{10}O_{28}$-Cu-pyno-NEt is plotted in

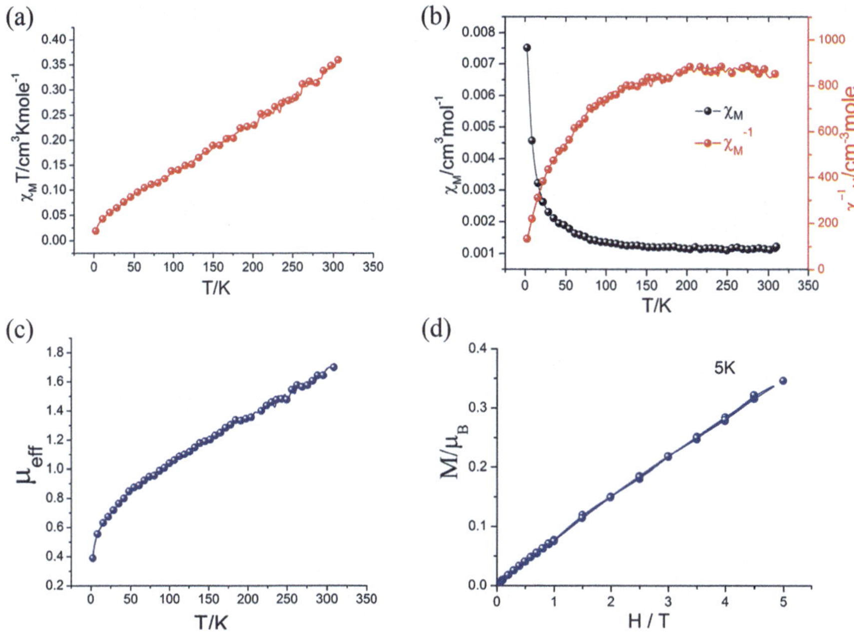

Fig. 7.14 (**a**) χ_MT vs. T plot, (**b**) χ_M vs. T and χ_M^{-1} vs. T plots, (**c**) μ_{eff} vs. T plot, and (**d**) (M/N$_\beta$) vs. H plot

Fig. 7.14. It is apparent from the plot that the susceptibility is positive throughout the temperature range and hardly varies with the temperature down to 75 K. Below this temperature, the susceptibility increases drastically, indicating a probable ferromagnetic ordering at a lower temperature. The inset in Fig. 7.14 shows χ_MT vs. T plot. At room temperature, the χ_MT value is 0.35 cm^3 Kmol^{-1}, which is a little bit lower than expected for the spin-only value for one isolated copper(II) ion with S = 1/2 and g = 2.00. The complex follows Curie-Weiss law with a Curie constant of C = 0.39 cm^3 K mol^{-1} and a Weiss constant of θ = −0.44 K (Sama et al. 2017b). The C value is in the expected range for an isolated Cu(II) ion and θ arises from the weak noncovalent (π···π and O···H) interactions present in the molecule.

Such behavior is indicative of a weak antiferromagnetic interaction. The magnetic moment calculated using the equation,

$$\mu_{eff} = 2.83\sqrt{\chi_M T} \tag{7.2}$$

At 300 K is 1.68 μ_B per formula unit indicating a Cu atom in +2 state (a spin only magnetic moment for copper = 1.73 μ_B) and all ten vanadium sites in the compound exist in the +5-oxidation state with no unpaired electrons. This is also further indicated by the colour of the crystal (dark brown) corresponding to the magnetic copper in +2 state and vanadium in +5 state. The steep increase in susceptibility suggests that the Cu^{2+} spins (s = 1/2) align along the applied magnetic field (Guillou et al. 1992).

The reduced magnetization (per Cu entity), expressed in terms of μ_B (M/N_β) vs. applied field (H) curves of complex increases linearly and progressively, which is not saturated upto 5 T. The magnetization value at the highest measured field (5 T) was found to be 0.35 μ_B. The experimental magnetization of the complex is lower than that of the isolated S = 1/2 the system further affirming the presence of weak antiferromagnetic interactions in monomeric entity (Sama et al. 2017b).

7.3.6 Luminescent Property and Detection of PA and Pd^{2+}

Since, $V_{10}O_{28}$-Cu-pyno-NEt exhibit luminescent properties, emission of the complex is investigated in different solvents (Fig. 7.15). Among them, the intense emission is observed in water in comparative to others. As, water is green solvent, so we decided to investigate the luminescent properties of POV in water. Emission at 384 nm upon excitation at 320 nm in **1**, could be assigned to the luminescence from the intra-ligand transition. The bright-blue luminescence of **1** promotes us to investigate its potential application in selective sensing of explosive nitro aromatics (NACs) and some essential metal ions.

The photoluminescence spectra of **1** dispersed in water possess a strong fluorescence band at 384 nm. To examine the sensing ability of **1** for different nitroaromatics, fluorescence titration experiments were carried out by gradual addition of 2 mM stock solutions of various NACs, viz. NB (Nitrobenzene), MNP (3-Nitrophenol), PNP (4-Nitrophenol), 2,4-DNP (2,4-Dinitrophenylhydrazine) and

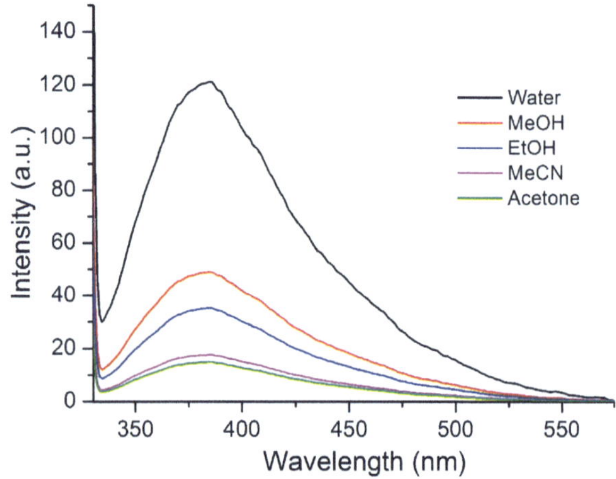

Fig. 7.15 Emission spectrum of **1** dispersed in different solvents upon excitation at 320 nm

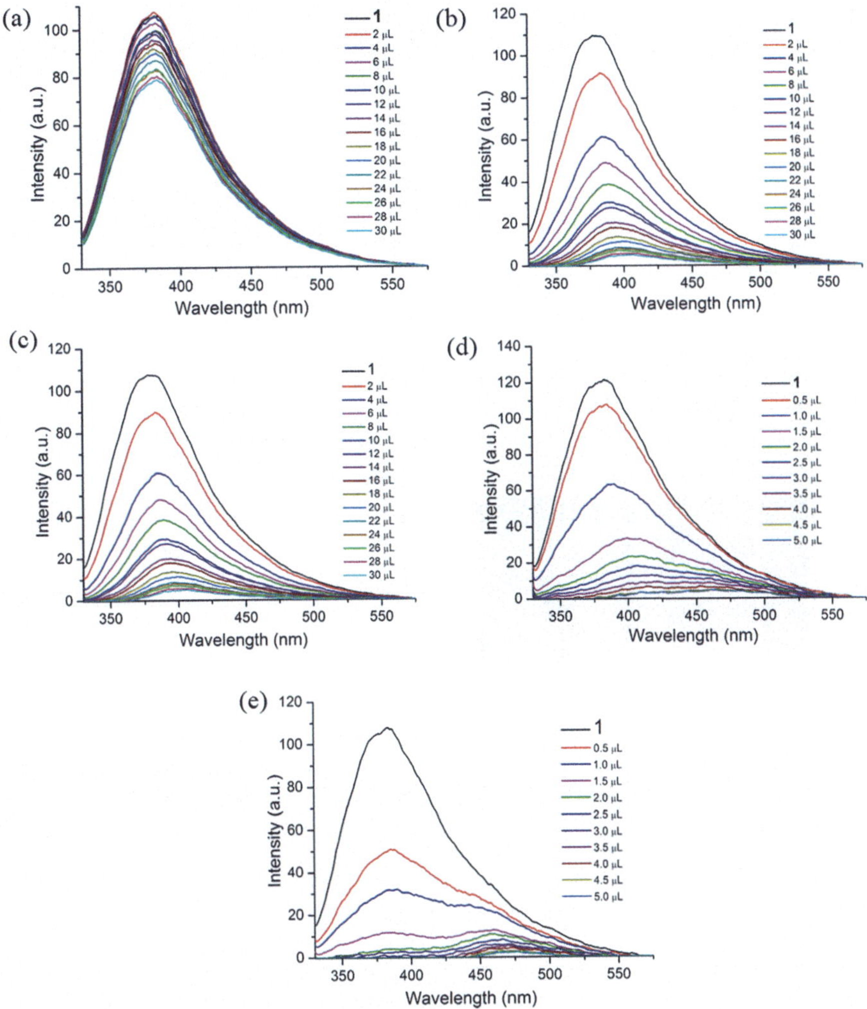

Fig. 7.16 The change in fluorescence intensity of **1** upon incremental addition of (**a**) NB, (**b**) MNP, (**c**) PNP, (**d**) 2,4-DNP and (**e**) PA solution in Water

PA (2,4,6- trinitrophenol). Incremental addition of NB, MNP, and PNP had a minor effect on the fluorescence intensity of **1** (Fig. 7.16).

Interestingly, **1** shows smashing sensing behaviour towards PA by quenching (Quenching $= (I_0 - I)/I_0 \times 100\%$, where, $I_0 =$ initial fluorescence intensity, $I =$ intensity of **1** containing PA solution), the initial fluorescence intensity by 91%, after adding 0.91 ppb of PA in aqueous phase (Fig. 7.17). The appreciable fluorescence quenching of 52% could be clearly observed even at a very low concentration of PA

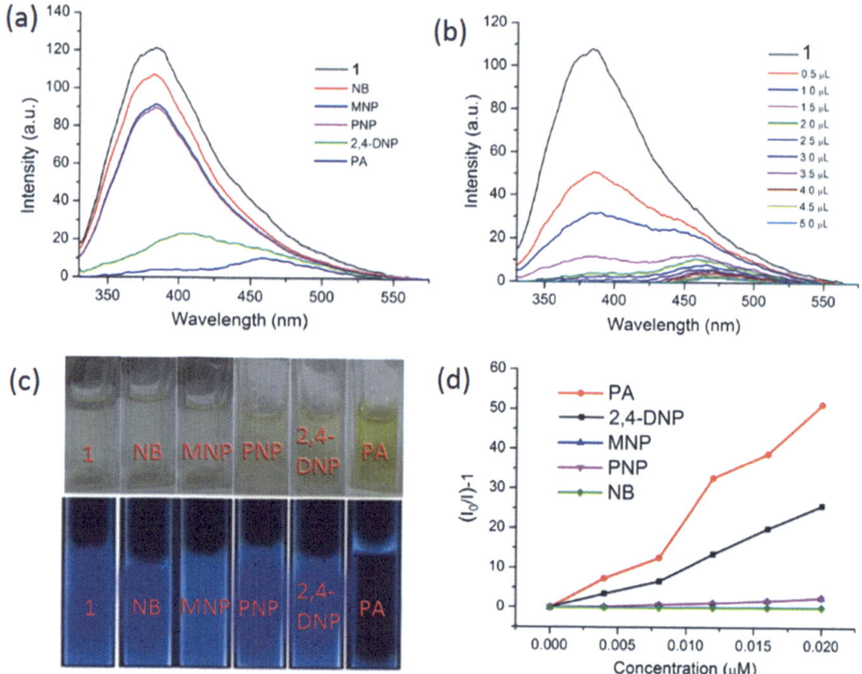

Fig. 7.17 (**a**) The change in fluorescence intensity of **1** ($\lambda_{ex} = 304$ nm) on addition of 0.91 ppb of different NACs. (**b**) The change in fluorescence intensity of **1** with incremental addition of PA. (**c**) Digital photographs of solutions of **1** in the presence of different nitroaromatic analytes under normal light (top) and under portable UV light (bottom). (**d**) The stern-volmer plot for various nitro analytes

with 0.18 ppb in the aqueous phase. The quenching efficiency of **1** is amongst the highest values reported for POMs and MOFs (Hu et al. 2014; Cui et al. 2012; O'Connor 1982; Cui et al. 2012; Zhang et al. 2016) and **1** is most sensitive among them (Fig. 7.18). The stability of the POV framework in a solution state is confirmed by the identical patterns of the solid-state and solution UV-visible spectra even at high temperatures (Fig. 7.19) (Sama et al. 2017b).

To confirm the fluorescence response of POV towards other NACs, we also performed the sensing experiments with NB, MNP, PNP, 2,4-DNP but negligible quenching was examined with NB, MNP and PNP except for 2,4-DNP, which shows appreciable quenching ability (Fig. 7.18). This clearly suggests that there is a strong interaction between PA with **1**, which helps show the highly selective sensing behavior towards PA with the record low, to the best of our knowledge, the detection limit of 0.18 ppb in the aqueous phase.

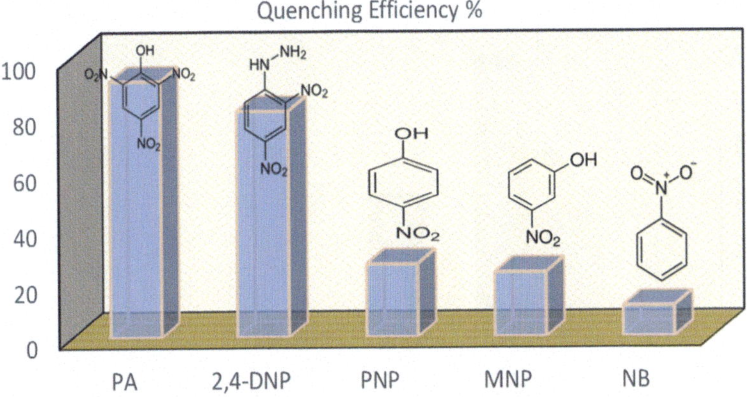

Fig. 7.18 The fluorescence quenching efficiencies of different analytes

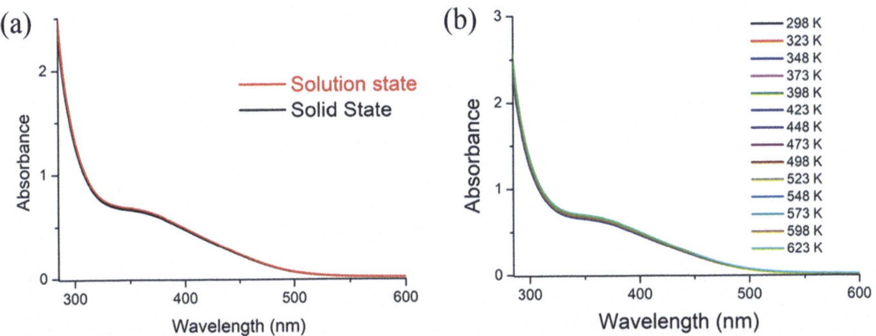

Fig. 7.19 (**a**) Solid and solution (water) state UV-Visible spectra of **1**. (**b**) UV-Visible spectra at different temperatures

To explore the reason behind the quenching mechanism for selective sensing of PA, we analyzed the quenching efficiencies of all the analytes by using the the Stern–Volmer equation (Barea et al. 2014):

$$I_0/I = K_{SV}[A] + 1 \tag{7.3}$$

Where, I_0 = fluorescent intensity of **1** before the addition of the analyte, I = fluorescent intensity after the addition of the respective analyte, K_{SV} = Stern-Volmer constant, [A] = molar concentration of the analyte (M^{-1}). The plots with PA and 2,4-DNP show a linear increase at low concentrations, which deviates from linearity and turns upwards direction with increasing concentrations, whereas the other nitro analytes showed only linear trends in S-V plots (Fig. 7.17d). 3D representation of

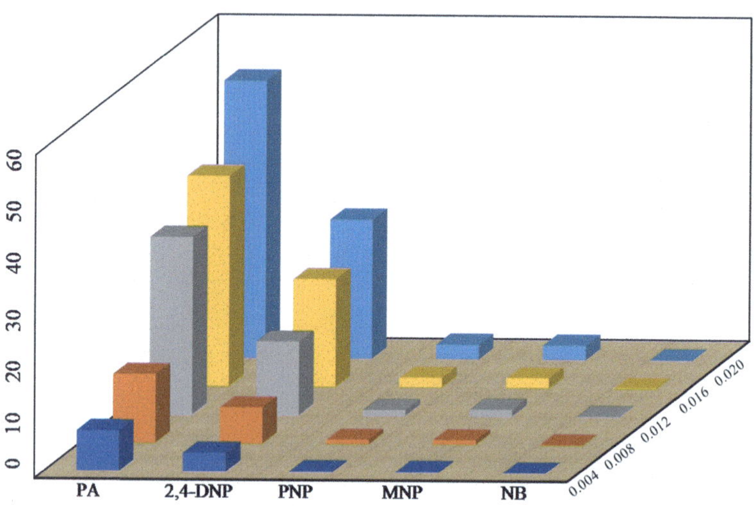

Fig. 7.20 3D representation of Stern-Volmer (SV) plots of **1** for various NACs

Stern-Volmer (SV) plots of **1** for various nitro analytes in the aqueous phase is shown in Fig. 7.20. From the linear fitting of the plot, the quenching constant for PA was found to be 2.4×10^9, which indicates the super quenching ability of **1** towards PA in water.

To the best of our knowledge, this is the highest quenching constant value among the reported POMs as well as MOFs for selective sensing of PA to date (Zhang et al. 2016; Ye et al. 2029). The behavior of interaction (whether static or dynamic) between POV and analytes, can be anticipated by (i) analyzing the absorption spectral changes of POV by the addition of analytes, (ii) time-resolved measurements of photoluminescence decays of the POV after addition of analytes. To explore the mechanism, we measured the fluorescence lifetime of **1** with and without the addition of PA. The average lifetime remained unchanged even after the addition of PA, suggesting that the quenching process follows a static mechanism (Fig. 7.21). To get a deeper insight, we recorded the UV-vis spectrum of **1** in the presence of PA (Fig. 7.22) and found that, upon gradual addition, a new band at 356 nm appears, which ascertained the formation of a non-emissive ground state complex and also supports the static mechanism (O'Connor 1982).

With the other NACs tested, no considerable change was observed in the spectral pattern, except for 2,4–DNP which shows a small peak at 366 (Fig. 7.23). The above spectroscopic results, it is confirmed that there is a strong interaction between PA and **1** in the ground state.

As PA possesses two types of functional groups and to verify whether the hydroxyl group plays any role in sensing, we performed some experiments with hydroxyl substituted aromatic and aliphatic molecules without nitro groups like catechol, 2,6-Bis(hydroxymethyl) p-cresol, di(trimethylolpropane), 1,1,1-Tris (hydroxymethyl) propane. From Fig. 7.24, it is observed that no significant

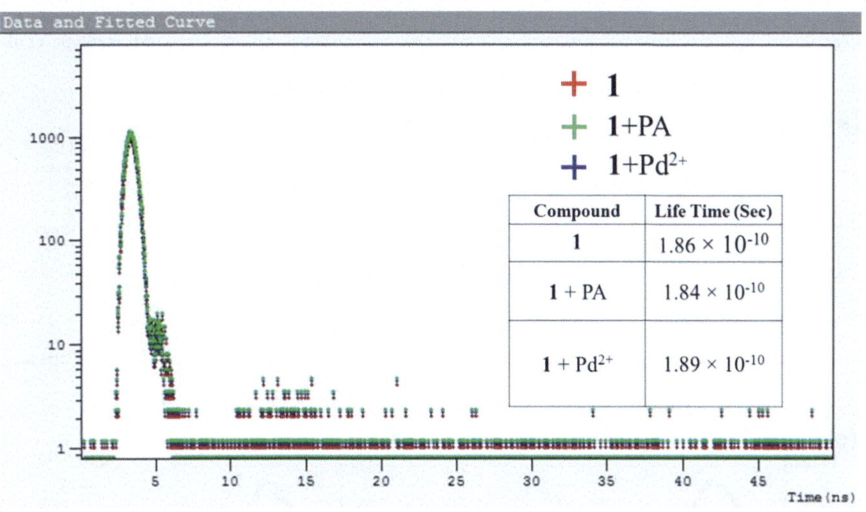

Compound	Life Time (Sec)
1	1.86×10^{-10}
1 + PA	1.84×10^{-10}
1 + Pd^{2+}	1.89×10^{-10}

Fig. 7.21 Fluorescence decay profile of **1** in the presence and absence of PA and Pd^{2+} solution

Fig. 7.22 UV-vis spectra of **1** upon gradual addition of PA showing spectral change with the appearance of the new band at 356 nm

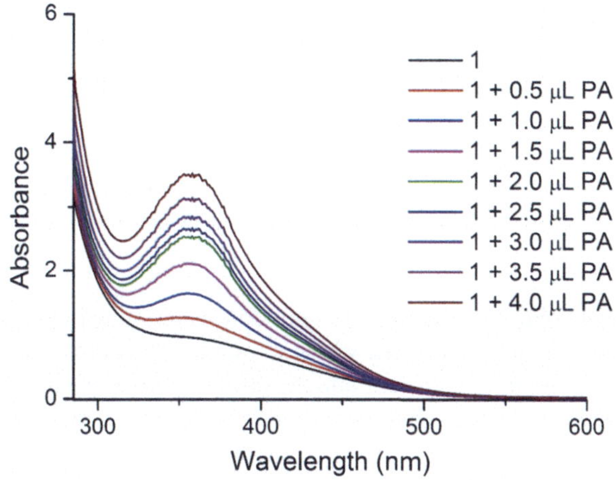

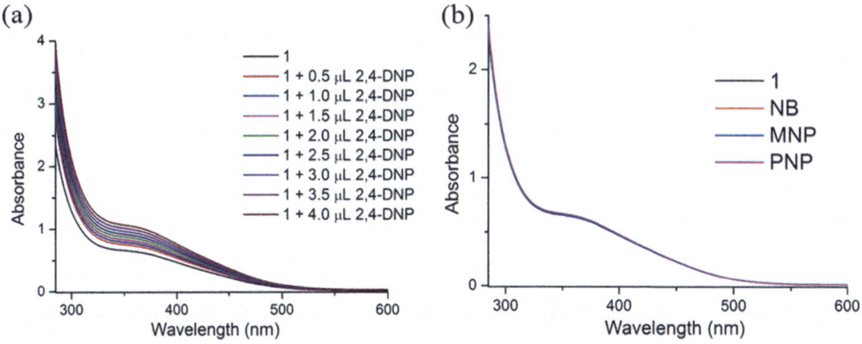

Fig. 7.23 (**a**) UV-vis spectra of **1** upon gradual addition of 2,4-DNP showing spectral change with the appearance of the new band at 366 nm. (**b**) UV-vis spectra of **1** in the presence of different nitro analytes

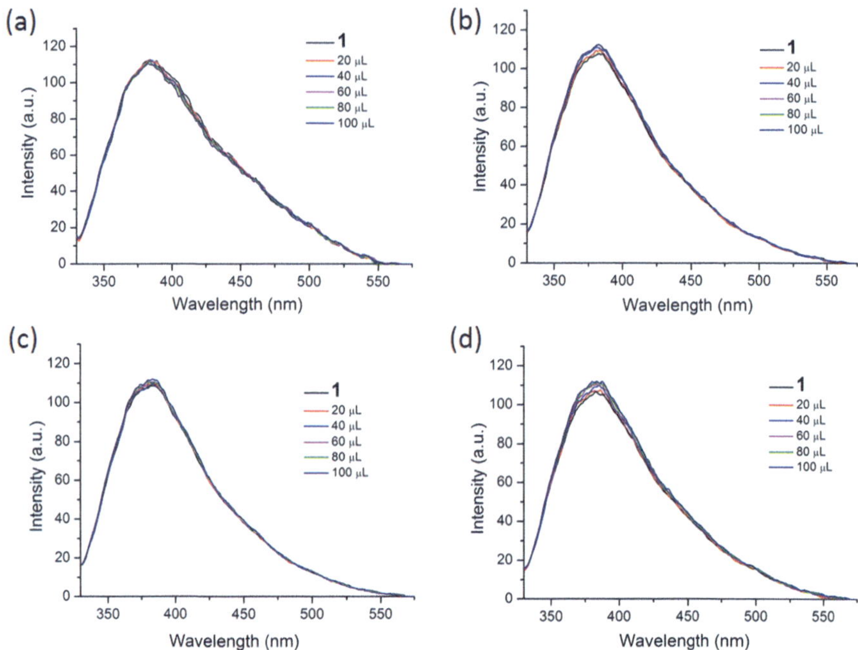

Fig. 7.24 The change in fluorescence intensity of **1** upon incremental addition of Catechol (**a**), 2,6 Bis(hydroxymethyl) p-cresol (**b**), di(trimethylolpropane) (**c**) and 1,1,1-Tris(hydroxymethyl) propane (**d**) (1 mM) solution in Water

quenching was found in the emission spectrum with the gradual addition of analytes, which confirms that the whole PA molecule is responsible for sensing.

To gain more insight into the discriminative sensing behavior of **1** towards PA in the presence of other interfering NACs, we performed some competition experiments by sequential addition of different NACs followed by PA into **1** and the corresponding emission spectra were recorded (Fig. 7.25). The initial addition of different NACs showed negligible intensity quenching, but effective quenching was observed after addition of the PA solution (Fig. 7.26), even in high concentrations, which demonstrates the exceptional selectivity of compounds **1** for PA. Further, from the crystal structure, it can be seen the pores are almost occupied and there is no chance for encapsulation of the nitro analytes into the pores. The mechanism behind the fluorescence quenching might also be photo induced electron transfer between the electron rich framework and the electron deficient analyte adsorbed on the surface (Chen et al. 2014) as **1** has electron rich framework of polyoxometalate.

Generally, POMs structures possess a band structure and exhibit definite band gap energy (O'Connor 1982; Cui et al. 2012; Zhang et al. 2016; Wu et al. 2012; Gole et al. 2014a, b; Salinas et al. 2012). As demonstrated by Li and co-workers, (Pramanik et al. 2011) calculation of electronic band structure is much more suitable compared to molecular orbital (MO) calculations. Indeed, the valence band (VB) and

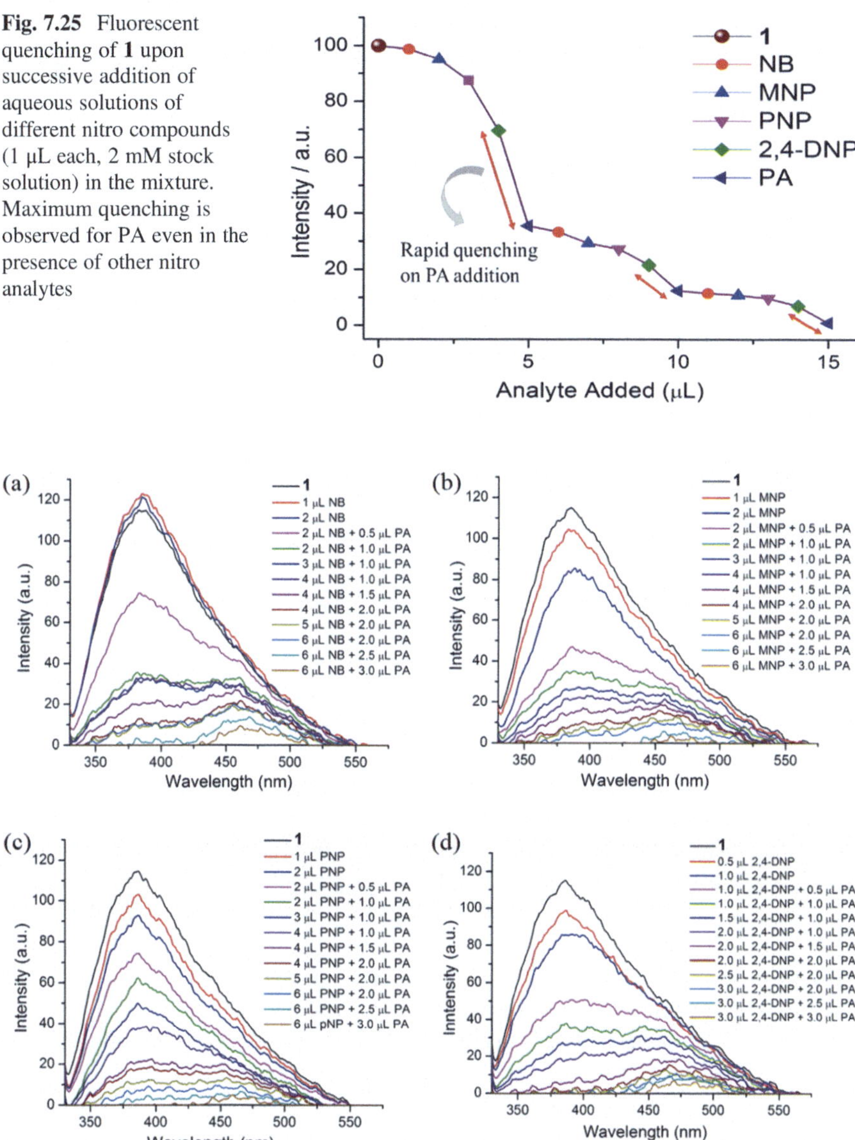

Fig. 7.25 Fluorescent quenching of **1** upon successive addition of aqueous solutions of different nitro compounds (1 μL each, 2 mM stock solution) in the mixture. Maximum quenching is observed for PA even in the presence of other nitro analytes

Fig. 7.26 The change in fluorescence intensity of **1** upon addition of NB (**a**), MNP (**b**), PNP (**c**) and 2,4-DNP (**d**) solution followed by PA solution respectively

conduction band (CB) energies can be defined in a similar manner to the MOs of discrete molecules. One can envisage that electron-deficient NACs will quench the luminescence of the POMs if the lowest unoccupied MOs (LUMOs) of the analytes, which are π^*-type orbitals, reside between the VB and CB of the luminescent POMs.

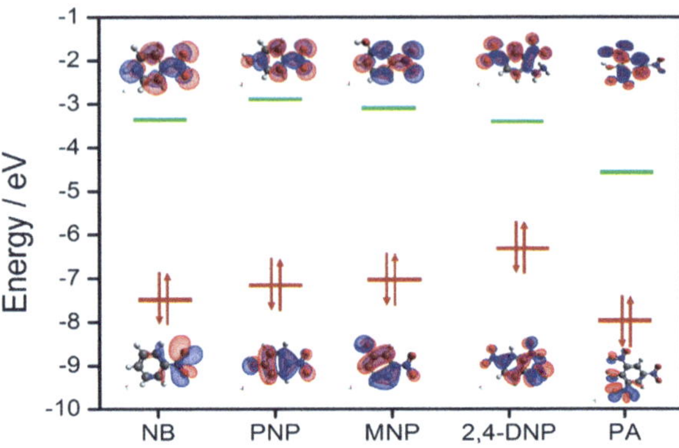

Fig. 7.27 HOMO and LUMO energy levels of the molecular orbitals considered for nitro analytes

Table 7.5 HOMO and LUMO energies calculated for nitroanalytes and ligand at B3LYP/6- 31G* level of theory

Analytes	HOMO (ev)	LUMO (eV)	Band gap (eV)
NB	−7.752	−3.023	4.729
PNP	−7.236	−2.722	4.514
MNP	−7.029	−2.984	4.045
2,4-DNP	−6.408	−3.014	3.394
PA	−8.205	−4.384	3.821

Upon excitation, effective charge transfer can take place from the CB of the POMs to the LUMO of the NACs, thereby quenching the fluorescence intensity. Hence, for more electron-deficient analytes with more stable LUMOs, electron transfer from the CB of a POMs to the LUMOs of the analytes becomes thermodynamically more favorable. However, with electron-rich analytes, for which the LUMO is located above the CB of the POMs, excited-state electrons from the LUMO of the analytes will be transferred to the CB of the POMs, leading to enhancement of the luminescence intensity. To get a deeper insight into this mechanism, we calculated the HOMO–LUMO energies of the employed NACs by DFT at the B3LYP/6-31G* level, as shown in Fig. 7.27, and from the results it is ascertained that the observed maximum fluorescence intensity quenching is due to easy electron transfer from the framework to the lowest LUMO energy level of PA compared to the other NACs (Table 7.5) (Hu et al. 2014; Cui et al. 2012; O'Connor 1982; Cui et al. 2012).

However, the observed order of quenching efficiency was not in full agreement with the corresponding LUMO energies of the other nitro analytes, which indicates that electron transfer is not the sole mechanism for the intensity quenching (O'Connor 1982; Cui et al. 2012; Ye et al. 2029). To explore this, we recorded the UV-vis spectra for all the NACs. Only PA showed the maximum spectral overlap with the emission spectrum of **1** whereas it was negligible for the other nitro analytes (Fig. 7.28), which confirms that the resonance energy transfer mechanism along with the electron transfer mechanism is responsible for the selective fluorescence

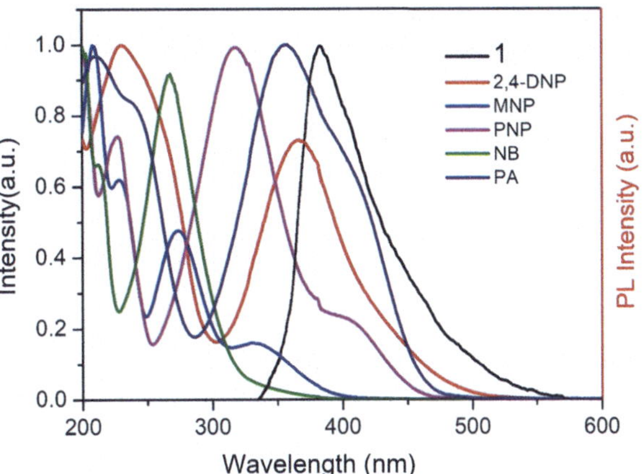

Fig. 7.28 Spectral overlap between normalized emission spectra of **1** ($\lambda_{ex} = 320$ nm) and normalized absorbance spectra of the nitro analytes

quenching with PA, while for other NACs fluorescence quenching happens only through the electron transfer mechanism (Hu et al. 2014; Cui et al. 2012; Nagarkar et al. 2014; Lin et al. 2012; Hendon et al. 2013). Since energy transfer is a long-range process, the emission quenching can pass on to the neighboring fluorophores.

Hence, when the energy-transfer process gets involved in quenching, the quenching efficiency could undergo a radical upsurge, amplifying the detection sensitivity range, as well as selectivity limits. Furthermore, the detection ability of POV can be restored, and it can be recycled for a significant number of cycles by centrifuging the dispersed solution after use (Fig. 7.29). Remarkably, the initial fluorescence intensity was almost regained even after five cycles suggesting a high photostability and reversibility of **1** for detection applications.

Moreover, the sensing ability of the present POV was also examined for some light and heavy metal ions in water. Expectedly, Pd^{2+} has been found to be detected selectively by **1**. The crystal structure of **1** possess pyno connected with a secondary moiety of copper, and thus it is expected that the aromatic (–CH=CH–) moiety pyno could facilitate the selective sensing behavior towards pd^{2+} ions. In a typical experiment, the fluorescence spectra of **1** was recorded upon gradual addition of Pd^{2+} in water solution (1 mM). As expected, the gradual addition of Pd^{2+} solution into **1** resulted in fast and high fluorescence quenching (78%) with a significant spectroscopic variation upon addition of 11 ppb of Pd^{2+} (Fig. 7.30).

1 also displays selective chromogenic behavior towards Pd^{2+} with a color change from colorless to greyish yellow, which was observed by the naked eye (Fig. 7.30b). Specifically, **1** could detect Pd^{2+} even at concentrations as low as 0.8 ppb. It is worth mentioning here that the calculated detection limit of 0.8 ppb is far below the permissible limit of 5–10 ppm set by the World Health Organization (WHO) for Pd^{2+} in drug chemicals and this value falls in the range of reported fluorescence probes for selective Pd^{2+} sensing via fluorescence quenching mechanism (Kaur et al.

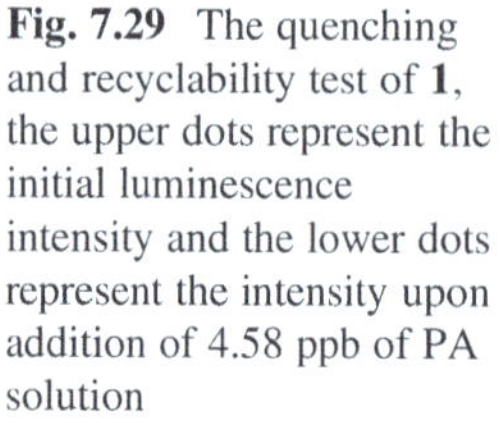

Fig. 7.29 The quenching and recyclability test of **1**, the upper dots represent the initial luminescence intensity and the lower dots represent the intensity upon addition of 4.58 ppb of PA solution

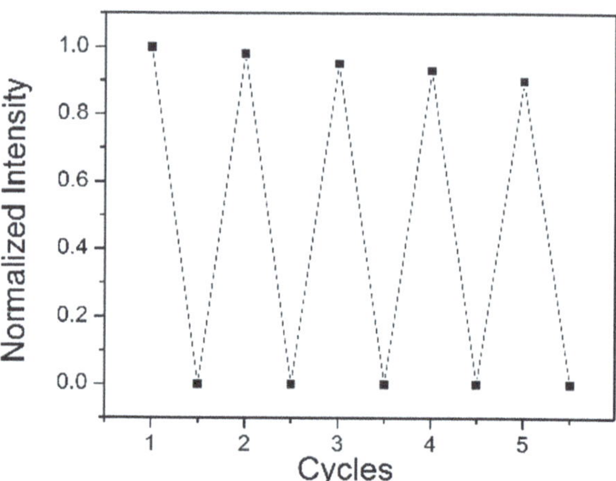

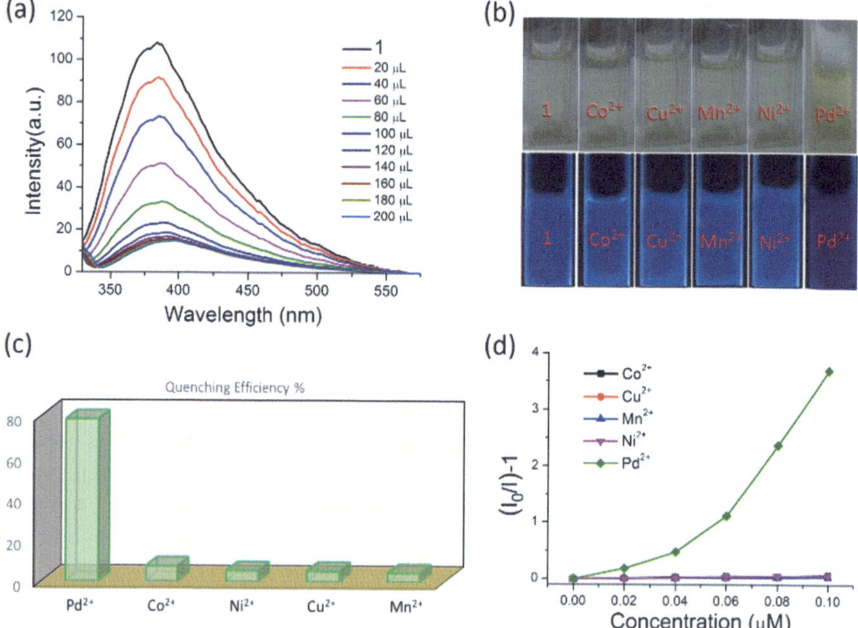

Fig. 7.30 (**a**) The change in fluorescence intensity of **1** ($\lambda_{ex} = 320$ nm) with incremental addition of Pd^{2+}. (**b**) Digital photographs of solutions of **1** in the presence of different metal ions under normal light (top) and under portable UV light (bottom). (**c**) The fluorescence quenching efficiencies of different analytes upon addition of 11 ppb. (**d**) The Stern-Volmer plot for various analytes

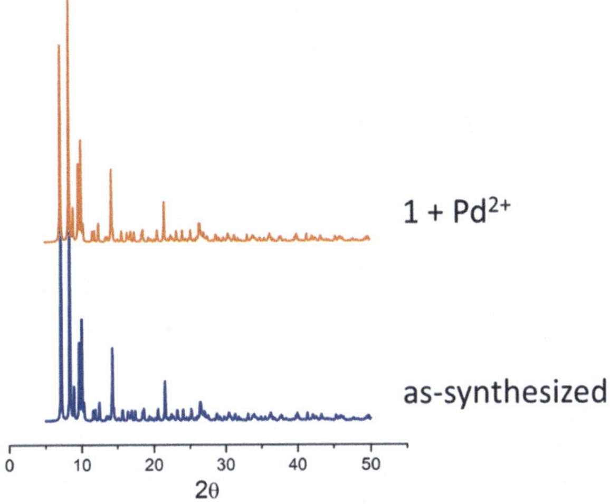

Fig. 7.31 PXRD patterns of **1**: as-synthesized (blue) and after immersion in $Pd2^{+}$ solution for 12 hrs (orange)

2014; Panchompoo et al. 2012; Cai et al. 2013; Garrett and Prasad 2004; Li et al. 2010). Moreover, the stability of the framework was further confirmed by the PXRD patterns of **1**, before and after soaking in the solution of Pd^{2+} for 12 h (Fig. 7.31).

The Stern-Volmer plot (Stern-Volmer constant, KSV $= 4.45 \times 10^{1}\ M^{-1}$) obtained from the fluorescence quenching of **1** with increasing Pd^{2+} concentrations showed an excellent linear response at low concentrations (Fig. 7.30d).

As the concentration increased, the plot deviates from linearity, demonstrating the simultaneous presence of both static and dynamic mechanisms. 3D representation of Stern-Volmer (SV) plots of **1** for various analytes in the aqueous phase is shown in Fig. 7.32. Furthermore, the fluorescence lifetime measurement proved that, in the case of Pd^{2+}, the fluorescence quenching also follows a static mechanism (Fig. 7.21).

To check the selective sensing behavior of **1** towards Pd^{2+}, we performed the fluorescence experiments with some light metal ions, such as Mn^{2+}, Co^{2+}, Ni^{2+} and Cu^{2+} (Fig. 7.33) as well as with some heavy metal ions like Cd^{2+}, Hg^{2+}, Pb^{2+} and Pt^{2+} (Fig. 7.33). All these metal ions had no significant effect on the fluorescence intensity under identical experimental conditions (Figs. 7.30c and 7.34), which suggests the selective sensing behavior of **1** towards Pd^{2+} (Fig. 7.30a).

To examine the selectivity of Pd^{2+} in the presence of other metal ions, we also performed some competition experiments. When we added different metal ions to **1**, the fluorescence intensities were not affected much, even at high concentrations, but effective quenching was observed after adding Pd^{2+} solution, which confirms the remarkable selectivity (Figs. 7.35 and 7.36).

Emission and absorption spectra analyzed the mechanism behind the Pd^{2+} selectivity. In the emission spectrum of **1** with Pd^{2+}. In addition to the intensity quenching, a slight red shift was observed in the band (~7 nm) upon adding an

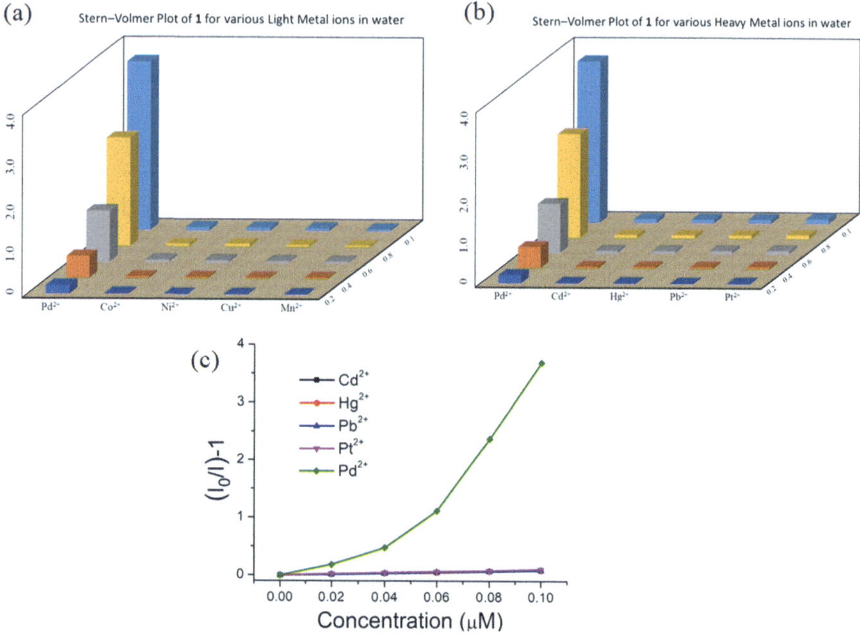

Fig. 7.32 3D representation of Stern-Volmer (SV) plots of **1** for various (**a**) light metal ions, (**b**) heavy metal ions, (**c**) The Stern–Volmer plot for different heavy metal ions

appropriate quantity of Pd^{2+} solution, which suggests the presence of significant interactions between the POV unit and Pd^{2+} (Fig. 7.30a) (Huang et al. 2004). From the selective chromogenic behaviour towards Pd^{2+} and the considerable variation in the fluorescence spectrum, we expect that the observed selective sensing behaviour of **1** for Pd^{2+} over other metal ions is due to the interaction between the unsaturated moieties (–CH=CH–) of the ligand pyno of copper moiety and Pd^{2+} as it is a well-known fact that the π-philic metal ions like Pd^{2+} could co-ordinate with (–CH=CH–) through the (π → d) transitions (Jia et al. 2000; Duan et al. 2008).

For calculating detection limit, PA (0.5–3 μL, 2 mM stock solution) and Pd^{2+} (20–100 μL, 1 mM stock solution) were added to **1**, and corresponding fluorescent intensities were recorded. By plotting fluorescence intensity with increasing concentration of the analyte, slope (*m*) of the respective graph was calculated (Figs. 7.37 and 7.38). Standard deviation (σ) was calculated from five blank measurements of **1** (Table 7.6).

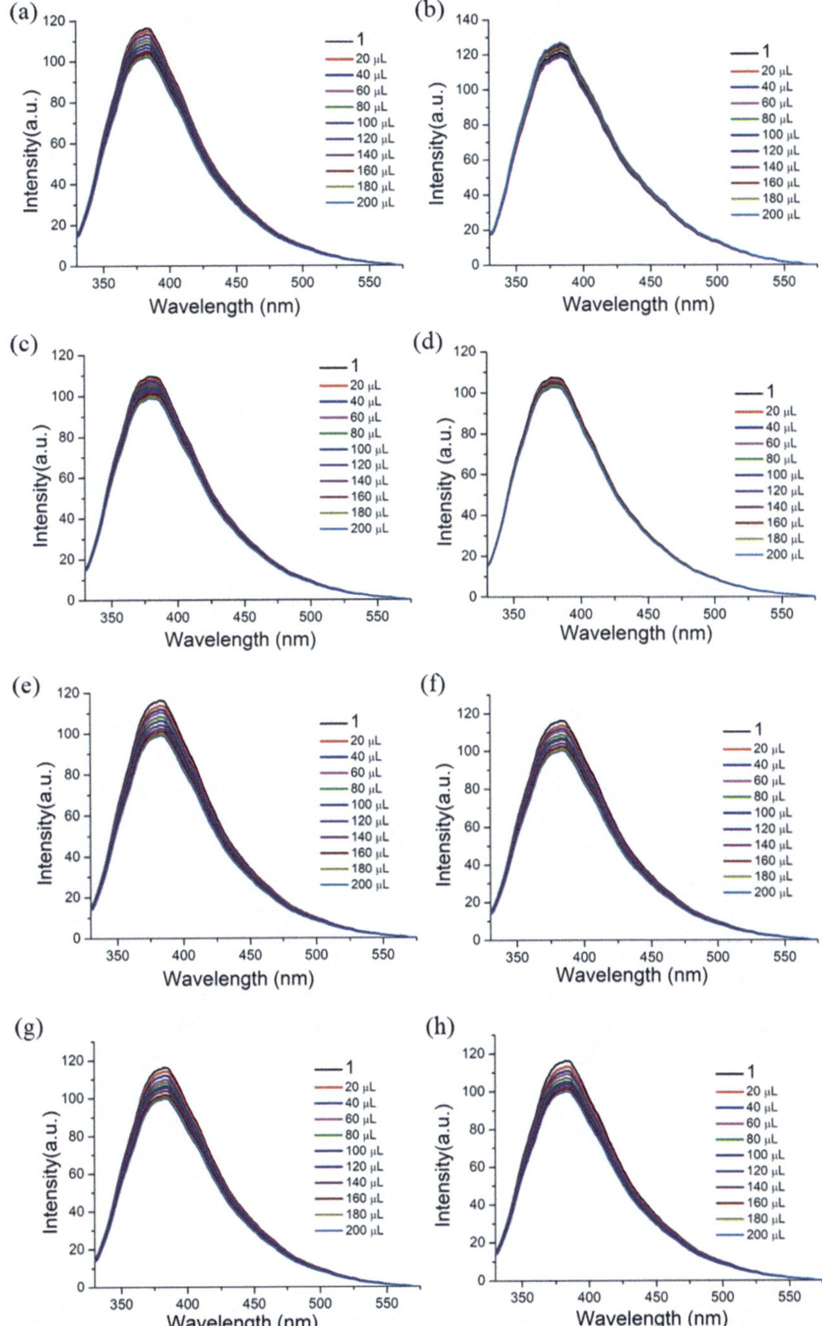

Fig. 7.33 Emission spectrum of **1** upon incremental addition of (**a**) Co^{2+}, (**b**) Ni^{2+}, (**c**) Cu^{2+}, (**d**) Mn^{2+}, (**e**) Cd^{2+}, (**f**) Hg^{2+}, (**g**) Pb^{2+}, (**h**) Pt^{2+} (1 mM) solution in Water

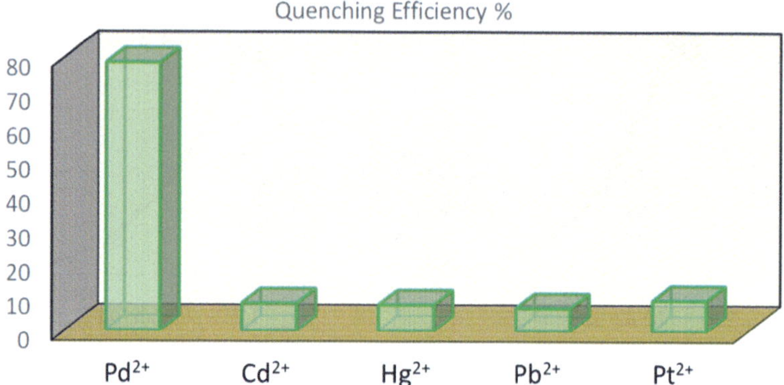

Fig. 7.34 The fluorescence quenching efficiencies of different analytes upon addition of 11 ppb

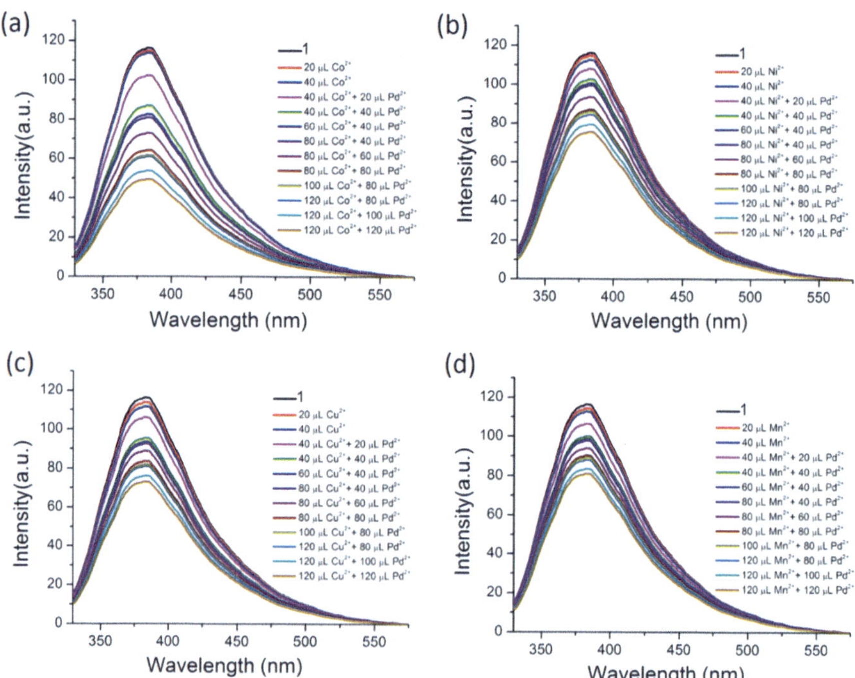

Fig. 7.35 The change in fluorescence intensity of **1** upon addition of Co^{2+}(**a**), Ni^{2+}(**b**), Cu^{2+}(**c**) and Mn^{2+}(**d**) solution followed by Pb^{2+} solution respectively

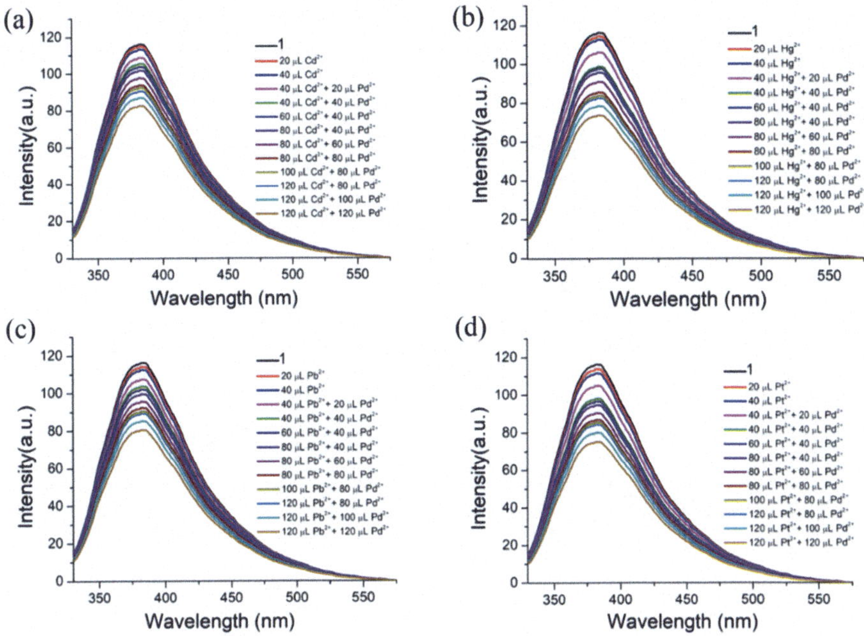

Fig. 7.36 The change in fluorescence intensity of **1** upon addition of Cd^{2+}(**a**), Hg^{2+}(**b**), Pb^{2+}(**c**) and Pt^{2+}(**d**) solution followed by Pb^{2+} solution respectively

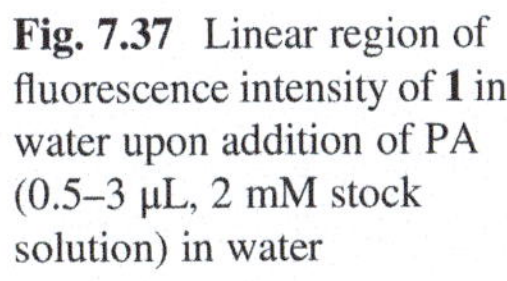

Fig. 7.37 Linear region of fluorescence intensity of **1** in water upon addition of PA (0.5–3 μL, 2 mM stock solution) in water

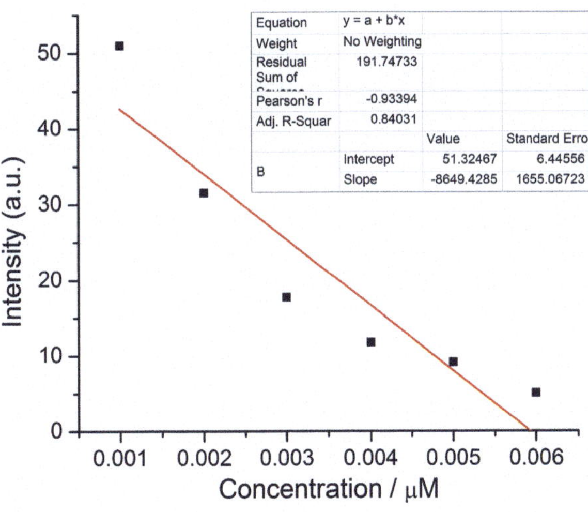

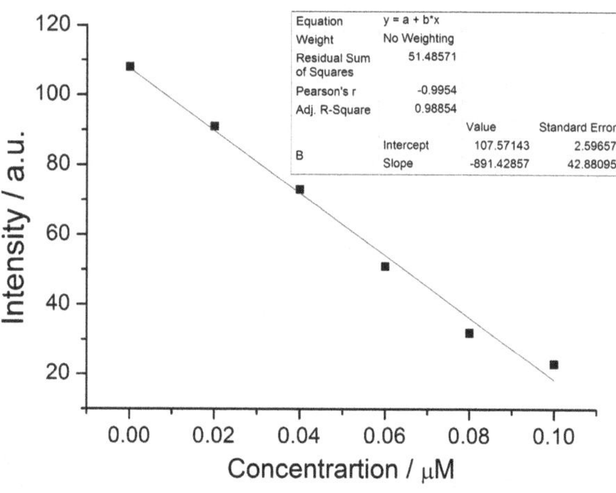

Fig. 7.38 Linear region of fluorescence intensity of **1** in water upon addition of Pd^{2+} (20–100 μL, 1 mM stock solution) in water

Table 7.6 Standard deviation for **1**

Blank readings (only probe)	FL intensity of **1**
Reading 1	111.03
Reading 2	110.26
Reading 3	109.04
Reading 4	107.50
Reading 5	114.49
Standard deviation (σ)	**2.34**

Table 7.7 Detection limit calculation of **1** for PA

Complex	Slope from graph (m)	Detection limit (3σ/*m*)	
1	8649.43	Mμ	ppb
		8.12E-04	~0.18

Table 7.8 Detection limit calculation of **1** for Pd^{2+}

Complex	Slope from graph (m)	Detection limit (3σ/*m*)	
1	891.42	μM	ppb
		7.88E-03	~0. 80

Calculation of Standard Deviation

The detection limit is calculated according to the formula: Detection limit = $(3\sigma/m)$. The detection limit of **1** was found to be ~0.18 for PA and ~0.8 ppb for Pd^{2+} (Tables 7.7 and 7.8).

Calculation of Detection Limit

7.4 Conclusion

Polyoxovanadate (POV) -based inorganic-organic hybrid compound showing sensing of hazardous species was successfully synthesized. The synthesized compound when characterized by FTIR, time decay and magnetic studies confirmed the presence of the Cu^{2+} state and intermolecular V···H, lp···π, V–O···H, π···π and C–H···H interactions thereby confirming the generation of supramolecular framework. The complex shows highly sensitive, discriminative and selective sensing behavior for the said species and is the first example of its type in the discrete molecule category. The sensing pathways are investigated by spectral titrations, time decay, and DFT (B3LYP/def2–SVP) studies. The lowest detection limit has been discovered for the present POV towards the sensing of both PA and Pd^{2+} ions with ~0.18 and ~0.80 ppb, respectively.

References

Ahmed I, Farha R, Goldmann M, Ruhlmann L (2013) Chem Commun 49:496–498
An H-Y, Wang E-B, Xiao D-R, Li Y-G, Su Z-M, Xu L (2006) Angew Chem Int Ed 45:904–908
An H, Hu Y, Wang L, Zhou E, Fei F, Su Z (2015) Cryst Growth Des 15:164–175
Ansari IA, Sama F, Raizada M, Shahid M, Ahmad M, Siddiqi ZA (2016) New J Chem 40:9840–9852
Armstrong DR, Davidson MG, Moncrief D (1995) Angew Chem Int Ed Eng 34:478
Aronica C, Chastanet G, Zueva E, Borshch SA, Clemente-Juan JM, Luneau D (2008) J Am Chem Soc 130:2365–2371
Barea E, Montoro C, Navarro JAR (2014) Chem Soc Rev 43:5419–5430
Basler R, Chaboussant G, Sieber A, Andres H, Murrie M, Kögerler P, Bögge H, Crans DC, Krickemeyer E, Janssen S, Mutka H, Müller A, Güdel H-U (2002) Inorg Chem 41:5675–5685
Braga D, Grepioni F, Biradha K, Desiraju GR (1996) J Chem Soc Dalton Trans:3925–3930
Breen JM, Schmitt W (2008) Angew Chem 120:7010–7014
Brookhart M, Green MLH, Parkin G (2007) Proc Natl Acad Sci U S A 104:6908
Brown ID (ed) (1981) Structure and bonding in crystals. Academic, New York
Brown ID, Altermatt D (1985) Acta Crystallogr Sect B 41:244
Cai S, Lu Y, He S, Wei F, Zhaoab L, Zeng X (2013) Chem Commun 49:822
Cameron JM, Newton GN, Busche C, Long D-L, Oshio H, Cronin L (2013) Chem Commun 49:3395–3397
Chen B, Huang X, Wang B, Lin Z, Hu J, Chi Y, Hu C (2013) Chem Eur J 19:4408–4413
Chen W-C, Qin C, Wang X-L, Li Y-G, Zang H-Y, Jiao Y-Q, Huang P, Shao K-Z, Su Z-M, Wang E-B (2014) Chem Commun 50:13265–13267
Cho W, Lee HJ, Choi G, Choi S, Oh M (2014) J Am Chem Soc 136:12201–12204
Chowdhury A, Mukherjee PS (2015) J Organomet Chem 80:4064–4075
Corigliano F, di Pasquale S (1975) Inorg Chim Acta 12:99–101
Cronin L, Long D-L (2012) Dalton Trans 41(thematic issue):33
Cronin L, Müller A (2012) Chem Soc Rev 41(thematic issue):24
Cui Y, Yue Y, Qian G, Chen B (2012) Chem Rev 112:1126
Desiraju GR (2005) Chem Commun 28:2995
Dey T, Chatterjee P, Bhattacharya A, Pal S, Mukherjee AK (2016) Cryst Growth Des 16:1442

Dimitrou K, Brown AD, Concolino TE, Rheingold AL, Christou G (2001) Chem Commun 14:1284–1285
Dolbecq A, Dumas E, Mayer CR, Mialane P (2010) Chem Rev 110:6009–6048
Du D-Y, Qin J-S, Li S-L, Su Z-M, Lan Y-Q (2014) Chem Soc Rev 43:4615–4632
Duan L, Xu Y, Qian X (2008) Chem Commun 47:6339
Eguchi R, Uchida S, Mizuno N (2012a) J Phys Chem C 116:16105–16110
Eguchi R, Uchida S, Mizuno N (2012b) Angew Chem Int Ed 51:1635–1639
El-Maali NA, Osman AH, Aly AAM, Al-Hazmi GAA (2005) Bioelectrochemistry 65:95–104
Evans HT (1966) Inorg Chem 5:967–977
Fernández de Luis R, Urtiaga MK, Mesa JL, Larrea ES, Iglesias M, Rojo T, Arriortua MI (2013) Inorg Chem 52:2615–2626
Fu H, Qin C, Lu Y, Zhang Z-M, Li Y-G, Su Z-M, Li W-L, Wang E-B (2012) Angew Chem Int Ed 51:7985–7989
Fuchs J, Mahjour S, Palm R, Naturforsch Z (1976) Anorg Chem Org Chem 31B:544
Garrett CE, Prasad K (2004) Adv Synth Catal 346:889
Gatteshi D, Ganeschi A, Pardi L, Sessoli R (1994) Science 265:1054
Germain ME, Knapp MJ (2009) Chem Soc Rev 38:2543–2555
Gole B, Bar AK, Mukherjee PS (2014a) Chem Eur J 20:2276–2291
Gole B, Song W, Lackinger M, Mukherjee PS (2014b) Chem Eur J 20:13662
Griffith WP, Lesniak PJB (1969) J Chem Soc A 7:1066–1071
Grimme S, Antony J, Ehrlich S, Krieg H (2010) J Chem Phys 132:154104
Guillou O, Bergerat P, Kahn O, Bakalbassis E, Boubekeur K, Batail P, Guillot M (1992) Inorg Chem 31:110–114
Han JW, Hill CL (2007) J Am Chem Soc 129:15094–15095
Hayashi Y (2011) Coord Chem Rev 255:2270–2280
He G, Peng H, Liu T, Yang M, Zhang Y, Fang Y (2009) J Mater Chem 19:7347
He Y-C, Yang J, Kan W-Q, Zhang H-M, Liu Y-Y, Ma J-F (2015) J Mater Chem A 3:1675–1681
Hendon CH, Tiana D, Fontecave M, Sanchez C, D'arras L, Sassoye C, Rozes L, Mellot-Draznieks-C, Walsh A (2013) J Am Chem Soc 135:10942
Heng S, Mak AM, Stubing DB, Monro TM, Abell A d (2014) Anal Chem 86:3268–3272
Holaday MGD, Tarafdar G, Kumar A, Reddy MLP, Srinivasan A (2014) Dalton Trans 43:7699
Hu Z, Deibert BJ, Li J (2014) Chem Soc Rev 43:5815
Hu X, Qin C, Wang X, Shao K, Su Z (2015) New J Chem 39:7858–7862
Huang H, Wang K, Tan W, An D, Yang X, Huang S, Zhai Q, Zhou L, Jin Y (2004) Angew Chem Int Ed 43:5635
Huynh HV, Wong LR, Ng PS (2008) Organometallics 27:2231
Ibers JA, Hamilton WC (1974) International tables for x-ray crystallography, vol IV. Kynoch Press, Birmingham
Ito T, Yamase T (2010) Materials 3:158
Ito T, Taira M, Fukumoto K, Yamamoto K, Naruke H, Tomita K (2012) Bull Chem Soc Jpn 85:1222–1224
Jia C, Piao D, Oyamada J, Lu W, Kitamura T, Fujiwara Y (2000) Science 287:1992
Kaiser CR, Pais KC, de Souza MVN, Wardell JL, Wardell SMSV, Tiekink ERT (2009) CrystEngComm 11:1133
Kanoo P, Ghosh AC, Maji TK (2011) Inorg Chem 50:5145–5152
Kaur P, Kaur N, Kaur M, Dhuna V, Singh J, Singh K (2014) RSC Adv 4:16104
Kawahara R, Uchida S, Mizuno N (2014) Inorg Chem 53:3655–3661
Kögerler P, Tsukerblat B, Müller A (2010) Dalton Trans 39(21):36
Lan Y-Q, Li S-L, Wang X-L, Shao K-Z, Du D-Y, Su Z-M, Wang E-B (2008) Chem Eur J 14:9999–10006
Lee C, Yang W, Parr RG (1988) Phys Rev B 37:785
Li J-R, Yu Q, Sañudo C, Tao Y, Song W-C, Bu X-H (2008) Chem Mater 20:1218–1220
Li H, Fan J, Du J, Guo K, Sun S, Liu X, Peng X (2010) Chem Commun 46:1079

Li S, Liu S, Liu S, Liu Y, Tang Q, Shi Z, Ouyang S, Ye J (2012) J Am Chem Soc 134:19716–19721
Li S, Sun W, Wang K, Ma H, Pang H, Liu H, Zhang J (2014) Inorg Chem 53:4541–4547
Li J, Huang X, Yang S, Xu Y, Hu C (2015a) Cryst Growth Des 15:1907–1914
Li J-K, Huang X-Q, Yang S, Ma H-W, Chi Y-N, Hu C-W (2015b) Inorg Chem 54:1454–1461
Lin C-K, Zhao D, Gao W-Y, Yang Z, Ye J, Xu T, Ge Q, Ma S, Liu D-J (2012) Inorg Chem 51:9039
Liu W, Thorp HH (1993) Inorg Chem 32:4102
Liu D, Lu Y, Tan H-Q, Chen W-L, Zhang Z-M, Li Y-G, Wang E-B (2013a) Chem Commun 49:3673–3675
Liu B, Yang J, Yang G-C, Ma J-F (2013b) Inorg Chem 52:84–94
Liu D, Lu Y, Tan H-Q, Wang T-T, Wang E-B (2015) Cryst Growth Des 15:103–114
Long D-L, Burkholder E, Cronin L (2007) Chem Soc Rev 36:105–121
Lv H, Guo W, Wu K, Chen Z, Bacsa J, Musaev DG, Geletii YV, Lauinger SM, Lian T, Hill CL (2014) J Am Chem Soc 136:14015–14018
Mahimaidoss MB, Krasnikov SA, Reck L, Onet CI, Breen JM, Zhu N, Marzec B, Shvets IV, Schmitt W (2014) Chem Commun 50:2265–2267
Martin JW, Newman PWG, Robinson BW, White AH (1972) J Chem Soc Dalton Trans 1:2233
McGlone T, Thiel J, Streb C, Long D-L, Cronin L (2012) Chem Commun 48:359–361
Meaney MS, McGuffin VL (2008) Anal Bioanal Chem 391:2557–2576
Miras HN, Yan J, Long D-L, Cronin L (2012) Chem Soc Rev 41:7403–7430
Miras HN, Vilà-Nadal L, Cronin L (2014) Chem Soc Rev 43:5679–5699
Mukhopadhyay A, Pal S (2006) Eur J Inorg Chem 23:4879–4887
Müller A, Peters F, Pope MT, Gatteschi D (1998) Chem Rev 98:239–271
Murphy G, Nagle P, Murphy B, Hathaway B (1997) J Chem Soc Dalton Trans 15:2645–2652
Nagarkar SS, Desai AV, Ghosh SK (2014) Chem Commun 50:8915
Nakamura S, Ozeki T (2008) Dalton Trans 0:6135–6140
Neelakantana MA, Rusalraja F, Dharmarajaa J, Johnsonrajaa S, Jeyakumarb T, Pillaic MS (2008) Spectrochim Acta Part A 71:1599–1609
Neese F (2009) Orca. An ab Initio, Density functional and semiempirical program package version
Neese F (2012) WIREs Comput Mol Sci 2:73
Newman PWG, White AH (1972a) J Chem Soc Dalton Trans 1:1460
Newman PWG, White AH (1972b) J Chem Soc Dalton Trans 1:2239
Nohra B, El Moll H, Rodriguez Albelo LM, Mialane P, Marrot J, Mellot-Draznieks C, O'Keeffe M, Ngo Biboum R, Lemaire J, Keita B, Nadjo L, Dolbecq A (2011) J Am Chem Soc 133:13363–13374
Nyman M, Ingersoll D, Singh S, Bonhomme F, Alam TM, Brinker CJ, Rodriguez MA (2005) Chem Mater 17:2885
Nyman M, Rodriguez MA, Anderson TM, Ingersoll D (2009) Cryst Growth Des 9:3590
O'Connor CJ (1982) Prog Inorg Chem 29:203
Onodera S, Ikegami Y (1980) Inorg Chem 19:615–618
Panchompoo J, Aldous L, Baker M, Wallace MI, Compton RG (2012) Analyst 137:2054
Parrot A, Izzet G, Chamoreau L-M, Proust A, Oms O, Dolbecq A, Hakouk K, El Bekkachi H, Deniard P, Dessapt R, Mialane P (2013) Inorg Chem 52:11156–11163
Pope MT (1983) Heteropoly and isopoly oxometalates. Springer, Berlin
Pope MT, Müller A (eds) (1994) Polyoxometalates: from platonic solids to anti-retroviral activity. Kluwer, Dordrecht
Pope MT, Müller A (eds) (2001) Polyoxometalate chemistry: from topology via self-assembly to applications. Kluwer, Dordrecht
Pramanik S, Zheng C, Zhang X, Emge TJ, Li J (2011) J Am Chem Soc 133:4153–4155
Qin J-S, Du D-Y, Guan W, Bo X-J, Li Y-F, Guo L-P, Su Z-M, Wang Y-Y, Lan Y-Q, Zhou H-C (2015) J Am Chem Soc 137:7169–7177
Rakovsky E, Joniakova D, Gyepes R, Schwendt P, Micka Z (2005) Cryst Res Technol 40:719–722

Saha S, Mahato P, Reddy U, Suresh GE, Chakrabarty A, Baidya M, Ghosh SK, Das A (2012) Inorg Chem 51:336-–345

Salinas Y, Manez RM, Marcos MD, Sancenon F, Castero AM, Parra M, Gil S (2012) Chem Soc Rev 41:1261–1296

Sama F, Ansari IA, Raizada M, Ahmad M, Nagaraja CM, Shahid M, Kumar A, Khan K, Siddiqi ZA (2017a) New J Chem 41:1959–1972

Sama F, Dhara AK, Akhtar MN, Chen Y, Tong M, Ansari IA, Raizada M, Ahmad M, Shahid M, Siddiqi ZA (2017b) Dalton Trans 46:9801

Samara CD, Janakoudakis PD, Kessissaglou DP, Manoyussakis GE, Mentzapfos D, Terzis A (1992) J Chem Soc Dalton Trans 1:3259–3264

Saßmannshausen J (2012) Dalton Trans 41:1919

Schaefer A, Horn H, Ahlrichs R (1992) J Chem Phys 97:2571

Schaefer A, Huber C, Ahlrichs R (1994) J Chem Phys 100:5829

Schindler M, Hawthorne FC, Baur WH (2000) Chem Mater 12:1248–1259

Schöler S, Wahl MH, Wurster NIC, Puls A, Hättig C, Dyker G (2014) Chem Commun 50:5909

Seth SK (2014) J Mol Struct 1070:65–74

Seth SK, Sarkar D, Royd A, Kar T (2011a) CrystEngComm 13:6728–6741

Seth SK, Sarkar D, Kar T (2011b) CrystEngComm 13:4528–4535

Seth SK, Lee VS, Yana J, Zain SM, Cunha AC, Ferreira VF, Jordão AK, de Souza MCBV, Wardell SMSV, Wardellf JL, Tiekink ERT (2015) CrystEngComm 17:2255–2266

Shanmugaraju S, Mukherjee PS (2015) Chem Eur J 21:6656–6666

Sheldon RA, Kochi JA (1981) Metal Catalysed Oxidations of Organic Compounds. Academic, New York

Sheldrick GM (2002) SADABS, software for empirical absorption correction, Ver. 2.05. University of Göttingen, Göttingen

Sheldrick GM (2008) SHELXL97, Program for crystal structure refinement. University of Göttingen, Göttingen

Sheldrick GM (2015) Acta Cryst C 71:3–8

Shiraishi Y, Matsunaga Y, Hongpitakpong P, Hirai T (2013) Chem Commun 49:3434–3436

Shivaiah V, Das SK (2005) Inorg Chem 44:8846–8854

Siddiqi ZA, Mathew VJ (1994) Polyhedron 13:799–805

Siddiqui KA, Tiekink ERT (2013) Chem Commun 49:8501

SMART & SAINT (2003) Software reference manuals, version 6.45. Bruker Analytical X-Ray Systems, Inc., Madison

Sokolov AN, Friscic T, Blais S, Ripmeester JA, MacGillivray LR (2006) Cryst Growth Des 6:2427

Spackman MA, Jayatilaka D (2009) CrystEngComm 11:19

Stavila V, Talin AA, Allendorf MD (2014) Chem Soc Rev 43:5994–6010

Steed JW, Atwood JL (2000) Supramolecular Chemistry. VCH, New York

Steffen C, Thomas K, Huniar U, Hellweg A, Rubner O, Schroer A (2010) J Comput Chem 31:2967

Tahmasebi E, Masoomi MY, Yamini Y, Morsali A (2015) Inorg Chem 54:425–433

Takahashi S, Jukurogi T, Katagiri T, Uneyama K (2006) CrystEngComm 8:320

Tan Y-X, Zhang Y, He Y-P, Zheng Y-J, Zhang J (2014) Inorg Chem 53:12973–12976

Tanaka S, Annaka M, Sakai K (2012) Chem Commun 48:1653–1655

Tiekink ERT, Schpector J-Z (2011) J Chem Soc ChemCommun 47:6623

Toal SJ, Trogler WC (2006) J Mater Chem 16:2871–2883

Truflandier LA, Boucher F, Payen C, Hajjar R, Millot Y, Bonhomme C, Steunou N (2010) J Am Chem Soc 132:4653–4668

Uchida S, Mizuno N (2007) Coord Chem Rev 251:2537–2546

Venegas-Yazigi D, Brown KA, Vega A, Calvo R, Aliaga C, Santana RC, Cardoso-Gil R, Kniep R, Schnelle W, Spodine E (2011) Inorg Chem 50:11461–11471

Wang Y, Ye L, Wang T-G, Cui X-B, Shi S-Y, Wang G-W, Xu JQ (2010a) Dalton Trans 39:1916–1919

Wang X, Hu H, Liu G, Lin H, Tian A (2010b) Chem Commun 46:6485–6487

Wang J, Li Y, Patel NG, Zhang G, Zhou D, Pang Y (2014) Chem Commun 50:12258–12261
Weigend F, Ahlrichs R (2005) Phys Chem Chem Phys 7:3297
Wery ASJ, Gutierrez-Zorrila JM, Luque A, Roman P, Martinez-Ripoll M (1996) Polyhedron 15:4555–4564
Wolff SK, Grimwood DJ, McKinnon JJ, Turner MJ, Jayatilaka D, Spackman MA (2012) CrystalExplorer, version 3.0. University of Western Australia, Crawley
Wu W, Ye S, Yu G, Liu Y, Qin J, Li Z (2012) Macromol Rapid Commun 33:164
Wutkowski A, Näther C, Kögerler P, Bensch W (2013) Inorg Chem 52:3280–3284
Xi Y-P, Mak TCW (2012) Chem Commun 48:1123–1125
XPREP, version 5.1. Siemens Industrial Automation Inc., Madison (1995)
Xu B, Wu X, Li H, Tong H, Wang L (2011) Macromolecules 44:5089
Yadav R, Trivedi M, Kociok-Köhn G, Prasad R, Kumar A (2015) CrystEngComm 17:9175
Yao W, Eisenstein O, Crabtree RH (1997) Inorg Chim Acta 254:105
Ye J, Zhao L, Bogale RF, Gao Y, Wang X, Qian X, Guo S, Zhao J, Ning G (2029) Chem Eur J 2015:21
Yin P, Wu P, Xiao Z, Li D, Bitterlich E, Zhang J, Cheng P, Vezenov DV, Liu T, Wei Y (2011) Angew Chem 123:2569–2573
Zhang L, Schmitt W (2011) J Am Chem Soc 133:11240–11248
Zhang Y, Lewis JC, Bergman RG, Ellman JA, Oldfield E (2006) Organometallics 25:3515–3519
Zhang Z, Sadakane M, Murayama T, Izumi S, Yasuda N, Sakaguchi N, Ueda W (2014) Inorg Chem 53:903–911
Zhang Z, Sadakane M, Noro S, Murayama T, Kamachi T, Yoshizawa K, Ueda W (2015) J Mater Chem A 3:746–755
Zhang H, Yang J, Kan W, Liu Y, Ma J (2016) Cryst Growth Des 16:265–276
Zheng S-T, Zhang J, Yang G-Y (2008) Angew Chem Int Ed 47:3909–3913
Zhong YR, Cao ML, Mo HJ, Ye BH (2008) Cryst Growth Des 8:2282
Zhou J, Zhao J-W, Wei Q, Zhang J, Yang G-Y (2014) J Am Chem Soc 136:5065–5071

Chapter 8
Photocatalytic Decontamination of Organic Pollutants Using Advanced Materials

Krishnasamy Lakshmi, Venkatramanan Varadharajan, and Krishna Gounder Kadirvelu

Abstract Organic pollutants released into the natural environment such as a river, air and land pose a high threat to organisms that thrive in these environment. Industrial effluents are considered the major source of organic pollutants such as dyes, pesticides, and drugs such as antibiotics. Even with extensive regulations, numerous studies have found the presence of such organic pollutants in natural environments. Currently, the use of semiconductor materials for the photocatalytic destruction of organic pollutants got increased owing to their advantages such as high efficiency, low cost, and less toxicity. Among these materials, TiO_2 based materials have shown superior pollutant degrading capability coupled with lower secondary pollution. This chapter reviews advanced photocatalytic materials available to degrade organic pollutants.

Keywords Organic pollutants · Environment · Dye · Photocatalysis · TiO_2

8.1 Introduction

Organic pollutants generated by various industries pose a significant threat to public health and the environment. Effluents produced by industries such as paper and pulp, textile, pharmaceutical, petroleum, pesticides, fertilizers, etc. contain harmful chemicals that require extensive treatments prior to their release into the natural environment such as a river. The common procedures employed for the removal of toxic chemicals from industrial wastes are classified into physical, chemical and biological treatments. Physical treatments include adsorption, ion exchange, sedimentation, irradiation and membrane filtration. Chemical methods used for the treatment of effluents like dyes include oxidation, photochemical destruction, electrochemical degradation, use of chemical reagents such as Fenton's reagent and ozonation (Robinson et al. 2001). Biological treatments recruit living or dead

K. Lakshmi · V. Varadharajan · K. G. Kadirvelu (✉)
DRDO-BU Center for Life Sciences, Bharathiar University, Coimbatore, India

M. Oves et al. (eds.), *Modern Age Waste Water Problems*,
https://doi.org/10.1007/978-3-030-08283-3_8

biological agents such as bacteria, fungi, and algae to degrade or remove harmful chemicals from industrial wastes.

A myriad of organic pollutants is released by different industries. These pollutants belong to either one of the following categories, including organic dyes, pesticides, fertilizers, phenols, biphenyls, pharmaceuticals, plasticizers, oils, hydrocarbons, proteins, carbohydrates, microbes, etc. (Wang et al. 2014). Textile industries consume a large volume of organic dyes, and it was estimated that around half a million metric tons of dyes were produced annually to meet the demand. Commercially more than 1,00,000dyes were available for various purposes such as fabric and food coloring (Clarke and Anliker 2005; Lazar 2005). Some of the commercially available organic dyes include orange G, methyl orange, congo red, cresol red, cotton blue, methylene blue, rhodamine blue, methyl violet, methyl red, reactive red X3B, Bismarck brown R, rhodamine B, thiophene and so on.

If organicdyes containing industrial effluents reach water bodies such as a river, it will change the color of the water and can give rise to immediate public complaints. Hence, it is considered an aesthetic problem in addition to the eco-toxic hazard. The presence of dye in water bodies can adversely affect aquatic ecosystem by reducing the penetration of sunlight into the water. This depletion of sunlight restricts photosynthesis in phytoplankton that thrive on the deepest part of river or lake. Since phytoplankton are the primary producers, this effect can create a chain of changes that can disrupt entire aquatic eco-system. Apart from this, organic dyes are known to have detrimental effects on living organisms. Certain azo dyes and their breakdown products posses' significant biological toxicity by inducing hemoglobin adduct formation, which in turn causes disturbances in the normal physiology of blood. Azo dyes such as direct black 38 can induce DNA damages that lead to the development of malignant tumors in human (Carmen and Daniela 2012). Dyes used for food coloring also causes progressive damages to a human nervous system including behavioral and cognitive defects in children.

Pesticides are chemical compounds used to annihilate pests. Pesticides are classified into insecticides, rodenticides, herbicides, fungicides, and fumigants. Some of the common pesticides include organophosphates, organochlorines, carbamates, paraquats, diquates, 2,4-di-chlorophenoxyacetic acid, captan, dithiocarbamates, ethylene dibromide and methyl bromide (Abdollahi et al. 2004).

A list ofpesticides and their use is provided in Table 8.1. Pesticides are widely used in agriculture to protect the crops from pests and therefore to improve yield. As agricultural practices and food production increased drastically during the last few decades, the production and usage of pesticides got increased. In 2000, the global production of pesticide had reached a peak of around three million tons (Tilman et al. 2002). Even though natural pesticides are exploited for their potency, synthetic pesticides are still used in large quantities owing to their ready availability, cheap cost, and efficiency.

As these chemicals become an integral part of the eco-system, the hazards associated with this chemical should be explored extensively to determine its long-term effects on the environment and human health. The wide-scale uses of pesticides are linked with severe environmental pollution and health hazard. Pesticides are capable of causing acute and chronic illness in humans. Pesticides can able to induce

Table 8.1 A list of pesticides available in market

Pesticide	Structure	Description
Atrazine	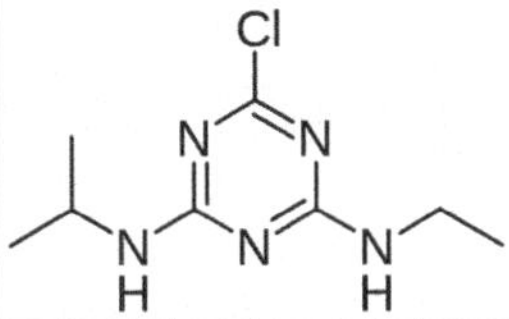	A commonly used herbicide to eliminate weeds in major crops.
Acephate	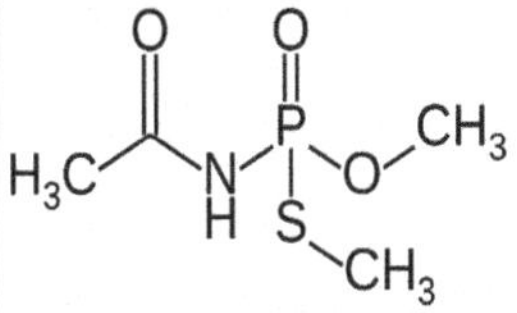	An insecticide used commonly for fruits, vegetables, and vines.
Acetamiprid	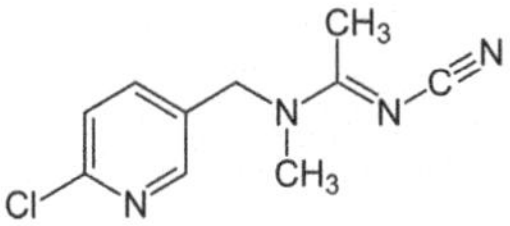	An insecticide targeted to sucking insects that infect plants such as tomatoes and citrus fruits.
Boscalid	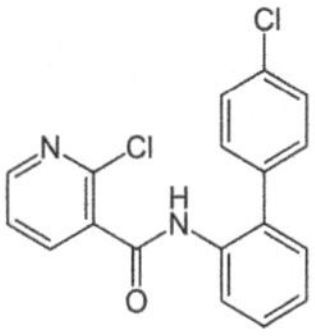	A fungicide used to control fungal pathogens in specialty crops like a bean, grapes, sunflower,etc.
Bifenthrin	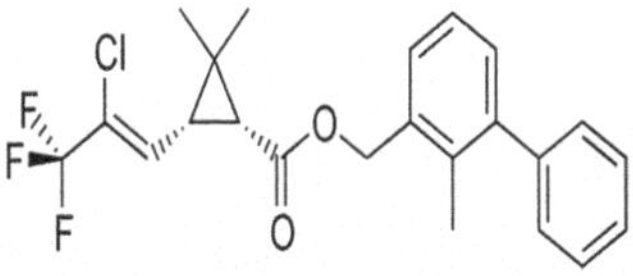	An insecticide primarily targeted to imported red ants, and it is widely used in major crops such as corn.
Carbendazim	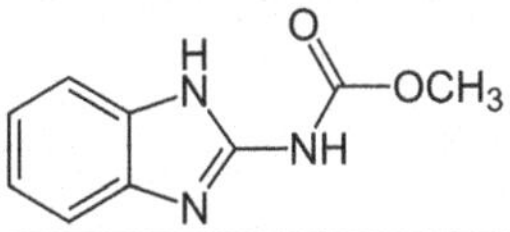	A broad spectrum fungicide used to control fungal infestations in crops such as banana, strawberry, and pineapple.
Cyprodinil	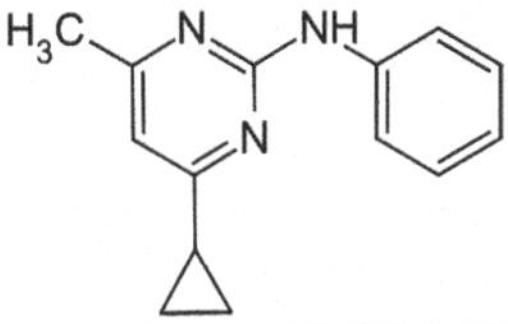	A fungicide used for fruits and ornamental crops. It is also used as seed dressing in barley
Chlorpyrifos	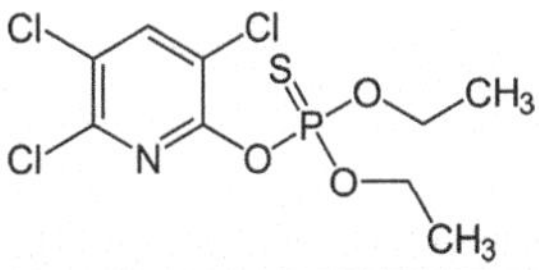	An insecticide against various types of worms such as cutworms and corn rootworm. It is widely used for corn, almonds, and fruits like apple and orange.

(continued)

Table 8.1 (continued)

Pesticide	Structure	Description
Diflubenzuron	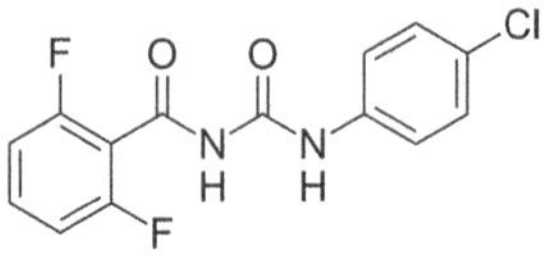	An insecticide targeted against plant-eating larvae of insects that affect ornamental plants
Endosulfan	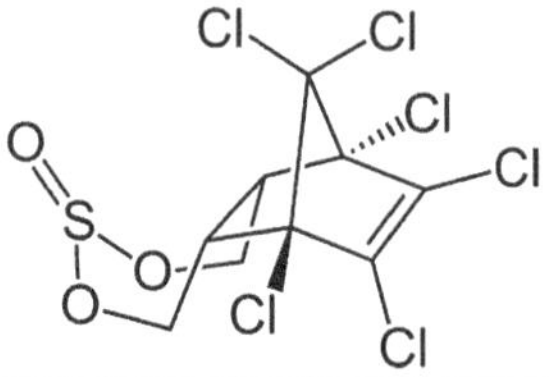	A widely used insecticide is targeting whiteflies, aphids, beetles, worms, and grasshoppers.
Fluridone	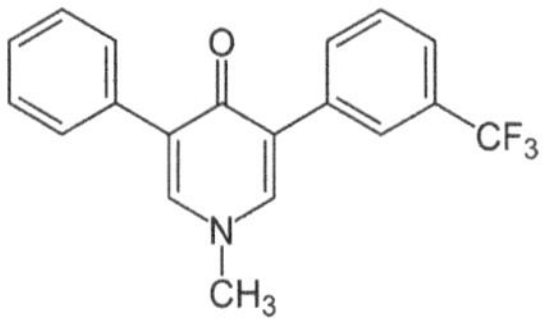	Aherbicide is targeting aquatic plants such as hydrilla.
Imidacloprid	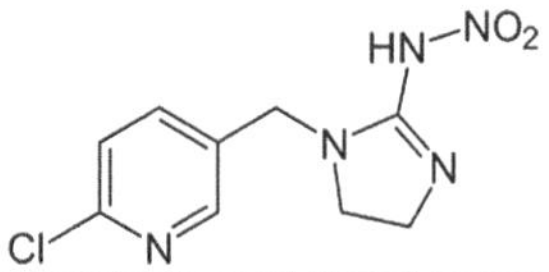	A most widely used insecticide in the world and it can control aphids, beetles, and locusts.
Methomyl	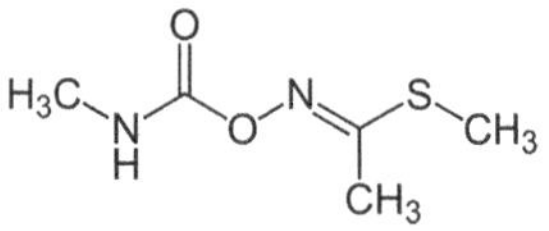	A broad spectrum insecticide used to kill various insects that infest vegetables and orchids.
Parathion	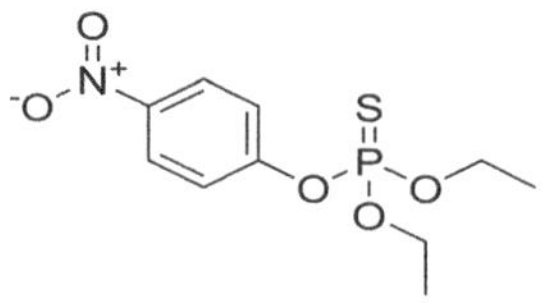	An organophosphate insecticide often applied to cotton, rice, and fruits. It is banned in many countries.

oxidative stress in mammals by the formation of free radicals. Free radical accumulation in living organisms was proven to cause damage to biomolecules such as proteins, lipids, and nucleic acids. Hence, pesticides were long exploited for their ability to cause carcinogenesis in humans.

Pharmaceutical compounds found in wastewater can adversely affect the health of public and certain pharmaceuticals present in industrial effluents can change due to chemical processes such as oxidation to produce compounds that can be detrimental to the environment. A meta-study by Deblonde, 2011 concluded the presence of significant quantities of pharmaceutical compounds in wastewater (influent and effluent) of wastewater treatment plants (Deblonde et al. 2011). Among the compounds studied, caffeine (psychostimulant) was observed to be found in higher concentration in influent, and the rate of removal was found to be around 97%. After removal, the concentration of caffeine in the effluent was 1.77 μg/L. The other pharmaceutical compounds present in the wastewater includes antibiotics such as ciprofloxacin, clarithromycin, doxycycline, erythromycin, norfloxacin, ofloxacin, roxithromycin, sulfapyridine, tetracycline and trimethoprim, antiepileptics such as carbamazepine, 4-aminoantipyrine, antipyrin, codeine and diclofenac, anti-inflammatory and analgesics such as ibuprofen, indomethacin, ketoprofen, ketorolac, naproxen and clofibric acid, lipid regulators such as fenofibric acid, bezafibrate, gemfibrozil, acebutolol and atenolol, beta blockers such as celiprolol, metoprolol, propranolol and sotalol, diuretics such as furosemide, hydrochlorothiazide, amidotrizoic acid and diatrizoate, contrast media such as iotalamic acid, iopromide, iomeprol, iohexol and iopamidol and cosmetics such as galaxolide and tonalide. Apart from influents and effluents, pharmaceutically active substances were reported to found in several surface waters present in downstream of wastewater treatment plants indicating an inefficient treatment (Heberer 2002). The presence of pharmaceutical wastes such as clofibric acid, carbamazepine, primidone or iodinated contrast agents in groundwater samples of Germany showed a significant leach of these compounds through subsoil in that region. The presence of pharmaceutically active substances in drinking water samples were also detected in fewer parts of the world.

Biological pollutants also play a significant role in determining the safeness of the environment. Many biological entities present in different environments can cause severe damage to human and animal health. Biological agents such as proteins, bacteria, fungi, and the virus can be easily found in anyenvironments. The most common mode of entry of these agents into human or animal occurs via contaminated water, food or air. Some of the diseases that transmit via contaminated water include cholera, amoebic dysentery, typhoid, parathyroid, gastroenteritis, tularemia and so on (Behnam et al. 2016). The release of protein containing effluents into water streams from industries such as dairy and milk processing can increase the nutrient content of water bodies such as a lake, support the growth of the microorganism and lead to adverse conditions such as eutrophication of ponds and lakes.

The alarming rise in the number of organic pollutants in natural environments encouraged the genesis of novel technologies for effective decontamination of organic pollutants and their break down products. These technologies aim to convert harmful pollutants into less toxic compounds that can be easily degradable under less harsh or gentle conditions. One of the important developments in the field of effluent treatment is advanced oxidation processes (AOPs). The AOPs decontaminates organic pollutants by mineralizing the pollutants with highly reactive and short living chemical species such as H_2O_2, OH^*, $O_2^*$and O_3. The process of decontamination contains 3 steps: (i) generation of hydroxyl radicals, (ii) attack on a pollutant by OH^* that can lead to fragmentation of pollutant and (iii) subsequent attack of OH^* on fragments till complete mineralization to CO_2 and H_2O. There are numerous methods available to generate hydroxyl radicals, and the commontechniques employ UV/H_2O_2, ozone, Fenton's reagent or photocatalysts. Upon excitation with UV radiation of wavelengths less than 280 nm, hemolytic bond cleavage of O-O bonds in H_2O_2 occurs and releases OH^* (Eq. 8.1) (Baxendale and Wilson 1957).

$$H_2O_2 + UV \rightarrow 2OH^* \quad (8.1)$$

Degradation of organic pollutants by ozone occurs in multiple steps. Usually, an initiator is required for the decomposition of ozone to form OH radicals. OH^- ions act as initiators for decomposition of ozone. Addition of
H_2O_2 improves the formation of OH radicals.

$$\begin{array}{c} H_2O_2 \\ H^+ \uparrow\downarrow \\ HO^- + O_3 \rightarrow O_2 + HO_2^- \end{array} \quad (8.2)$$

$$HO_2^- + O_3 \rightarrow HO_2^* + O_3^{*-} \quad (8.3)$$

$$HO_2^* \rightleftarrows H^+ + O_2^{*-} \quad (8.4)$$

$$O_2^{*-} + O_3 \rightarrow O_2 + O_3^{*-} \quad (8.5)$$

$$O_3^{*-} + H^+ \rightarrow HO_3^* \quad (8.6)$$

$$HO_3^* \rightarrow OH^* + O_2 \quad (8.7)$$

$$OH^* + O_3 \rightarrow HO_2^* + O_2 \quad (8.8)$$

It was also observed that an increase in pH favors the formation of hydroxyl radicals. As the concentration of conjugate base HO_2^* depends on pH, the increase in pH increases the concentration of HO_2^*. Apart from H_2O_2, UV radiation can also be used along with ozone to produce hydroxyl radicals. Upon excitation of ozone with UV radiation of 254 nm, hydroxyl radicals are generated based on the following equations (Andreozzi et al. 1999),

$$O_3 \xrightarrow{hv} O^1(D) + O_2 \quad (8.9)$$

$$O^1(D) + H_2O \rightarrow H_2O_2 \quad (8.10)$$

$$H_2O_2 \xrightarrow{hv} 2OH^* \quad (8.11)$$

Use of Fenton's reagent for the generation of hydroxyl radicals is one of the traditional and cheap methods for wastewater treatments. Fenton's reagent contains ferrous (Fe^{2+}) salts that can react with hydrogen peroxide to form hydroxyl radicals (Eq. 8.12). Fenton's reagent was reported to degrade different type of organic pollutants in wastewater.

$$Fe^{2+} + H_2O_2 \rightarrow Fe^{3+} + HO^- + OH^* \quad (8.12)$$

Heterogeneous semiconductors are currently researched for their ability to degrade a wide range of organic pesticides. Some of them include TiO_2, ZnO, ZnS, Fe_2O_3, CdS, and GaP. Among these, TiO_2 is used extensively as it does not undergo auto degradation to produce toxic products.

Apart from that, TiO_2 also exhibits superior photocatalytic properties at the excitation range of 300–390 nm along with higher stability for repeated use. The use of heterogeneous photocatalytic superconductors comes with a series of advantages including, provision to use under ambient temperature and pressure, economy and availability of raw materials. One of the importantadvantages is that photocatalytic degradation of pollutants by heterogeneous semiconductors does not retain any secondary pollutants from degradation of parent compounds. Currently, nanoscale semiconductor photocatalysts are explored extensively for their decontamination properties. Nanoscale materials possess high surface area to volume ratio that allows it to react with organic pollutants rapidly than in macro or micro scale. But the separation of nanoscale semiconductors from the reaction mixture after the completion of degradation is quite tedious. The agglomeration of nanoscale materials during the process also restricts the materials capabilities to decontaminate organic pollutants (Chong et al. 2010).

This chapter is a brief review of various novel materials that were explored for their ability to decontaminate organic pollutants like dyes and pesticides in the presence of photocatalyst. It also outlines the mechanisms involved in the process of decontamination of pollutants by TiO2nanoscale materials with emphasis on possible strategies to overcome problems associated with the use of photocatalytic materials for degradation of organic pollutant.

8.2 Advanced Photocatalytic Materials

8.2.1 TiO_2 Based Materials

Titanium dioxide is considered one of the best materials for photocatalytic degradation of organic compounds including pesticides. Before 1980, research in TiO_2 solely consisted of using $TiO_2$for water photolysis. As TiO_2 can only absorb UV portion of solar light which constituted about 3%, TiO_2 was found inefficient for hydrogen production. Instead, the researchers stated to use TiO_2 for its strong oxidizing potential. In 1977, the first demonstration of using TiO_2suspension for complete degradation of cyanide was reported (Frank and Bard 1976). Ever since $TiO_2$is widely researched for its photocatalytic potential. Currently, composite materials consisting of TiO_2 and other materials are studied briefly for the degradation of harmful compounds.

TiO_2 (rutile), upon excitation with light of wavelength shorter than its band gap (3.0 eV) can generate an electron-hole pair. The hole (h^+) generated during this process have higher oxidation potential and can oxidize organic compounds entirely by oxidizing all the component products of oxidized organic compounds. Apart from this, numerous reactive oxygen species (ROS) are also produced along the course of the process, and they aid in degrading the organic compound (Hashimoto et al. 2005). A typical method is described below in the form of equations.

$$TiO_2 \xrightarrow{hv} e^- + h^+ \tag{8.13}$$

$$e^- + O_2 \rightarrow O_2^- \tag{8.14}$$

$$O_2^- + H^+ \rightarrow HO_2^* \tag{8.15}$$

$$h^+ + H_2O \rightarrow OH^* + H^+ \tag{8.16}$$

$$h^+ + O_2^- \rightarrow 2O^* \tag{8.17}$$

8.2.1.1 Advantages

Cheap and ease of synthesis.
High photostability in solutions.
The holes produced by TiO_2 are highly oxidizing and redox selective (Gupta and Tripathi 2011).

8.2.1.2 Disadvantages

TiO_2 nanoparticles tend to agglomerate in aqueous a suspension, which in turn reduces photocatalytic potential.
The band gap energies of TiO_2 (rutile and anatase) is high. Hence, it can absorb a small portion of visible light.

8.2.2 *Strategies to Overcome the Limitations of TiO_2*

Developing new TiO_2 based materials that can have band gap energies lower than 3.0 eV and flexibility of charged carriers within conduction and valence bands.
Agglomeration of TiO_2 nanoparticles can be prevented by stabilizing nanoparticles using steric or electrostatic stabilization. In many cases, electrostatic stabilization can be attained by adding polyelectrolytes such as polyacrylic acid (Othman et al. 2012)

8.2.2.1 Doping and Coupling of TiO_2

In order to overcome the limitations associated with the use of TiO_2 for photocatalysis, TiO_2 can be doped with other metals or coupled with other semiconductors. TiO_2 can be doped with large number of metal cations and anions such as rare earth metals including scandium, lanthanum, yttrium etc., noble metals like gold, silver, platinum and rhodium, transition metals such as nickel, cobalt, iron and copper, post-transition metals such as gallium, tin, aluminum and lead and non-metals such as carbon, nitrogen and sulfur. Apart from this, spectral absorption can be extended to the visible region by coupling TiO_2 with other semiconductors such as CdS, WO_3 and SnO_2. Upon doping with metals such as Fe, the electronic structure of TiO_2 gets modified and the absorption shifts from UV to the visible region. Various forms of Fe ions (Fe^{2+}, Fe^{3+}, and Fe^{4+}) can effectively inhibit recombination of the electron-hole pair by capturing the electrons from TiO_2 valence band (Fig. 8.2).

Coupling of TiO_2 with other semiconductors can also induce similar effects. When semiconductors of small band gap are coupled with TiO_2 of the broad band gap, transfer in charge carriers from one to another takes place and allows separation of electrons from holes (Daghrir and Drogui 2017).

8.2.3 Example of TiO_2 Based Photocatalysis

Different examples of TiO_2 based photocatalysis of organic pollutants are described below. A list of recent TiO_2 based materials used for photocatalysis of organic pollutantsis tabulated in Table 8.2.

8.2.4 Photocatalytic Degradation of Methyl Parathion

Methyl parathion is one of the commonly used pesticides worldwide. It is an organophosphorus pesticide and known to cause acute toxicity in humans by inhibiting acetylcholinesterase enzyme in nerve cells. Hence, the use of methyl parathion is banned from using in agriculture by several countries. Under ambient conditions, the degradation of methyl parathion is a slow process that can be accelerated by the use of photocatalyst. Typically, TiO_2 degrades methyl parathion under ambient condition and UV irradiation within 45 min. A scheme of degradation of methyl parathion and its break down products is illustrated in Fig. 8.1. Initially, the ROS produced by TiO_2 attacks P=S bond of methyl parathion and forms paraoxon derivatives. The subsequent attack of ROS causes the destruction of the P-Obond and results in the formation of phenol and dialkyl phosphates. Apart from this, the degraded products also showed interchange of methyl and methoxy groups (Evgenidou et al. 2007). These products further undergo mineralization by subsequent attacks of ROS to form CO_2 and H_2O (Fig. 8.2).

Table 8.2 Novel TiO_2 based photocatalysts for degradation of organic pollutants

Material	Organic pollutant	References
TiO_2/Zeolite	Formaldehyde, phenol and methyl orange	Zhang et al. (2018)
TiO_2-C	Phenol, imidacloprid and dichloroacetic acid	Matos et al. (2007)
N-doped TiO_2	Ciprofloxacin, naproxen and paracetamol	Shetty et al. (2014)
TiO_2/C_3N_4	Methyl orange	Shen et al. (2017)
Zr-doped TiO_2	Formaldehyde	Huang et al. (2017)
TiO_2/graphene	Rhodamine 6G	Pu et al. (2017)
Au, Ag and Cu nanoparticles on TiO_2	Sulfamethoxazole	Zanella et al. (2017)
TiO_2/Hydroxyapatite	Methylene blue	Sahibed-dine et al. (2017)
TiO_2-AgI	Acid orange 7	Tavakoli et al. (2017)
BiOI/TiO_2	Reactive brilliant blue and tetrabromobisphenol A	Ao et al. (2017)
TiO_2-WO_3	Bisphenol A	Dominguez et al. (2017)
CeO_2–TiO_2	Methylene blue	Moongraksathum and Chen (2017)
CuPc-TiO_2	Methylene blue	Cabir et al. (2017)
Polyaniline- TiO_2	Bisphenol A	Chen et al. (2017)
Au/TiO_2	Phenoxyacetic acid	Lannoy et al. (2017)
Fe-doped TiO_2	Diazinon	Tabasideh et al. (2017)
W-doped TiO_2	Diuron	Foura et al. (2016)
WO_3/ TiO_2-N	Diclofenac	Cordero-García et al. (2017)
Cr(III)-doped TiO_2	4-chloro-2-methylphenoxyacetic acid	Mendiola-Alvarez et al. (2017)

8.2.5 Photocatalytic Decontamination of Acid Blue 40

Anthraquinonedyes are used widely next to the azo dyes. Upon treatment with different chemicals that confer amino or hydroxyl group to the parent compound, these dyes can give a palate of colors. As these dyes pose a series environmental hazard, photocatalytic break down of dye to harmless compounds is encouraged. A study by Tang, 1995 (Tang and An 1995) gave a complete overview of the photocatalytic degradation of acid blue 40 by TiO_2/UV. Briefly, upon treatment with an aqueous suspension of TiO_2 and irradiation with 350 nm source, acid blue 40 turned to a yellow colored intermediated after 20 min at pH 5. The intermediate has a maximum absorbance at 456 nm, and it slowly started to disappear after 5 h. During this period, an absorbance peak at 290 nm was also observed along with the major peak. It showed the presence of main anthraquinone chromophore in the intermediate, and it is usually hard to oxidize. The reaction mechanism indicated an initial attack of hydroxyl radical on a C-N bond of the acid blue 40. The second

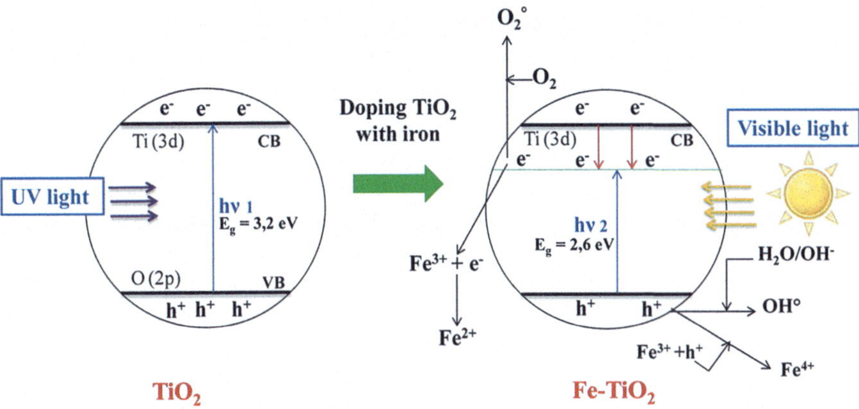

Fig. 8.1 Energy level diagram of TiO_2 and Fe doped TiO_2. (Reprinted with permission from Daghrir, 2013 (Daghrir and Drogui 2017). Copyright (1993) American chemical society)

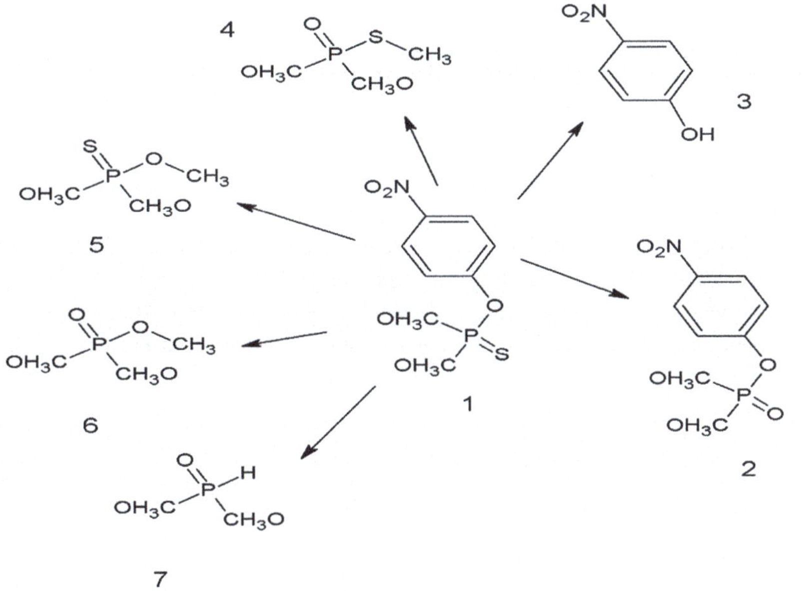

Fig. 8.2 Photocatalytic degradation of methyl parathion using TiO_2. 1. Methyl parathion, 2–3. Paraoxon derivatives and 4–7 phenols and dialkyl phosphates

attack by hydroxyl radical replaced amine group of main anthraquinone ring to hydroxyl group. Subsequent attacks broke anthraquinone ring and resulted further in complete mineralization.

8.2.6 Photocatalytic Degradation of Sulfamethoxazole

Sulfamethoxazole is a synthetic antimicrobial agent used to treat urinary tract infection. It is a sulfonamide drug extensively used in the veterinary field. Current data suggested the presence of sulfonamide drugs in wastewater (Haller et al. 2002; Hartig et al. 1999; Jen et al. 1998). Abellen and his colleagues (Abellán et al. 2007) used TiO_2 for photocatalytic degradation of sulfamethoxazole in aqueous solution. The LC/MS study indicated the presence of four intermediates during degradation. A scheme of the mechanism of photocatalytic degradation of sulfamethoxazole is illustrated in Fig. 8.3. Initially, the inclusion of hydroxyl radical to isoxazole ring of Sulfamethoxazole occurs and results in the formation of hydroxylatedanalog. The aniline ring of this analog can be attacked by hydroxyl radical at two different positions such as meta-directing NH_2 or ortho-directing SO_2 group to form different compounds. These compounds undergo subsequent oxidation to form SO_4^{2-} and NH_4^+ ions and mineral acids (Fig. 8.4).

8.3 Non TiO_2 Materials for Photocatalysis

Apart from TiO_2, many semiconductor materials were employed for photocatalysis. It includes ZnO, ZnS, Fe_2O_3, CdS, GaP, and composites of these materials. The band gap energies of these materials are lower than TiO_2 (Fig. 8.5) Even though these materials show moderate photocatalysis compared to TiO_2, they are suffered from a series of disadvantages includinglower photostability. Materials such as ZnO showed lower photostability and could undergo photocorrosion. Materials such as ZnO, CdS, and GaP can lead to secondary pollution, i.e.; it is hard to remove these

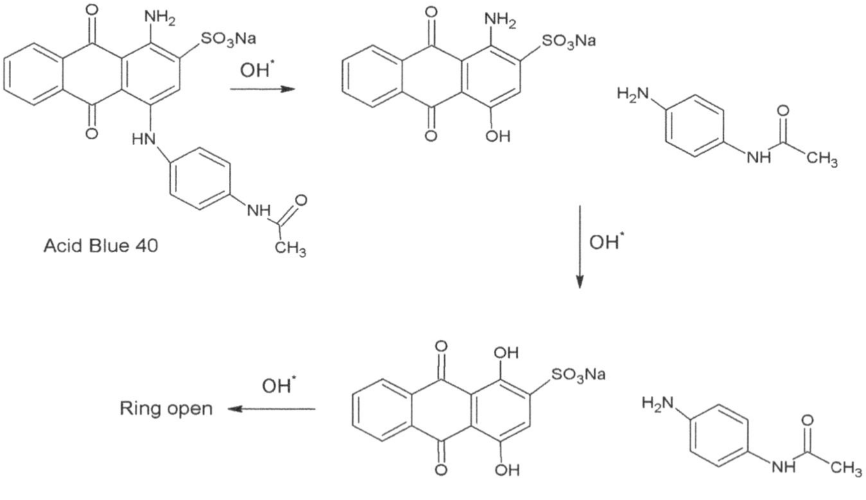

Fig. 8.3 Photocatalytic degradation pathway of acid blue 40

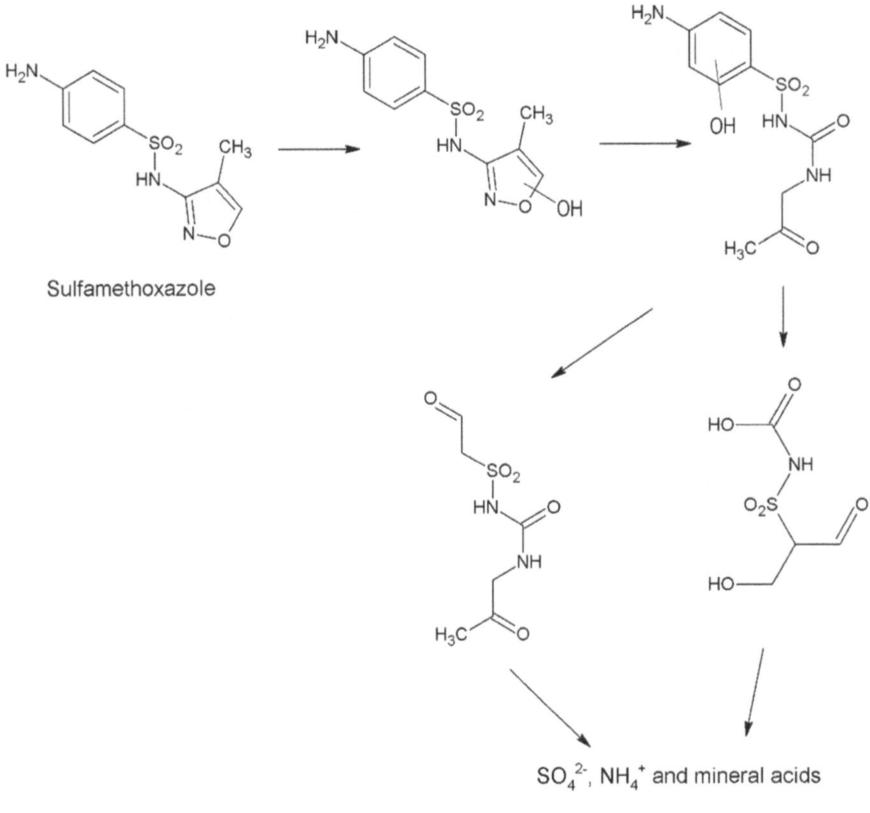

Fig. 8.4 Photocatalytic degradation pathway of antibiotic drug sulfamethoxazole

nanoscale materials once photocatalysis got completed. In order to solve these disadvantages, numerous composite materials were developed. A list of non-TiO_2 based materials is given in Table 8.3.

8.4 Advanced Photocatalytic Materials for Wastewater Treatment

Photocatalytic materials are widely explored for their role in degradation of organic compounds present in industrial and domestic wastewaters. Almost all the materials that are employed for wastewater treatment using photocatalysis make use of solar energy for degradation of organic compounds. The efficient TiO_2 nanomaterial loaded volcanic soil was used by Borges, 2016 (Borges et al. 2017) to degrade organic pollutant (methylene blue) in wastewater. This material depends on sun light as a light source to degrade almost 95% of organic material from the wastewater.

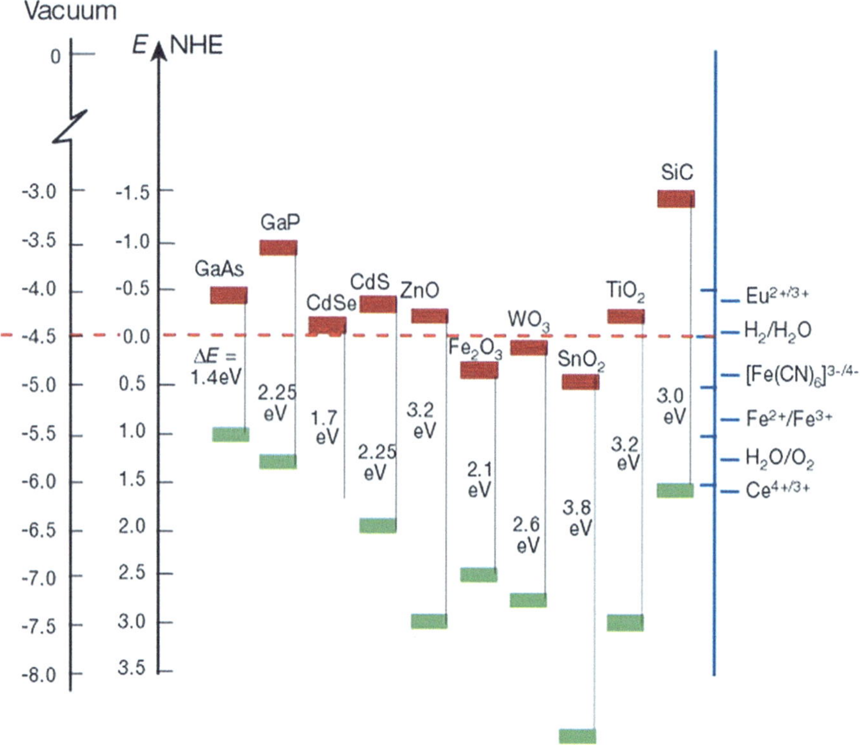

Fig. 8.5 Energy band levels of various semiconductors. (Reprinted with permission from Wenderich, 2016 (Wenderich and Mul 2016). Copyright (2016) American chemical society)

Table 8.3 Novel non-TiO_2 materials for photocatalytic decontamination of organic pollutants

Materials	Pollutants	References
Fe^{3+} and Eu^{3+} doped ZnO	Methyl orange	Yin et al. (2014)
N doped ZnO	Azure B	Rathore et al. (2015)
Fe and Cu promoted ZnO	Methylene blue	Mardani et al. (2015)
Zrcodoped Ag-ZnO	Naphthol blue black	Subash et al. (2013)
Ce doped ZnO	Methylene blue	Rezaei and Habibi-Yangjeh (2013)
ZnO/graphene	Methylene blue	Saranya et al. (2013)
α Fe_2O_3/graphene	Phenol	Pradhan et al. (2013)
BiOI/ZnO	Methyl orange	Jiang et al. (2011)
g-C_3N_4/CdS/graphene oxide	Rhodamine B and congo red	Pawar et al. (2014)
Ga_2O_3	Volatile pollutants in the air	Hou et al. (2007)
Ag_2CrO_4-graphene oxide	Methylene blue	Xu et al. (2015)

An ideal photocatalyst for wastewater treatment should be degrade wide variety of organic pollutants in wastewater within negligible period of time. Wang, 2015 had synthesized a novel bifunctional photocatalyst BiOI/Ag_3VO_4 to degrade multiple dyes in textile wastewater (S Wang et al. 2015). After 1 h of adsorbtion and 2 h of visible light exposure, the photocatalyst (30 wt%) degraded 97, 92 and 85% of basic dyes such as fuchsin, malachite green and crystal violet present in the wastewater.

8.5 Conclusion

Heterogeneous photocatalysis using semiconductors remains superior compared to other traditional methods of photocatalysis including UV, UV/H_2O_2, and Fenton's reagent. Even though a wide array of materials are available for photocatalysis of organic pollutants, titania-based materials still dominate the research owing to its photocatalytic properties. Current research indicated the use of TiO_2 nanoparticles for degradation of major organic pollutants including dyes, pesticides, and antibiotics. As TiO_2 has absorbance only in UV side of the spectra, composite materials are researched extensively. Apart from TiO_2 composites, composites of ZnO, Fe_2O_3 and CdS are also employed for degradation of organic pollutants.

References

Abdollahi M, Ranjbar A, Shadnia S, Nikfar S, Rezaiee A (2004) Pesticde and oxidative stress: a review. Med Sci Monit 10(6):RA141–RA147

Abellán MN, Bayarri B, Giménez J, Costa J(2007) Photocatalytic degradation of sulfamethoxazole in aqueous suspension of TiO2. Appl Catal B Environ 74(3):233–241

Andreozzi R, Caprio V, Insola A, Marotta R (1999) Advanced oxidation processes (AOP) for water purification and recovery. Catal Today 53(1):51–59

Ao Y, Xu J, Ashraf MA, Wang P, Wang C (2017) Bio-TiO2 nanobelt PN heterojunction with enhanced photocatalytic activity for degradation of reactive brilliant red and tetra bromobisphenol A under visible light. Environ Conserv Clean Water Air Soil:192

Baxendale JH, Wilson JA(1957) The photolysis of hydrogen peroxide at high light intensities. Trans Faraday Soc 53:344

Behnam B, Vali F, Hooman N (2016) Genetic study of nephrotic syndrome in Iranian children-systematic review. J Ped Nephrol 42:51–55

Borges ME, Sierra M, Cuevas E, García RD, Esparza P (2017) Photocatalytic Removal of the Antibiotic Cefotaxime on TiO2 and ZnO Suspensions Under Simulated Sunlight Radiation. Sol Energy 135:527–535

Cabir B, Yurderi M, Caner N, Agirtas MS, Zahmakiran M, Kaya M(2017) Methylene blue photocatalytic degradation under visible light irradiation on copper phthalocyanine-sensitized TiO2 nanopowders. Mater Sci Eng B 224:9–17

Carmen Z, Daniela S (2012) Textile Organic Dyes – Characteristics, Polluting Effects and Separation/Elimination Procedures from Industrial Effluents – A Critical Overview. Environmental and Analytical Update. Environmental and analytical update. InTech

Chen F, An W, Li Y, Liang Y, Cui W (2017) Fabricating 3D porous PANI/TiO2-graphene hydrogel for the enhanced UV-light photocatalytic degradation of BPA. Appl Surf Sci 427:123–132

Chong MN, Jin B, Chow CW, Saint C (2010) Recent developments in photocatalytic water treatment technology: A review. Water Res 44(10):2997–3027

Clarke EA, Anliker R (2005) , Color Chemistry: Synthesis, Properties, and Applications of Organic Dyes and Pigments, 3rd revised edition, The handbook of environmental chemistry. vol 3(3A). Springer, Berlin/Heidelberg

Cordero-García A, Palomino GT, Hinojosa-Reyes L, Guzmán-Mar JL, Maya-Teviño L, Hernández-Ramírez A (2017) Photocatalytic behaviour of WO3/TiO2-N for diclofenacdegradation using simulated solar radiation as an activation source. Environ Sci Pollut Res 24(5):4613–4624

Daghrir R, Drogui P (2017) Advances in technologies for pharmaceuticals and personal care products removal. Robert D IEC Res 52:3581–3599

Deblonde T, Cossu-Leguille C, Hartemann P (2011) Emerging pollutants in wastewater: A review of the literature. Int J Hyg Environ Health 214(6):442–448

Dominguez S, Huebra M, Han C, Campo P, Nadagouda MN, Rivero MJ, Ortiz I, Dionysiou DD (2017) Magnetically recoverable TiO2-WO3 photocatalyst to oxidize bisphenol A from model wastewater under simulated solar light. Environ Sci Pollut Res 24(14):12589–12598

Evgenidou E, Konstantinou I, Fytianos K, Poulios I, Albanis T (2007) Photocatalytic oxidation of methyl parathion over TiO 2 and ZnO suspensions. Catal Today 124(3):156–162

Foura G, Soualah A, Robert D (2016) Effect of W doping level on TiO2 on the photocatalytic degradation of Diuron. Water Sci Technol 75(1):20–27

Frank SN, Bard AJ (1976) Heterogeneous Photocatalytic Oxidation of Cyanide Ion in Aqueous Solutions at T1O2 Powder. J Am Chem Soc 99:303

Gupta SM, Tripathi M Chin (2011) A review of TiO2 nanoparticles. Sci Bull 56:1639

Haller MY, SR M"l, McArdell CS, Alder AC, Suter MJF (2002) Quantification of veterinary antibiotics (sulfonamides and trimethoprim) in animal manure by liquid chromatography–mass spectrometry. J Chromatogr A 952:111–120

Hartig C, Storm T, Jekel M (1999) Detection and identification of sulphonamide drugs in municipal waste water by liquid chromatography coupled with electrospray ionisation tandem mass spectrometry. J Chromatogr A 854:163–173

Hashimoto K, Irie H, Fujishima A Jpn (2005) TiO2 Photocatalysis: A Historical Overview and Future Prospects. J Appl Phys 44(12R):8269

Heberer T (2002) Occurrence, fate, and removal of pharmaceutical residues in the aquatic environment: a review of recent research data. Toxicol Lett 131(1):5–17

Hou Y, Wu L, Wang X, Ding Z, Z Li, Fu X (2007) Photocatalytic performance of α-, β-, and γ - Ga2O3 for the destruction of volatile aromatic pollutants in air. J Catal 250(1):12–18

Huang C, Ding Y, Chen Y, Li P, Zhu S, Shen S (2017) Highly efficient Zr doped-TiO2/glass fiber photocatalyst and its performance in formaldehyde removal under visible light. J Environ Sci 60:61–69

Jen JF, Lee HL, Lee BN (1998) Simultaneous determination of seven sulfonamide residues in swine waste water by high-performance liquid chromatography. J Chromatogr A 793:378–382

Jiang J, Zhang X, Sun P, Zhang L (2011) ZnO/BiOI Heterostructures: Photoinduced Charge-Transfer Property and Enhanced Visible-Light Photocatalytic Activity. J Phys Chem C 115 (42):20555–20564

Lannoy A, Bleta R, Machut-Binkowski C, Addad A, Monflier E, Ponchel A (2017) Cyclodextrin-Directed Synthesis of Gold-Modified TiO2 Materials and Evaluation of Their Photocatalytic Activity in the Removal of a Pesticide from Water: Effect of Porosity and Particle Size. ACS Sustain Chem Eng 5(5):3623–3630

Lazar T (2005) Color Chemistry: Synthesis, Properties, and Applications of Organic Dyes and Pigments Color. Res Appl 30:313–314

Mardani HR, Forouzani M, Ziari M, Biparva P (2015) Visible light photo-degradation of methylene blue over Fe or Cu promoted ZnO nanoparticles. Spectrochim Acta Part A Mol Biomol Spectrosc 141:27–33

Matos J, Miralles-Cuevas S, Ruíz-Delgado A, Oller I, Malato S (2007) Development of TiO2-C photocatalysts for solar treatment of polluted water. Carbon 122:361–373

Mendiola-Alvarez SY, Guzmán-Mar JL, Turnes-Palomino G, Maya-Alejandro F, Hernández-Ramírez A, Hinojosa-Reyes L (2017) UVand visible activation of Cr(III)-doped TiO2 catalyst prepared by a microwave-assisted sol–gel method during MCPA degradation. Environ Sci Pollut Res 24(14):12673–12682

Moongraksathum B, Chen YW (2017) CeO2–TiO2 mixed oxide thin films with enhanced photocatalytic degradation of organic pollutants. J Sol-Gel Sci Technol 82(3):772–782

Othman SH, Rashid SA, Ghazi TIM, Abdullah N (2012) Dispersion and stabilization of photocatalytic TiO2 nanoparticles in aqueous suspension for coatings applications. J Nanomater 2012:2

Pawar RC, Khare V, Lee CS (2014)Hybrid photocatalysts using graphitic carbon nitride/cadmium sulfide/reduced graphene oxide (g-C3N4/CdS/RGO) for superior photodegradation of organic pollutants under UV and visible light. Dalton Trans 43(33):12514–12527

Pradhan GK, Padhi DK, Parida KM (2013) Fabrication of α-Fe2O3 nanorod/RGO composite: A novel hybrid photocatalyst for phenol Degradation. ACS Appl Mater Interfaces 5(18):9101–9110

Pu S, Zhu R, Ma H, Deng D, Pei X, Qi F, Chu W (2017) Facile In-situ Design Strategy to Disperse TiO2 Nanoparticles on Graphene for the Enhanced Photocatalytic Degradation of Rhodamine 6G. Appl Catal B Environ 218:208–219

Rathore P, Chittora AK, Ameta R, Sharma S (2015) Enhancement of Photocatalytic Activity of Zinc Oxide by Doping with Nitrogen. Sci Rev Chem Commun 5(4)

Rezaei M, Habibi-Yangjeh A (2013) Simple and large scale refluxing method for preparation of Ce-doped ZnO nanostructures as highly efficient photocatalyst. Appl Surf Sci 265:591–596

Robinson T, McMullan G, Marchant R, Nigam P (2001) Remediation of dyes in textile effluent: a critical review on current treatment technologies with a proposed alternative . Bioresour Technol 77(3):247–255

Sahibed-dine A, Bentiss F, Bensitel M (2017) The photocatalytic degradation of methylene bleu over TiO2 catalysts supported on hydroxyapatite. J Mater 8(4):1301–1311

Saranya M, Garg S, Singh I, Ramachandran R, Santhosh C, Harish C, Vanchinathan TM, Chandra MB, Grace AN (2013) Solvothermal Preparation of ZnO/Graphene Nanocomposites and Its Photocatalytic Properties . Nanosci Nanotechnol Lett 5(3):349–354

Shen L, Xing Z, Zou J, Z Li XW, Zhang Y, Zhu Q, Yang S, Zhou W (2017) Black TiO2 nanobelts/g-C3N4 nanosheets Laminated Heterojunctions with Efficient Visible-Light-Driven Photocatalytic Performance. Sci Rep 7

Shetty R, Chavan VB, Kulkarni PS, Kulkarni BD, Kamble SP (2014) Photocatalytic Degradation of Pharmaceuticals Pollutants Using N-Doped TiO2 Photocatalyst: Identification of CFX Degradation. Indian Chem Eng 59(3):177–199

Subash B, Krishnakumar B, Swaminathan M, Shanthi M (2013) Solar-light-assisted photocatalytic degradation of NBB dye on Zr-codoped Ag–ZnO catalyst. Res Chem Intermed 39(7):3181–3197

Tabasideh S, Maleki A, Shahmoradi B, Ghahremani E, McKay G Sep (2017) Sono photocatalytic degradation of diazinon in aqueous solution using iron-doped TiO2 nanoparticles. Purif Technol 189:186–192

Tang WZ, An H (1995) Photocatalytic Degradation Kinetics and Mechanism of Acid Blue 40 by TiO_2/UV in Aqueous Solution. Chemosphere 31(9):4171–4183

Tavakoli F, Badiei A, Ziarani GM, Tarighi S (2017) Photocatalytic Application of TiO2–AgI Hybrid for Degradation of Organic Pollutants in Water. Int J Environ Res:1–8

Tilman D, Cassman KG, Matson PA, Naylor R, Polasky S (2002) Agricultural sustainability and intensive production practices. Nature 418(6898):671–677

Wang CC, Li JR, Lv XL, Zhang YQ (2014) Photocatalytic organic pollutants degradation in metal–organic frameworks. Guo G Energy Environ Sci 7(9):2831–2867

Wang S, Yuan G, Liping W, Wei Z, Huan H, Jun X, Shaogui Y, Cheng S (2015) Carbon Dots Sensitized BiOI with Dominant {001} Facets for Superior Photocatalytic Performance. Appl Catal Environ 168:448–457

Wenderich K, Mul G (2016) Methods, Mechanism, and Applications of Photodeposition in Photocatalysis: A Review. Chem Rev 116(23):14587–14619

Xu D, Cheng B, Cao S, Yu (2015) Enhanced photocatalytic activity and stability of Z-schemeAg2CrO4-GO composite photocatalysts for organic pollutantdegradation. J Appl Catal B Environ 164:380–388

Yin D, Zhang L, K Song YO, Wang C, Liu B, Wu M (2014) ZnO Nanoparticles Co-Doped with Fe3+ and Eu3+ Ions for Solar Assisted Photocatalysis. J Nanosci Nanotechnol 14(8):6077–6083

Zanella R, Avella E, Ramírez-Zamora RM, Castillón-Barraza F, Durán-Álvarez JC (2017) Enhanced photocatalytic degradation of sulfamethoxazole by deposition of Au, Ag and Cu metallic nanoparticles on TiO2. Environ Technol:1–12

Zhang G, Song A, Duan Y, Zheng S (2018) Enhanced photocatalytic activity of TiO2/zeolite composite for abatement of pollutants. Microporous Mesoporous Mater 255:61–68

Chapter 9
Metal Oxide Nanostructured Materials for Water Treatment: Prospectives and Challenges

Sayfa Bano, Saima Sultana, and Suhail Sabir

Abstract Water being scarce is worth saving. The water contamination and pollution are increasing at greater pace posing serious threats to the ecosystem. However, the problem of water contamination has been solved upto much extent with the development and advancement in field of nanotechnology. Several modified and doped metal oxide nanostructured materials are used in water purification. These materials are synthesized by various techniques like hydrothermal synthesis, chemical vapour deposition method, sol gel method to obtain various types of materials possessing different properties which are immensely useful at nanoscale level due to their large surface area to volume ratio, band gap tunability along with high catalytic activity. Various characterisation techniques like X-ray Diffraction, Fourier Transform –Infra-Red spectroscopy, Scanning electron microscopy and Energy-dispersive X-ray spectroscopy, Transmission electron microscopy are needed to ensure the formation of metal oxide nanostructured materials. These metal oxide nanostructured materials are inevitably helpful in the form of catalyst for removal and purification of wastewater (like dyes and organic pollutants, inorganic materials) discharged from different industries by methods like photocatalysis, adsorption techniques, ion exchange process, disinfection process and electrochemical techniques for simultaneous heavy metal separation and detection.

Keywords Nanostructured metal oxides · Hydrothermal method · Photocatalysis · Adsorption method · Disinfection process · Sensors

9.1 Introduction

Water is the necessity of life. Around 50% of world's population is dependent on ground water and surface water for drinking and domestic purposes (World Water Assessment Programme (WWAP) 2015). Due to urbanization, growth in population

S. Bano (✉) · S. Sultana · S. Sabir
Environmental Research Laboratory, Department of Chemistry, Aligarh Muslim University, Aligarh, Uttar Pradesh, India

M. Oves et al. (eds.), *Modern Age Waste Water Problems*,
https://doi.org/10.1007/978-3-030-08283-3_9

and industrialisation the quality of water has been deteriorated to a greater extent making limited access to potable water (Tan et al. 2011). Poor quality of water is due to the discharges from industrial wastewater of various organic, inorganic pollutants including dye discharge from textiles industry, effluents from pulp and paper industry, and heavy metal ions from chemical manufacturing units (Subba Rao and Venkatarangaiah 2014; Xu et al. 2012; Sara 2014). Domestic wastewater consist of traces of pharmaceutical products, soap and detergents which also pollute water bodies due to accumulation and eutrophication (Subba Rao and Venkatarangaiah 2014; Teja and Koh 2009). As a result, various health issues have arisen. To avoid and reduce the number of diseases, various preventive measures were taken into consideration by government officials. The traditional materials and treatment technologies like activated carbon, oxidation, reverse osmosis (RO) membranes and activated sludge are not efficient for wastewater treatment consisting of pharmaceuticals, surfactants, various industrial additives, however methods like Oxidation, Coagulation, Adsorption techniques etc. have been previously used which have now become obsolete or it can only be applied at primary level for removal of bulk materials (Mondal 2017; Amin et al. 2014; Kyzas and Matis 2015). These methods are not effective and cannot be used entirely for removal of toxic materials. Various other water treatment process also include membrane technologies, chemical treatment method but these are very costly due to high operating cost and they also produce secondary by-products which are very toxic and cannot be used at large scale. However there are other water treatment processes like the advanced oxidation processes (AOPs), such as the sonolysis, Fenton reaction, ozonation, and photocatalysis have gained a considerable attention in the removal of pollutants, owing to their simplicity, low cost, high efficiency, easy handling, and good reproducibility (Byrne et al. 2017; Opoku et al. 2017; Chan et al. 2011). The advanced oxidation process produces highly reactive chemical oxidants like hydroxide radicals, superoxide radicals, peroxide based radicals and ozone molecule for removal of organic and inorganic toxic molecules into less toxic molecules (Opoku et al. 2017; Journal 2016; Deng and Zhao 2015). Since Advanced oxidation process produces sludge and fouling smell after treatment after chlorination, not much effective for removal of pollutants. Other techniques like ion exchange process, water disinfection treatment and electrochemical method for simultaneous determination of heavy metal ions were used for selective and efficient treatment when used with modified nanocomposites as catalysts (Tan et al. 2014). However, photocatalysis is considered as one of the most effective and efficient approaches for treatment of wastewater using photocatalyst, sunlight and other light sources as driving force (Opoku et al. 2017; Chan et al. 2011). This technique can degrade several harmful organic pollutants into less toxic smaller molecules. Photocatalysis is mainly dependent on the photocatalyst and light source. Photocatalyst is mainly metal oxides nanoparticles or nanocomposites with various attractive optical properties, high surface area to volume ratio, photostability and low cost, but there are certain crucial steps still needed to overcome the limitations and challenges in water treatment processes (Mondal 2017; Byrne et al. 2017; Mahlambi et al. 2015; Amin and Alazba 2014).

9.2 Metal Oxide Nanostructured Materials

Metal oxides are of considerable importance since they show enhanced structural tunability at nanometre scale (Fernández-garcia and Rodriguez 2007). Metal oxides therefore adopt several structural geometries with an electronic structure, which can be either semiconductor, metallic, or insulator depending on the nature of the structure. The unique properties of metal oxides which makes them the most important class of materials are optical, electronic, electrical, magnetic, catalytic, and photoelectronic properties, imparting wider range of applicability such as gas sensing, catalysis, superconductors, and energy storage devices (Fernández-garcia and Rodriguez 2007; Yang et al. 2013). Various metal oxides like TiO_2 (Chen et al. 2015), SnO_2 (Reddy et al. 2016), ZnO (Stankic et al. 2016), CeO_2 (Gurunathan and Ponnusamy 2017), ZrO_2 (Fernández-garcia and Rodriguez 2007) etc. can be used as effective photocatalyst in degradation of water treatment. These oxides have high band gap tunability, low transparency in visible light, increased surface area at nanoscale level making it as promising photocatalyst (Choudhury et al. 2017; Cheng et al. 2012). However there are some properties like low quantum yield, high recombination rate, high agglomeration, wide energy band gap limiting its applicability in water treatment process (Opoku et al. 2017).

9.3 Synthesis of Metal Oxide Nanoparticles

Metal oxide nanostructured materials synthesis can be broadly categorised into bottom up and top down approaches. In top down approaches, bulk solids are finely grounded where each component becomes part of nanostructure level (Rodríguez and Fernández-García 2006). Milling and attrition are examples of top down approaches. In bottom up approaches, very small nanoparticles are formed due to presence of chemical and physical forces operating while self-assembling process. Colloidal dispersion method is the best method for bottom up approaches (Alagarasi 2013; Rogach et al. 2004). Other methods which are useful in synthesizing nanostructured materials are as follows.

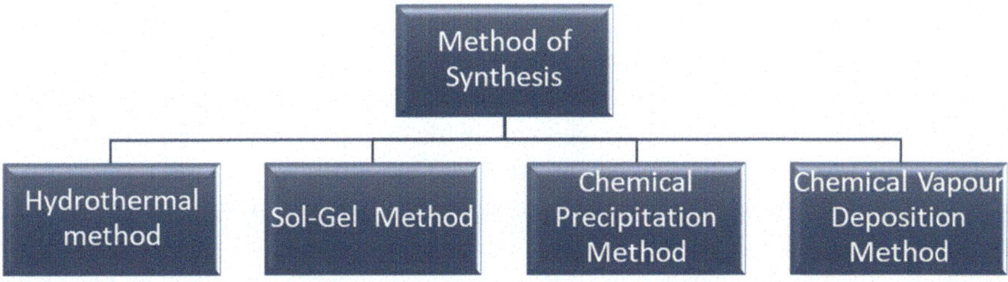

Fig. 9.1 Different method for synthesizing metal oxide nanoparticles

9.3.1 Chemical Precipitation Method

In this method metal salt (chlorides, nitrates) precursor solution are formed in water (any other solvent can also be used), upon addition of base there is formation of oxy-hydroxide precipitate followed by calcination. The size of nanomaterials are dependent on pH, reaction temperature, solvents, surfactants etc. however this method is very easy and simple for synthesizing in aqueous solutions (Alagarasi 2013; Rogach et al. 2004).

9.3.2 Sol –Gel Method

In this method formation of colloidal suspension takes place followed by gelation to form broad network of structure. The precursor solution (metals and metalloids element) are surrounded by reactive ligands. In this method removal of liquid and calcination temperature mainly controls the particle size and morphology (Mishra et al. 2015).

9.3.3 Hydrothermal Synthesis

In this method thermal treatment of reaction mixture is carried out in an autoclave above boiling temperature of water. Metal precursors are reduced to give metal oxides nanoparticles having wide variety of shape. If solvent other than water is used the process is termed as solvothermal method of synthesis. Fine size of metal oxides are obtained by controlling the rate of hydrolysis, reaction temperature and pressure, solvents. Usually a high temperature solvent is used (Akhir et al. 2016).

9.3.4 Chemical Vapour Deposition Techniques

This is a gas phase synthesis techniques and are very useful in controlling parameters like size, shape and chemical composition. In this method gaseous products are allowed to react either homogenously or heterogenously depending on the application of obtained product. In homogenous chemical vapour deposition method, the volatile precursors are introduced into the reacting chamber and get adsorbed on substrate. The particles formed in gaseous phase diffuses towards cold surface through thermophoretic effect. The formed nanopowders are scrapped. In heterogenous chemical vapour deposition method, the solid formed at substrate first catalyse the reaction forming dense film of nanomaterial. This method is excellent for obtaining highly pure material, excellent control size, shape, crystallinity and chemical composition (Subba Rao and Venkatarangaiah 2014; Pokropivny et al. 2007).

9.4 Characterization Techniques

The surface analysis techniques and conventional characterization techniques are designed to determine the composition, crystallinity and defects in structure, shape, size, topography and morphology of surfaces. Some of the techniques that are very helpful in fabrication and characterisation of nanostructured metal oxides are as follows.

9.4.1 Fourier Transform Infra-Red (FT-IR) Spectroscopy

This technique is very helpful in determining the functional group of the nanostructured material symmetric and asymmetric stretching vibration of molecules. It is also helpful in elucidating the structure of unknown compounds.

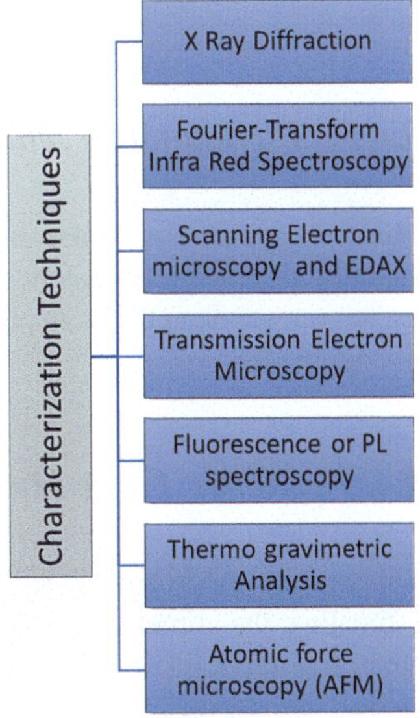

Fig. 9.2 Techniques for characterising metal oxide nanoparticles

9.4.2 X-ray Diffraction (XRD)

It is a non -destructive and widely used technique for characterization of metal oxide nanostructured materials. It reveals information about chemical composition, crystallite size, lattice plane, orientation and various defects in crystalline materials (Sharma et al. 2012).

9.4.3 Scanning Electron Microscopy (SEM) & EDAX

In this technique samples are scanned with a highly energetic beam of electron and reveal information about surface morphology, shape and other surface features. The surface of sample should be conductive, otherwise thin layer of carbon or gold coating is done in order to make it conductive. However EDAX reveal information about elemental and chemical composition of the given nanostructured materials (Buhr et al. 2009).

9.4.4 Transmission Electron Microscopy (TEM)

The images are produced by illuminating the sample with highly energetic beam of electron within vacuum region, detecting the electron transmitted through the sample and obtaining image on photographic film. It also reveal information about surface topography, structure, symmetry, orientation and size of materials.

9.4.5 Photoluminescence (PL) Technique

This technique is based on instantaneous emission of light from the given sample. Its signal is dependent on nature of material to be examined and also depends on the excitation wavelength. From photoluminescence spectra, band gap can be estimated.

9.4.6 Atomic Force Microscopy (AFM)

This is very helpful in directly visualizing the surface of materials in nanoscale, estimating the dimensions of the surface. A highly three dimensional image is created as probe is scanned across the surface. It is also a versatile technique, and creates the surface image of any type of material like polymer, metal oxides, ceramics, glass and biological samples. Surface image of both conducting and non-conducting are easily observed.

9.4.7 Thermo Gravimetric Analysis (TGA) and Differential Thermal Analysis (DTA) Technique

This technique is also very helpful in determining the thermal stability of nanostructured material over wider range of temperature. It also reveals information about various phenomenon (curing, endotherm and exotherm nature of decomposition, phase transformation etc.) occurring while heating a sample in an inert atmosphere.

9.5 Application in Water Treatment Technologies

9.5.1 Metal Oxide as Nanoadsorbents

Various methods like filteration, desalination, biological methods, sedimentation and adsorption are used in treatment of wastewater or water treatment process (Bora and Dutta 2014). Among these, adsorption method is most suitable for treatment having several advantages over others. It is a surface phenomenon where adsorbate (pollutants) molecules are adsorbed on the surface by chemisorption or physiosorption process (Hua et al. 2012). Pollutants mostly heavy metal ions are adsorbed via various interactions like electrostatic, hydrogen bonding, vander wall interaction onto the surface of catalyst. Metal oxides based nanoadsorbants must be porous, possess high surface area for effective and efficient catalytic activity (Gehrke et al. 2015). This is estimated by isotherm, used to describe the mechanism of adsorption based on Langmuir and Freundlich adsorption isotherm (Taman 2015). Although, it relies on several factors like temperature, pH, contact time, particle size and concentration of pollutants. Since nanoadsorbants show higher efficiency due to specific characteristics properties (high surface area) leading to higher rate of adsorption. Recently various nanoadsorbants like copper oxide nanoparticles, zinc oxide nanoparticles, titanium oxide nanoparticles, iron oxides/hydroxides nanoparticles (Nassar 2012) have been envisaged at different pH range for removal of different kind of heavy metal ions (Bora and Dutta 2014). Singh et al. reported ZnO to be nontoxic and works effectively in 5.8–6.8 pH range. Various authors (Bang et al., Pena et al.) reported TiO_2 nanoparticles with high chemical stability, nontoxicity, low cost in removal of arsenic at specified pH. Several metal oxides are doped with carbon based materials (like graphene, multiwalled carbon nanotubes) and conducting polymer like polyaniline, polypyrrole etc. for better efficiency due to synergistic effect (Hua et al. 2012). There are several other composites of metal oxides like acetate/PPy/TiO_2, succinic/PPy/TiO_2 and tartaric/PPy/TiO_2 (Opoku et al.

2017) which shows maximum adsorption capacity and require less interval of time and can also be reused. Somayeh et.al reported the effective removal of copper ions by using TiO_2, PPy/TiO_2, PPy/TiO_2/DHSNa composites as nano-adsorbents optimizing parameters at pH 3.0, catalyst loading- 50 mg/L and equilibrium time of 30 min (Opoku et al. 2017). Another most important metal oxides, iron oxide magnetic nanoparticles are employed as nanoadsorbents in removal of organic pollutants. Iron oxide magnetic nanoparticles possess high surface area, fast kinetics, adsorption capacity and magnetic properties. These nanoparticles showed excellent efficiency in removal of organic pollutants like polycyclic aromatic hydrocarbon (PAHs), polychlorinated biphenyls (PCBs) and pesticides (Gutierrez et al. 2018).

9.5.2 *Metal Oxide as Photocatalyst in Degradation of Pollutants*

The work in photocatalysis area is explored by Fujishima by describing splitting of water using TiO2 photoelctrode (Hashimoto et al. 2005). Various metal oxides have been explored in the area of photocatalysis for the degradation of various types of contaminants due to their attractive optical properties, ease in synthesis and photostability. However, there are major limitations like high recombination rate, delay in charge separation and wide band gap which reduces the overall efficiency of catalytic activity (Hanaor and Sorrell 2011).

Photocatalysis uses semiconductors such as metal oxides, nitrides, and sulphides etc. A metal oxide is photoactivated by the photon from the light source. Natural sunlight as a renewable and safe energy source as well as abundant and clean, has been the ideal source of energy for the activation process (Chan et al. 2011). Upon interaction with light electro-hole pairs are produced which on oxidation produces highly powerful oxidants, capable of reducing contaminants (Sultana et al. 2015). Heterogeneous catalysis is most widely used with the help of modification in metal oxide nanostructured material by doping (Chen et al. 2015). Doping can be achieved within metal oxide-metal oxide, metal-metal oxide to improve the optical properties of the derived composite material like reduction in band gap and making it suitable for visible light irradiation, reducing recombination rate of electron-hole pair thereby enhancing the binding with pollutant molecules and also increase the electron carrier mobility to enhance charge separation effect (Mahlambi et al. 2015; Qu and Alvarez 2013; Anjum et al. 2016).

9.5.2.1 Mechanism of Photocatalytic Degradation

When the metal oxide is irradiated by sunlight, there occur separation of electrons and holes from valence and conduction band respectively only if the energy of the incident photons is equal to or greater than ($\geq$) the metal oxide band gap. The

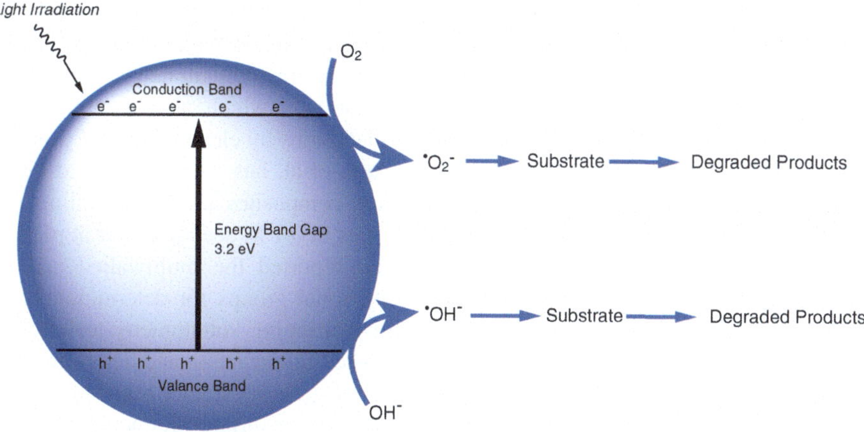

Fig. 9.3 Mechanism of photocatalysis of TiO_2

photogenerated electron then migrates to the metal oxide surface where separation and redox reaction occur. The electrons in conduction band are trapped by oxygen molecules dissolved in solution forming superoxide radicals O_2^{-} which subsequently transformed into. OH hydroxide radicals. Now the photogenerated holes in valence band react with water molecules also producing $^•OH$ radicals. These hydroxides and superoxides radicals are highly reactive and act as strong oxidising agents, thereby reacting with dye molecules thus decomposing it into smaller molecules, CO_2, H_2O (Chen et al. 2015; Sultana et al. 2015) (Fig. 9.3).

9.5.3 Metal Oxide Composites in Electrochemical Detection of Heavy Metal Ions

Heavy metal ions are essential for normal functioning of human body. However, excessive intake will cause damage to vital organs by interfering with normal functions and hindering metabolic system (Zhou et al. 2016). Detection and sensing of heavy metal ions relieve potential threats to the ecosystem. However, detection of heavy metal involving spectroscopic techniques like atomic absorption spectroscopy (AAS), flame photometry, inductively coupled plasma mass spectrometry (ICP-MS) and inductively coupled plasma atomic emission spectrometry (ICP-AES) are most widely used but there are certain limitations regarding detection time and tedious procedures (Lu et al. 2018). Exploiting nanotechnology in electrochemical method with the help of variety of modified electrodes based on nanostructured metal oxides, and different types of conducting and biodegradable polymer is reliable and accurate technique for detection of metal ions at lower concentration (Arduini et al. 2010; Chamjangali et al. 2015; Zhu et al. 2014). Besides voltammetry other electrochemical techniques like anodic stripping voltammetry (ASV), linear sweep voltammetry

(LSV), differential pulse voltammetry (DPV), square wave voltammetry (SWV) were also used due to high sensitivity, wide range of detection limits at lower concentration and high compatibility. Anodic stripping voltammetry consists of two step process, firstly accumulation of metal ions which also require better adsorption property followed by stripping of metal ions by electrochemical process (Lu et al. 2018; Promphet et al. 2015). M. Arab et al. has shown modified GCE coated with MWCNT poly (pyrocatechol violet) composites and bismuth films for simultaneous determination of Cd^{2+} and Pb^{2+} ions in 0.1 M acetate buffer at pH 5.0 (Chamjangali et al. 2015). Xueying Li et al. reported the multiwalled carbon nanotubes grafted with nitrogen and thiol functional groups for determination of very harmful cadmium and lead ion by modified stripping voltammetry techniques (Li et al. 2016). Modified biosensors were also used for detection and determination of heavy metal ion based on DNA as template for growing strand. It appears to be very feasible and compatible method in order to avoid further production of bye-products (Saidur et al. 2017).

9.5.4 *Water Disinfection Treatment Process*

One of the most critical step in water treatment plant, helpful in reducing generation of harmful by-products and makes water for potable use. Water disinfection process is also carried out by chlorination, however it does have some adverse effect of fouling. Various nanostructured metal oxide like TiO_2 (Li et al. 2008), ZnO (Prasanna and Vijayaraghavan 2015) and noble metals like Ag (Li et al. 2008), Au, Pt, Pd nanoparticles possess strong antibacterial, disinfectant, and antimicrobial properties (Anjum et al. 2016; Li et al. 2008). Besides, various other carbon based materials and biodegradable polymer like chitin, chitosan etc. are also shown to possess good properties towards different bacterial systems (Amin and Alazba 2014; Zhang et al. 2016). Disinfection process can be efficiently carried out by photocatalysis. Photocatalysis disinfection of wastewater was first reported by Matsunaga et al. In this method, the solution is exposed to UV or visible light, metal atom gets converted to metal ion, results in formation of highly reactive oxygen species. The highly reactive oxygen species and metal ions helps in lysis of bacterial cell wall and cause inactivation (Matsunaga et al. 1985; Upadhyay et al. 2014). Special feature of this method is that upon irradiation the metal ion gets released from inactivated bacteria and can be used further. Some metal oxides can be activated without light irradiation and this process is useful in removal of volatile organic compounds (Mondal 2017).

9.5.5 *Chemical Treatment of Wastewater*

Wastewater treatment via chemical method involve hydroxides, carbonates and sulphides as coagulants for removal of pollutants. In chemical method, oxiadation and ozonation is most widely used (Kao 2008). Further, chemical oxidation method is categorised into chemically and UV assisted method. Strong oxidizing agents like hydrogen peroxide, chlorine, potassium permanganate and fenton's reagent were used (Opoku et al. 2017). Bouyakoub. A.Z. et al. reported the significant removal of Levafix Brilliant Blue dye with the help of manganese chloride and magnesium chloride (Bouyakoub et al. 2011). Various other dye molecules like Direct Yellow R, Acid Orange II, Reactive Yellow K-6G etc. were also treated by white mud solution with efficient removal rate of 90% within time period of 90 sec (Janaki et al. 2012). Ozonation is also most widely used with the action of strong oxidizing agent like ozone (O_3). Ozone gets converted into free radicals (highly reactive species) which ultimately reacts with pollutant molecules decomposing it into the final by products. Conditions required for effective ozonation are optimum temperature, UV/peroxide and optimum pH (alkaline or acidic) depending on pollutants (Manuscript et al. n.d.; Mahmoodi 2011; Mehmood et al. 2013). Integrated chlorination process has also been used for suppressing the growth of bacteria. However, high dosage can also be effective for efficient removal of dye molecules but can cause adverse effect by the production of hazardous by-products with undesirable fouling odour (Gopal et al. 2007; Vincenzo Naddeo 2013).

9.5.6 *Treatment of Wastewater by Ion Exchange Technology*

Ion exchange process involve treatment of wastewater containing heavy metal pollutants on the basis of strong interaction between contaminants (charged functional groups) and resin used. Ion exchange process is divided into anionic and cationic ion exchange on the basis of different types of resin used. Cation exchange resins have negatively charged groups (sulphate, phosphate and carboxylates) attached to the backbone, whereas anion exchange resins have positively charged molecules like alkyl substituted phosphine (PR_3^+), alkyl substituted sulphide (SR_2^+) and alkyl substituted amine fixed to the backbone. Besides these, amphoteric exchangers are also used for simultaneous exchange of cation and anions with the help of special type of mixed bed reactor. Cationic and anionic exchangers are further categorized into strong and weak acid base ion exchange resin on the basis of functional groups (Kumar and Jain 2013; Katayama and Furuichi 1996; Khan et al. 2009). Ion exchange materials are available in many forms like synthetic, natural and organic materials. First organic material based resin was synthesized by copolymerization of styrene and divinyl benzene. These resins have greater exchange capacity and stability, mostly used as anion exchanger for removing carbonic acid, silicic acid and other anions for complete demineralization of water.

However these organic polymers do not withstand high temperature (Kumar and Jain 2013; Miller and Castagna 2017). To overcome various flaws, the existing exchangers are modified with metal oxides, polymer doped with polymetallic acids and organic-inorganic hybrid materials. These materials will possess various properties like high ion exchange capacity, selectivity, good reproducibility, mechanical and thermal stability (Opoku et al. 2017; Wawrzkiewicz et al. 2018). For removal of heavy metal ions like Arsensic (III) and Chromium (IV), commercial anion exchange resins like Amberlite IRA-400, Purolite A-505 and Relite A 490 were used. These ion exchange resins have shown higher affinity towards removal of divalent anions as compared to monovalent anions attaining maximum sorption at pH range of 6.0–9.0 (Rivas et al. 2015). A.A.Khan has reported the synthesis of modified cation exchange resin of poly-o-aniside Sn (IV) arsenophosphate composites for selective determination of Pb (II) in wastewater operating over wide range of pH with quick response time. The modified composites can also be used as ion selective metal electrode for determination of Pb by potentiometric titration (Khan et al. 2009). Another group has reported the removal of Acid Orange 7, Reactive Blue 5 and Direct Blue 71 dye present in textile wastewaters under strongly basic anion exchange resin (Lewatit Mono Plus MP 500) (Rivas et al. 2015). Presence of charged group on dye molecules aid in its strong interaction with resin thereby helps in its easy removal. These resins were most effective in treatment of wastewater dye solutions with monolayer sorption capacity ranging from 979 to 1004 mg/g with increasing temperature from 25 °C to 45 °C. Its regeneration process was easily carried out by using NaOH or NaOH-MeOH solution. Beside this cation exchange resins were also used for removal of wastewater dye solutions (Wawrzkiewicz et al. 2018) (Table 9.1).

9.6 Challenges in Water Treatment Technology

Rapid progress and development in nanotechnology has led to the exploitation of nanocomposites, conducting polymer and biodegradable polymer in treatment of water and wastewater by using different techniques. There are certain challenges faced while processing various treatment techniques on large scale due to high capital cost, energy input, man power and optimizing operating conditions. However traditional methods involving water treatment creates lots of hazardous by-products, toxic sludges which ultimately creates pollution. Some evolved technologies like photocatalysis, ion exchange technology, water disinfection treatment and electrochemical determination of heavy metal ions were being efficient for wastewater treatment because of various attractive properties like low production of hazardous by-products, no sludge production, simultaneous determination and detection of metal ions, degradation of harmful cationic and anionic dyes along with removal of microorganisms and pathogens. For operating these processes on large scale, development in synthesizing various novel hybrid nanocomposites based on

Table 9.1 Description of treatment of wastewater by various methods

S. No	Nanocomposites	Synthesis method	Treatment mode	Applications	References
1.	Polyamide grafted carbon microspheres	Interfacial polymerization technique	Adsorption	Rhodamine B dye and heavy metals (Pb, Cd, Hg, Cr, Ni, and Cu)	Saleh and Ali (2018)
2.	Magnetic graphene oxide (MGO) and sand	Reduction method	Adsorption	Removal of Cr(VI), Pb(II), As(III), and Cd(II)	Ghasemabadi et al. (2018)
3.	Polyaluminum chloride	Chemical method	Adsorption	Remazol brilliant blue R dye	Trivedi et al. (2009)
4.	TiO_2-Pt composite	Hydrothermal followed by chemical reduction method	Photocatalysis	Organic pollutants	Alamelu and Ali (2018)
5.	Pd-V_2O_5 Heterostructure Nanorods	Hydrothermal synthesis	Photocatalysis	Photodegradation of Rh-6G	Kumar and Thangadurai (2018)
6.	BSA-$ZnWO_4$ Nanocomposites	Hydrothermal method	Photocatalysis	Degradation of methylene blue	Singh et al. (2013)
7.	Bismuth modified electrodes	Electroplating	Electrochemical method	Detection of lead	Arduini et al. (2010)
8.	Gold nanoparticle-graphene- cysteine composite modified bismuth film electrode	Electrochemical synthesis	Square wave anodic stripping voltammetry (SWASV)	Determination of Cd(II) and Pb(II)	Zhu et al. (2014)
9.	Ruthenium(II)-textured graphene oxide nanocomposite	Modified Hummers' method	Electrochemical method	Detection of Cd(II), Pb(II), As(III) and Hg(II)	Gumpu et al. (2017)
10.	Ag-TiO_2 photocatalyst	Hydrothermal synthesis	Disinfection process via photocatalysis	Antimicrobial and antibacterial activity.	Castro et al. (2012)
11.	Zinc oxide nanoparticles	Precipitation method	Disinfection	Treatment of pathogenic bacteria, viruses and fungi by disinfection process	Li et al. (2008)

(continued)

Table 9.1 (continued)

S. No	Nanocomposites	Synthesis method	Treatment mode	Applications	References
12.	Ag- $CoFe_2O_4$-GO Nanocomposite	Solvothermal reaction	Disinfection	Simultaneous disinfection and Pb (II) determination	Ma et al. (2015)
13.	$MnCl_2$ and $MgCl_2$	Chemicals	Chemical method	Removal of reactive dye and Levafix Brilliant Blue EBRA	Bouyakoub et al. (2011)
14.	PAni/HA/ TiO_2/Fe_3O_4 Nanocomposite	Chemical oxidation method	Chemical treatment method	Can be used in degradation of pollutants	Barkoula et al. (2008)
15.	Copper ferrite nanoparticle	Co-precipitation method	Ozonation method	Removal of reactive Red 198 (RR198) and Reactive Red 120 (RR120)	Mahmoodi (2011)
16.	Composites of mixed tin(IV) oxide	Chemical precipitation method	Ion exchange process	Separation of Gallium from waste	Paganini and Felinto (2005)
17.	TiO_2 and Fe-WBAX	Chemical precipitation method	Ion exchange process	Simultaneous determination of Chromium and arsenic	Gifford (2016)
18.	Poly-o-aniside Sn(IV) arsenophosphate	Chemical precipitation method	Ion exchange process	Determination of lead	Khan et al. (2009)

nanostructured metal oxides had to be explored for better selectivity, sensitivity, detection, determination and treatment of wastewater.

9.7 Conclusion and Future Prospectives

Pertaining to current situation faced in water treatment process, this chapter has summarised various synthesis and characterisation techniques of metal oxide nanostructured materials, and their application in water treatment technologies which are efficient and very effective in removal of toxic pollutants like dye discharges and heavy metal ions by chemical oxidation method, photocatalysis, ion exchange process, electrochemical method, disinfection process and adsorption techniques using viable alternatives for degradation of different kinds of persistent contaminants. Since metal oxides nanostructured material must possess some specific properties like very high surface area to volume ratio, specific surface area for better adsorption, low band gap energy for high efficiency in visible region of light. In order to reach the desired goal on large scale these metal oxide nanostructured materials are modified by doping with various other composite materials like conducting polymers, zeolites, clay etc. and are needed to be revolutionized in these aspects to minimise hurdles in accomplishing tasks like adsorption capacity, improvement in efficiency and photostability in visible region.

Acknowledgement Authors are thankful to Aligarh Muslim University for providing necessary research facilities. SS thanked UPCST (CST/223) for providing financial support to carry out the work.

References

Akhir MAM, Mohamed K, Lee HL, Rezan SA (2016) Synthesis of tin oxide nanostructures using hydrothermal method and optimization of its crystal size by using statistical design of experiment. Procedia Chem 19:993–998. https://doi.org/10.1016/j.proche.2016.03.148

Alagarasi A (2013) Introduction to nanomaterials. 1–19

Alamelu K, Ali BMJ (2018) SC. Biochem Pharmacol. https://doi.org/10.1016/j.jece.2018.08.042

Amin MT, Alazba AA (2014) A review of nanomaterials based membranes for removal of contaminants from polluted waters. Membr Water Treat 5:123–146. https://doi.org/10.12989/mwt.2014.5.2.123

Amin MT, Alazba AA, Manzoor U (2014) A review of removal of pollutants from water/wastewater using different types of nanomaterials. Adv Mater Sci Eng 2014:1

Anjum M, Miandad R, Waqas M et al (2016) Remediation of wastewater using various nanomaterials. Arab J Chem. https://doi.org/10.1016/j.arabjc.2016.10.004

Arduini F, Calvo JQ, Palleschi G et al (2010) Bismuth-modified electrodes for lead detection. TrAC Trends Anal Chem 29:1295–1304. https://doi.org/10.1016/j.trac.2010.08.003

Barkoula NM, Alcock B, Cabrera NO, Peijs T (2008) Fatigue properties of highly oriented polypropylene tapes and all-polypropylene composites. Polym Polym Compos 16(2):101–113

Bora T, Dutta J (2014) Applications of nanotechnology in wastewater treatment–a review. J Nanosci Nanotechnol 14:613–626. https://doi.org/10.1166/jnn.2014.8898

Bouyakoub AZ, Lartiges BS, Ouhib R et al (2011) MnCl2and MgCl2for the removal of reactive dye Levafix Brilliant Blue EBRA from synthetic textile wastewaters: an adsorption/aggregation mechanism. J Hazard Mater 187:264–273. https://doi.org/10.1016/j.jhazmat.2011.01.008

Buhr E, Senftleben N, Klein T et al (2009) Characterization of nanoparticles by scanning electron microscopy in. Meas Sci Technol 20:1–9. https://doi.org/10.1088/0957-0233/20/8/084025

Byrne C, Subramanian G, Pillai SC (2017) Recent advances in photocatalysis for environmental applications. J Environ Chem Eng:0–1. https://doi.org/10.1016/j.jece.2017.07.080

Castro CA, Jurado A, Sissa D, Giraldo SA (2012) Performance of Ag-TiO2photocatalysts towards the photocatalytic disinfection of water under interior-lighting and solar-simulated light irradiations. Int J Photoenergy 2012:1. https://doi.org/10.1155/2012/261045

Chamjangali MA, Kouhestani H, Masdarolomoor F, Daneshinejad H (2015) A voltammetric sensor based on the glassy carbon electrode modified with multi-walled carbon nanotube/poly(pyrocatechol violet)/bismuth film for determination of cadmium and lead as environmental pollutants. Sens Actuators B Chem 216:384–393. https://doi.org/10.1016/j.snb.2015.04.058

Chan SHS, Wu TY, Juan JC, Teh CY (2011) Recent developments of metal oxide semiconductors as photocatalysts in advanced oxidation processes (AOPs) for treatment of dye waste-water. J Chem Technol Biotechnol 86:1130–1158. https://doi.org/10.1002/jctb.2636

Chen J, Qiu F, Xu W et al (2015) Recent progress in enhancing photocatalytic efficiency of TiO2-based materials. Appl Catal A Gen 495:131–140. https://doi.org/10.1016/j.apcata.2015.02.013

Cheng Z, Tan ALK, Tao Y et al (2012) Synthesis and characterization of iron oxide nanoparticles and applications in the removal of heavy metals from industrial wastewater. Int J Photoenergy 2012:5. https://doi.org/10.1155/2012/608298

Choudhury SP, Kumari N, Bhattacharjee A (2017) Study of structural, electrical and optical properties of Ni-doped SnO2 for device application: experimental and theoretical approach. J Mater Sci Mater Electron 28:18003–18014. https://doi.org/10.1007/s10854-017-7743-3

Deng Y, Zhao R (2015) Advanced oxidation processes (AOPs) in wastewater treatment. Curr Pollut Rep 1:167–176. https://doi.org/10.1007/s40726-015-0015-z

Fernández-garcia MJA, Rodriguez JA (2007) Metal oxide nanoparticles. Nanomater Inorg Bioinorg Perspect 60. https://doi.org/10.1002/0470862106.ia377

Gehrke I, Geiser A, Somborn-Schulz A (2015) Innovations in nanotechnology for water treatment. Nanotechnol Sci Appl 8:1–17. https://doi.org/10.2147/NSA.S43773

Ghasemabadi SM, Baghdadi M, Safari E (2018) Investigation of continuous adsorption of Pb (II), As (III), Cd (II), and Cr (VI) using a mixture of magnetic graphite oxide and sand as a medium in a fi xed- bed column. J Environ Chem Eng 6:4840–4849. https://doi.org/10.1016/j.jece.2018.07.014

Gifford JM (2016) Sustainable drinking water treatment: using weak base anion exchange sorbents embedded with metaloxide nanoparticles to simultaneously remove multiple oxoanions

Gopal K, Tripathy SS, Bersillon JL, Dubey SP (2007) Chlorination byproducts, their toxicodynamics and removal from drinking water. J Hazard Mater 140:1–6. https://doi.org/10.1016/j.jhazmat.2006.10.063

Gumpu MB, Veerapandian M, Krishnan UM, Rayappan JBB (2017) Simultaneous electrochemical detection of Cd(II), Pb(II), As(III) and Hg(II) ions using ruthenium(II)-textured graphene oxide nanocomposite. Talanta 162:574–582. https://doi.org/10.1016/j.talanta.2016.10.076

Gurunathan L, Ponnusamy V (2017) Photocatalytic effect of (TiO2/CeO2) with support of β-cyclodextrin for enhanced performance under solar light. J Mater Sci Mater Electron 28:18666–18674. https://doi.org/10.1007/s10854-017-7816-3

Gutierrez AM, Dziubla TD, Hilt JZ (2018) HHS Public Access 32:111–117. https://doi.org/10.1515/reveh-2016-0063.Recent

Hanaor DAH, Sorrell CC (2011) Review of the anatase to rutile phase transformation. J Mater Sci 46:855–874. https://doi.org/10.1007/s10853-010-5113-0

Hashimoto K, Irie H, Fujishima A (2005) TiO 2 photocatalysis: a historical overview and future prospects. Jpn J Appl Phys 44:8269–8285. https://doi.org/10.1143/JJAP.44.8269

Hua M, Zhang S, Pan B et al (2012) Heavy metal removal from water/wastewater by nanosized metal oxides: a review. J Hazard Mater 211–212:317–331. https://doi.org/10.1016/j.jhazmat.2011.10.016

Janaki V, Vijayaraghavan K, Ramasamy AK et al (2012) Competitive adsorption of reactive orange 16 and reactive brilliant blue R on polyaniline/bacterial extracellular polysaccharides composite-A novel eco-friendly polymer. J Hazard Mater 241–242:110–117. https://doi.org/10.1016/j.jhazmat.2012.09.019

Journal GN (2016) Use of selected advanced oxidation processes (AOPs) for wastewater treatment – a mini review

Kao CM (2008) Chemical oxidation of chlorinated solvents in contaminated groundwater: review. Pract Period Hazard Toxic Radioact Waste Manag 2:116. https://doi.org/10.1061/(ASCE)1090-025X(2008)12

Katayama N, Furuichi R (1996) Modeling of ion-exchange reactions on metal oxides with acid – base and charge characteristics of MnO 2, TiO 2, Fe 3 O 4, and Al 2 O 3 surfaces and adsorption affinity of alkali metal ions. Environ Sci Technol:30, 1198–1204. https://doi.org/10.1021/es9504404

Khan AA, Habiba U, Khan A (2009) Synthesis and characterization of organic-inorganic nanocomposite poly-o-anisidine Sn (IV) arsenophosphate: its analytical applications as Pb (II) ion-selective membrane electrode. 2009. https://doi.org/10.1155/2009/659215

Kumar S, Jain S (2013) History, introduction, and kinetics of ion exchange materials. J Chem 2013:13. https://doi.org/10.1155/2013/957647

Kumar JS, Thangadurai P (2018) SC. Biochem Pharmacol. https://doi.org/10.1016/j.jece.2018.08.028

Kyzas GZ, Matis KA (2015) Nanoadsorbents for pollutants removal: a review. J Mol Liq 203:159–168. https://doi.org/10.1016/j.molliq.2015.01.004

Li Q, Mahendra S, Lyon DY et al (2008) Antimicrobial nanomaterials for water disinfection and microbial control: potential applications and implications. Water Res 42:4591–4602. https://doi.org/10.1016/j.watres.2008.08.015

Li X, Zhou H, Fu C et al (2016) A novel design of engineered multi-walled carbon nanotubes material and its improved performance in simultaneous detection of Cd(II) and Pb(II) by square wave anodic stripping voltammetry. Sensors Actuators B Chem 236:144–152. https://doi.org/10.1016/j.snb.2016.05.149

Lu Y, Liang X, Niyungeko C et al (2018) A review of the identification and detection of heavy metal ions in the environment by voltammetry. Talanta 178:324–338. https://doi.org/10.1016/j.talanta.2017.08.033

Ma S, Zhan S, Jia Y, Zhou Q (2015) Highly efficient antibacterial and Pb(II) removal effects of Ag-CoFe<inf>2</inf>O<inf>4</inf>-GO nanocomposite. ACS Appl Mater Interfaces 7:10576–10586. https://doi.org/10.1021/acsami.5b02209

Mahlambi MM, Ngila CJ, Mamba BB (2015) Recent developments in environmental photocatalytic degradation of organic pollutants: the case of titanium dioxide nanoparticles – a review. J Nanomater 2015:1–29. https://doi.org/10.1155/2015/790173

Mahmoodi NM (2011) Photocatalytic ozonation of dyes using copper ferrite nanoparticle prepared by co-precipitation method. Desalination 279:332–337. https://doi.org/10.1016/j.desal.2011.06.027

Matsunaga T, Tomoda R, Nakajima T, Wake H (1985) Photoelectrochemical sterilization of microbial-cells by semiconductor powders. FEMS Microbiol Lett 29:211–214. https://doi.org/10.1111/j.1574-6968.1985.tb00864.x

Mehmood CT, Batool A, Qazi IA (2013) Combined biological and advanced oxidation treatment processes for COD and color removal of sewage water. Int J Environ Sci Dev 4:88–93. https://doi.org/10.7763/IJESD.2013.V4.311

Miller AWS, Castagna CJ (2017) Understanding ion-exchange resins for water treatment systems. Water Technol Solut Tech Pap

Mishra NK, Kumar C, Kumar A et al (2015) Structural and optical properties of SnO2–Al2O3 nanocomposite synthesized via sol-gel route. Mater Sci 33:714–718. https://doi.org/10.1515/msp-2015-0101

Mondal K (2017) Recent advances in the synthesis of metal oxide nanofibers and their environmental remediation applications. Inventions 2:9. https://doi.org/10.3390/inventions2020009

Nassar NN (2012) Iron oxide nanoadsorbents for removal of various pollutants from wastewater: an overview. Appl Adsorbents Water Pollut Control:81–118. https://doi.org/10.1073/pnas.0703993104

Opoku F, Kiarii EM, Govender PP, Mamo MA (2017) Metal oxide polymer nanocomposites in water treatments. Descr Inorg Chem Res Met Compd. https://doi.org/10.5772/67835

Paganini P, Felinto MCFC (2005) Inorganic ion exchanger based on tin oxide for heavy metals. Int Nucl Atl Conf -

Pokropivny V, Hussainova I, Vlassov S (2007) Introduction to nanomaterials and nanotechnology. Introd Nanomater Nanotechnol:1–138

Prasanna VL, Vijayaraghavan R (2015) Insight into the mechanism of antibacterial activity of ZnO: surface defects mediated reactive oxygen species even in the dark. Langmuir 31:9155–9162. https://doi.org/10.1021/acs.langmuir.5b02266

Promphet N, Rattanarat P, Rangkupan R et al (2015) An electrochemical sensor based on graphene/polyaniline/polystyrene nanoporous fibers modified electrode for simultaneous determination of lead and cadmium. Sens Actuators B Chem 207:526–534. https://doi.org/10.1016/j.snb.2014.10.126

Qu X, Alvarez PJJ (2013) Applications of nanotechnology in water and wastewater treatment. Water Res 47:3931–3946. https://doi.org/10.1016/j.watres.2012.09.058

Reddy CV, Babu B, Vattikuti SVP et al (2016) Structural and optical properties of vanadium doped SnO2nanoparticles with high photocatalytic activities. J Lumin 179:26–34. https://doi.org/10.1016/j.jlumin.2016.06.036

Rivas BL, Sánchez J, Urbano BF (2015) Polymers and nanocomposites: synthesis and metal ion pollutant uptake. Polym Int 65:255–267. https://doi.org/10.1002/pi.5035

Rodríguez JA, Fernández-García M (2006) Introduction: the world of oxide nanomaterials. Synth Prop Appl Oxide Nanomater:1–5. https://doi.org/10.1002/9780470108970.ch

Rogach AL, Talapin DV, Weller H (2004) Semiconductor nanoparticles. Colloids Colloid Assem Synth Modif Organ Util Colloid Part:52–95. https://doi.org/10.1002/3527602100

Saidur MR, Aziz ARA, Basirun WJ (2017) Recent advances in DNA-based electrochemical biosensors for heavy metal ion detection: a review. Biosens Bioelectron 90:125–139. https://doi.org/10.1016/j.bios.2016.11.039

Saleh TA, Ali I (2018) PT SC. Biochem Pharmacol. https://doi.org/10.1016/j.jece.2018.08.033

Sara M (2014) Magnetic nanocomposites for heavy metals removal from stormwater. 1–91

Sharma R, Bisen DP, Shukla U, Sharma BG (2012) X-ray diffraction: a powerful method of characterizing nanomaterials. Recent Res Sci Technol 4:77–79

Singh P, Raizada P, Pathania D et al (2013) Preparation of BSA-ZnWO4nanocomposites with enhanced adsorptional photocatalytic activity for methylene blue degradation. Int J Photoenergy 2013:1. https://doi.org/10.1155/2013/726250

Stankic S, Suman S, Haque F, Vidic J (2016) Pure and multi metal oxide nanoparticles: synthesis, antibacterial and cytotoxic properties. J Nanobiotechnol 14:1–20. https://doi.org/10.1186/s12951-016-0225-6

Subba Rao AN, Venkatarangaiah VT (2014) Metal oxide-coated anodes in wastewater treatment. Environ Sci Pollut Res 21:3197–3217. https://doi.org/10.1007/s11356-013-2313-6

Sultana S, Zain M, Umar K et al (2015) SnO 2 e SrO based nanocomposites and their photocatalytic activity for the treatment of organic pollutants. J Mol Struct 1098:393–399. https://doi.org/10.1016/j.molstruc.2015.06.032

Taman R (2015) Metal oxide nano-particles as an adsorbent for removal of heavy metals. J Adv Chem Eng 5:1–8. https://doi.org/10.4172/2090-4568.1000125

Tan YN, Wong CL, Mohamed AR (2011) an overview on the photocatalytic activity of nano-doped-TiO2 in the degradation of organic pollutants. ISRN Mater Sci 2011:1–18. https://doi.org/10.5402/2011/261219

Tan L, Xu J, Xue X, Lou Z, Zhu J, Baig SA, Xu X (2014) Multifunctional nanocomposite Fe_3O_4@SiO_2–mPD/SP for selective removal of Pb(II) and Cr(VI) from aqueous solutions, RSC Adv 4:45920–45929. https://doi.org/10.1039/C4RA08040H

Teja AS, Koh PY (2009) Synthesis, properties, and applications of magnetic iron oxide nanoparticles. Prog Cryst Growth Charact Mater 55:22–45. https://doi.org/10.1016/j.pcrysgrow.2008.08.003

Trivedi KN, Boricha AB, Bajaj HC, Jasra RV (2009) Adsorption of remazol brilliant blue R dye from water by polyaluminum chloride. Rasayan J Chemi 2:379–385

Upadhyay RK, Soin N, Roy SS (2014) Role of graphene/metal oxide composites as photocatalysts, adsorbents and disinfectants in water treatment: a review. RSC Adv 4:3823–3851. https://doi.org/10.1039/c3ra45013a

Vincenzo Naddeo AC (2013) Wastewater treatment by combination of advanced oxidation processes and conventional biological systems. J Bioremed Biodegr 04. https://doi.org/10.4172/2155-6199.1000208

Wawrzkiewicz M, Hubicki Z, Polska-adach E (2018) Strongly basic anion exchanger Lewatit MonoPlus SR-7 for acid, reactive, and direct dyes removal from wastewaters. Sep Sci Technol 53:1065–1075. https://doi.org/10.1080/01496395.2017.1293098

World Water Assessment Programme (WWAP) (2015) The United Nations world water development report 2015: water for a sustainable world, facts and figures. UN Water Rep 138. https://doi.org/10.1016/S1366-7017(02)00004-1

Xu P, Zeng GM, Huang DL et al (2012) Use of iron oxide nanomaterials in wastewater treatment: A review. Sci Total Environ 424:1–10. https://doi.org/10.1016/j.scitotenv.2012.02.023

Yang Y, Zhang C, Hu Z (2013) Impact of metallic and metal oxide nanoparticles on wastewater treatment and anaerobic digestion. Environ Sci Process Impacts 15:39–48. https://doi.org/10.1039/C2EM30655G

Zhang Y, Wu B, Xu H et al (2016) Nanomaterials-enabled water and wastewater treatment. Nano Impact 3–4:22–39. https://doi.org/10.1016/j.impact.2016.09.004

Zhou Y, Tang L, Zeng G et al (2016) Current progress in biosensors for heavy metal ions based on DNAzymes/DNA molecules functionalized nanostructures: a review. Sens Actuators B Chem 223:280–294. https://doi.org/10.1016/j.snb.2015.09.090

Zhu L, Xu L, Huang B et al (2014) Simultaneous determination of Cd(II) and Pb(II) using square wave anodic stripping voltammetry at a gold nanoparticle-graphene-cysteine composite modified bismuth film electrode. Electrochim Acta 115:471–477. https://doi.org/10.1016/j.electacta.2013.10.209

Chapter 10
Heavy Metal Remediation by Natural Adsorbents

Neha Dhingra, Ngangbam Sarat Singh, Talat Parween, and Ranju Sharma

Abstract Global industrialization and urbanization have led to serious, alarming levels of environmental pollution. Due to the property of high solubility in the aqueous solutions, heavy metals can quickly be absorbed by all living organisms. Once they enter the food chain, it is challenging to detoxify them. Metals are a part of the biological systems, but up to a certain permissible limit, beyond that limit, it becomes hazardous. The physical and chemical technologies require special equipment, it is also labor intensive as well as very costly. Whereas biological technologies of remediation are gaining popularity in order to solve the increasing levels of contamination in the environment. During the recent studies, it is clear that lime precipitation proves to be as one of the effective technique in order to treat inorganic effluent having a concentration of metal higher than 1000 mg/L; usage of new adsorbents, as well as the technique of membrane filtration, are frequently studied and is used for the remediation of the heavy metal-contaminated wastewater. Various techniques have been used for the remediation of contaminated wastewater, it is important to select the most effective method for remediation of metal-contaminated wastewater based on criteria of pH, initial metal concentration, the overall result after the treatment when compared with other technologies along with environmental impact and economics parameter including the capital investment and costs of operation. Finally, the technical applicability along with the simplicity of the plant and cost-effectiveness are major key factors that play an important role in the selection of the suitable treatment system for contaminated wastewater.

N. Dhingra
Department of Zoology, University of Delhi, New Delhi, India

N. S. Singh
University of Delhi to Department of Zoology, Dr. SRK Government Arts College, Yanam, Puducherry (UT), India

T. Parween
Department of Bioscience, Jamia Millia Islamia, New Delhi, India

R. Sharma (✉)
Indian Institute of Technology, New Delhi, India

M. Oves et al. (eds.), *Modern Age Waste Water Problems*,
https://doi.org/10.1007/978-3-030-08283-3_10

Keywords Heavy metals · Natural adsorbents · Industrialization · Wastewater treatments

10.1 Introduction

Due to the huge amount of discharges of wastewater laden with heavy metals such as Cd, Cr, Cu, Ni, As, Pb, and Zn, comprises of the most hazardous effluents among the chemical-intensive industries when compared to other effluents such as that are also emitted by the industries. Due to the property of high solubility in the aqueous solutions, heavy metals can easily be absorbed by all living organisms. Then once heavy metals enter the food chain, large concentrations of them can accumulate in the human body. If heavy metals are ingested by a human being beyond the certain permissible limit, thus they then lead to serious health issues (Babel and Kurniawan 2004). Thus making it very important for the remediation of metal contaminated wastewater before its discharge from the industries and different sources into the environment. Removal of heavy metal from the inorganic contaminated effluents can be easily achieved by some of the conventional treatment processes such as chemical precipitation, ion exchange, and electrochemical removal. These processes have significant advantages as well as disadvantages, which comprises of, incomplete removal of the contaminants, high-energy requirements for the process to carry on, and the most important being the production of toxic sludge as the by-products (Eccles 1999). During recent studies, various type of different approaches is under study for the development of cheaper and more efficient remedial technologies, both in order to decrease the quantity of wastewater produced and as well as to improve the quality standards of the treated effluents. Adsorption has become as one of the most efficient alternative treatments, according to various studies in recent years, the necessity for the low-cost adsorbents that have a good quantity of metal-binding capabilities has intensified (Leung et al. 2000). The adsorbents use in this technique may be of mineral, organic or biological origin, zeolites, industrial by-products, agricultural wastes, biomass, and use of polymeric materials were suggested by Kurniawan et al. (2005). The technique of membrane separation has been increasingly used as well as preferred for the treatment of inorganic contaminated effluent because of its convenient operation for the process. There are various types of different membrane filtrations such as ultrafiltration (UF), nanofiltration (NF) and reverse osmosis (RO), these were some of the techniques reported. Electrotreatments such as electrodialysis (Pedersen 2003) has also played an important role in the protection of the environment. The photocatalytic process is an upcoming innovative as well as a very promising new technique for the efficient remediation of pollutants in the wastewater (Skubal et al. 2002). Although many techniques can be used for the remediation processes of the inorganic effluent, the ideal treatment for the contaminants should be not only suitable, appropriate and applicable for the local conditions, but also should be able to meet the maximum contaminant level (MCL) standards which are established. In this chapter, we present an outline of various biological treatments of wastewater, which are heavily contaminated with heavy metals. The advantages along with their shortcomings are also discussed in this chapter. Conditions such as pH, temperature and treatment efficiency on which these technologies work are also presented in this chapter.

10.2 Heavy Metals in Industrial Wastewater

10.2.1 Definition and Toxicity

Heavy metals are the group of metals and metalloids having an atomic density greater than 5 g/cm^3 which is also 5 times more as compared to water (Garbarino 1995). Some of the heavy metals act as an essential micronutrient for every living being, but at higher concentrations, these metals lead to severe poisoning (Lenntech 2004).

Serious health effects are caused by heavy metals, such as reduction in growth and development of the organisms, cancer, organ disrupture, damages of the nervous system, and also causes death in extreme cases. Exposure to certain metals, those as mercury and lead, also promotes the development of autoimmunity in the organisms, where a person's immune system attacks their cells in the body. Autoimmunity leads to the development of joint diseases in the body such as rheumatoid arthritis, kidney diseases, failure of the circulatory system and nervous system, along with damaging of the fetal brain. Organisations have established wastewater regulations to minimize the exposure to human and environmental from hazardous chemicals.

The regulations include limitations on the different types and concentration of heavy metals that might be present in the discharged wastewater from the industries and other contaminating sources (Table 10.1).

Table 10.1 The MCL standards for the most hazardous heavy metals

Heavy metal	Toxicities	MCL (mg/L)
Arsenic	Skin manifestations, visceral cancers, vascular disease	0.050
Cadmium	Kidney damage, renal disorder, human carcinogen	0.01
Chromium	Headache, diarrhea, nausea, vomiting, carcinogenic	0.05
Copper	Liver damage, Wilson disease, insomnia	0.25
Nickel	Dermatitis, nausea, chronic asthma, coughing, a human carcinogen	0.20
Zinc	Depression, lethargy, neurological signs and increased thirst	0.80
Lead	Damage the fetal brain, diseases of the kidneys, circulatory system, and nervous system	0.006
Mercury	Rheumatoid arthritis, and diseases of the kidneys, circulatory system, and nervous system	0.00003

Babel and Kurniawan (2003a, b)

10.2.2 Industrial Wastewater Sources

Inorganic pigment manufacturing producing pigments that contain chromium compounds and cadmium sulfide; petroleum refining which generates conversion catalysts contaminated with nickel, vanadium, and chromium; and photographic operations producing the film with high concentrations of silver and ferrocyanide. All of these generators produce a lot of wastewaters. Industrial wastewater streams containing alarming amounts of heavy metals are produced from different industries. The process of electroplating along with other metal surface treatment generates a significant amount of heavy metals in the wastewaters (such as cadmium, zinc, lead, chromium, nickel, copper, vanadium, platinum, silver, and titanium). These include electroplating, electroless depositions, conversion-coating, anodizing-cleaning, milling, and etching. Another significant source of heavy metals wastes results from printed circuit board (PCB) manufacturing. Tin, lead, and nickel solder plates are the most widely used resistant over plates.

10.3 Physicochemical Methods of Remediation

Generally, the physicochemical technology of remediation provides a various number of advantages in the process of remediation like their rapid process, ease of operation and control, flexibility to change of temperature. Unlike in a biological system, physicochemical treatment can accommodate variable input loads and flow such as seasonal flows and complex discharge. Whenever it is required, chemical plants can be modified. In addition, the treatment system requires a lower space and installation cost. Their benefits, however, are outweighed by some drawbacks such as their high operational costs due to the chemicals used, high-energy consumption and handling costs for sludge disposal. However, with reduced chemical costs (such as utilizing of low-cost adsorbents) and feasible sludge disposal, physicochemical treatments have been found as one of the most suitable treatments for inorganic effluent (Kurniawan et al. 2006) (Table 10.2).

10.4 Bioremediation

In order to counterfeit the environmental contamination, bioremediation proves to be the most effective technology for remediation. It utilizes the use of microbes in order to remediate the contaminants. Bioremediation is an eco-friendly, a proficient and competent and most importantly cost effective. Bioremediation involves various *in situ* initiatives such as bioventing, bio-augmentation, and bio-sparging. The *ex situ* initiatives involve land farming, biopilling, and bioreactors. Bioremediation being economically feasible and effective can be employed anywhere in the environment based on the study. The process can be effectively carried out through the proper

Table 10.2 The main advantages and disadvantages of the various physicochemical methods for treatment of heavy metal in wastewater

#	Treatment method	Advantages	Disadvantages	References
1.	Chemical precipitation	Low capital cost, simple operation	Sludge generation, the extra operational cost for sludge disposal	Kurniawan et al. (2006)
2.	Adsorption with new adsorbents	Low-cost, easy operating conditions, having wide pH range, high metal-binding capacities	Low selectivity, production of waste products	Babel and Kurniawan (2003a, b), Aklil et al. (2004)
3.	Membrane filtration	Small space requirement, low pressure, high separation selectivity	High operational cost due to membrane fouling	Kurniawan et al. (2006)
4.	Electrodialysis	High separation selectivity	High operational cost due to membrane fouling and energy consumption	Mohammadi et al. (2005)
5.	Photocatalysis	Removal of metals and organic pollutant simultaneously, less harmful by-products	Long duration time, limited applications	Barakat et al. (2004), Kajitvichyanukula et al. (2005)

study of the parameters such as the type of pollutants, pH, the supply of nutrients, moisture content, diversity of the microbes at that particular place, temperature (Dua et al. 2002).

10.4.1 Techniques of Bioremediation

Bioremediation can be classified into two types: Ex-situ bioremediation and In-situ bioremediation. The efficiency of bioremediation is correlated to the contaminant and thus is site-specific.

10.4.1.1 *In situ* and *Ex situ* Bioremediation

This technique involves the use of living microbes having the capability to detoxify the pollutants at the contaminated site. This technique is considered to be most economical when compared to other physical and chemical methods. It is supposed to be the most advantageous alternative because it causes less interruption at the contaminated sites, by employing less harmful microbiome for the detoxification of organic contaminants. During the process of *Ex situ* technique, the place of contamination is relocated to another site.

- **Bioventing**:

This technique involves the process of bioremediation is maintained by inoculating oxygen or any other nutrients into the soil (Shanahan 2004). The two nutrients that are commonly added together are nitrogen and phosphorus as suggested by Rockne and Reddy (2003). Both of these nutrients are distributed adequately through the complete soil matrix, soil such as clay show lower permeability to the process of bioventing thus avoiding the distribution of nutrients through the soil matrix. This technique of venting works well along with well-drained, medium and coarse-textured soils therefore it is difficult for clayey soils to work under these conditions due to the presence of tiny pore and large surface area. These soils prove to be difficult in providing oxygen to the microbes in the soil to help process bioremediation; these soils are fine textured thus exhibiting low water retaining capacity. the process of bioventing declined the concentration of phenanthrene by 93% after 7 months of trial.

- **Bio-augmentation**:

This process involves amplifying the process of degradation by translocating the microbes to the contaminated sites. There are several drawbacks in this process of bioremediation, first being microbes introduced at the contaminated sites seldom participates with the native population of the contaminated site to extend as well as maintain effectively functional population ranks. Secondly, soils that are exposed to extreme amounts of biodegradables wastes also sustain a certain amount of degraders among them. This technique is correlated to biostimulation where sufficient amount of water, food resources along with oxygen are introduced at the contaminated sites to improve the process of degradation by microbes (Couto et al. 2010).

However, the technique of **in situ bioremediation** involves certain **drawbacks**:

1. It cannot treat a major portion of soil types.
2. Getting the microbes to act inclusively is very rare.
3. The required temperature for amplifying the process of degradation cannot be achieved easily.

- **Landfarming**:

In this technology, the contaminated soil is mixed with nutrients in order to design treatment beds which are shifted over or dug regularly to expose the soil matrix. Parameters of soil like moisture content, aeration, pH an addition of nutrients and bulking agents play the most important role in evaluating the efficacy in terms of pollutants biodegradation. It is an effective technique used for bioremediation of long chain hydrocarbons ranging from heavy oils to long-chain carbon fuels and creosote. the process of land farming majorly treats pesticide-contaminated soil.

- **Biopiling:**

The contaminated soil is filled out and then combined with organic products added to compost piles and then enclosed for treatment. The process of biopile comprises of treatment bed, ventilation frame, medium matrix and leachate compilation methods. All the soil parameters are kept under regulation to amplify the detoxification process. The nutrient matrix is kept under the soil matrix up to 20 ft.

deep to surpass gases and nutrients through the soil. A soil pile of about 20 ft. is enclosed in a plastic sheet to manage the overflow, evaporation, and volatilization in addition of enhanced solar heating leading to evaporation of volatile organic compounds from the contaminated soil and thus the air containing VOCs can be easily disposed of into the environment. The entire process of clean up is off 3–6 months after which the 20 ft. soil matrix is either replaced to its actual site or is cleared off (Wu and Crapper 2009).

10.4.2 Conventional Processes for Removal

The conventional techniques that are being used for the removal of heavy metals from contaminated wastewater include various processes such as **chemical precipitation, flotation, adsorption, ion exchange,** and **electrochemical deposition**.

- **Chemical Precipitation**: The process of chemical precipitation is the most commonly used for removing heavy metal from inorganic effluents. Adjustment of the pH under the basic conditions from the range of (pH 9–11) play as the major parameter that improves removal of heavy metal by chemical precipitation reaction significantly. Lime and limestone are the most preferred precipitating agents due to their easy availability and low-cost in many countries (Mirbagherp and Hosseini 2004; Aziz et al. 2008). Lime precipitation technique can be used to effectively remediate inorganic effluents having a metal concentration higher than 1000 mg/L in the wastewater. There are some other advantages of using lime precipitation technique which includes the simplicity of the process involved, inexpensive equipment required for the technique, and convenient along with safe operations of the process. Some other drawbacks are the excess sludge production that furthers needs treatment, slower metal precipitation, poor settling of heavy metals, the aggregation of metal precipitates, and the long-term impact on environmental after sludge disposal (Aziz et al. 2008).
- **Ion exchange:** It is one of the methods successfully used for the removal of heavy metals from the contaminated wastewaters in the industry. An ion exchanger works as a solid which is having the capacity of exchanging either cations or anions from the contaminated materials. Synthetic organic ion exchange resins are one of the most commonly used matrices for the process of ion exchanges. Ion exchange method has certain disadvantages also such as, that is incapable of handling highly concentrated solutions of heavy metal, as the matrix present in them gets fouled very easily by the presence of organics and other waste solids in the contaminated wastewater. Ion exchange is a nonselective technique and is extremely sensitive against the pH of the solution under the process of remediation.
- **Adsorption on a new variety of adsorbents:** The process of sorption involves the transfer of ions from the water to the soil, i.e., from a solution phase to the solid phase. During recent studies, the process of adsorption has now become one of the most effective alternative treatment technique for remediating the

wastewater which is laden with heavy metals ion. The process of adsorption is a mass transfer through which a substance is displaced from the liquid phase to the surface of a solid, and then becomes bound by physical and/or chemical interactions on that solid surface (Kurniawan and Babel 2003). Various kinds of low-cost adsorbents were derived from the agricultural waste, as well as an industrial by-product, some of the natural material, or modified biopolymers; they have been recently developed and employed to remove the heavy metals from metal-contaminated wastewaters.

- **Mechanisms of pollutant sorption onto solid sorbent:**
  - (i) The transfer of the contaminants from the bulk contaminated solution to the sorbent surface.
  - (ii) Adsorption of the contaminants on the surface of the particle.
  - (iii) Transport of the contaminant within the particle of the sorbent.

Technical applicability of such processes as well as the cost-effectiveness of these techniques are the major key factors that play an important role in the selection criteria of the most suitable as well as an efficient adsorbent in order to treat inorganic contaminated effluents.

10.5 Adsorption on Modified Natural Materials

Natural zeolites are among the most frequently studied. They have gained significant interest, mostly due to their valuable properties such as ion exchange capabilities. Some of the most common known natural zeolites, clinoptilolite was known to have the high selectivity for some of the heavy metal ions, such as Pb(II), Cd(II), Zn(II), and Cu(II). It was studied that the cation-exchange capabilities of clinoptilolite entirely depends on the method of pre-treatment and that the conditioning of clinoptilolite thus improves its ion exchange capacity and the heavy metal removal efficiencies (Babel and Kurniawan 2003a, b; Bose et al. 2002). The different capabilities of various another type of synthetic zeolites for the removal of heavy metals were also investigated. The pH plays a very important role in the selective adsorption of different types of heavy metal ions (Basaldella et al. 2007; Ríos et al. 2008; Barakat 2008). Basaldella et al. (2007) they studied the use of NaA zeolite for the removal of Cr(III) at a neutral pH, while Barakat (2008) used 4A zeolite, which was prepared by the dehydroxylation of low-grade kaolin. Barakat reported that these metal ions such as Cu(II) and Zn(II) were adsorbed at a neutral as well as at an alkaline pH, whereas Cr(VI) was only adsorbed at acidic pH as compared to the adsorption of Mn(IV) which was achieved at high alkaline pH values. Nah et al. (2006) prepared a variety of synthetic zeolite which was magnetically modified with the help of iron oxide (MMZ). MMZ respectively showed high adsorption capabilities for metal ions of Pb(II) and provides an excellent chemical resistance in a wide range of pH from 5 to 11 (Table 10.3).

Table 10.3 Adsorption capacities of modified natural materials for heavy metals

	Adsorption capacity (mg/g)						
Adsorbent	Pb^{2+}	Cd^{2+}	Zn^{2+}	Cu^{2+}	Cr^{6+}	Ni^{2+}	References
Zeolite, clinoptilolite	1.6	2.4	0.5	1.64		0.4	Babel and Kurniawan (2003a, b)
Modified zeolite, MMZ	123					8	Nah et al. (2006)
HCl-treated clay			63.2	83.3			Vengris et al. (2001)
Clay/poly(methoxyethyl) acrylamide	81.02		20.6	29.8		80.9	Şölener et al. (2008)
	85.6						Aklil et al. (2004)
Calcined phosphate	155.0						Mouflih et al. (2005)
Activated phosphate	4						Pan et al. (2007)
Zirconium phosphate	398						

10.6 Adsorption on Industrial By-Products

Fly ash, waste iron, iron slags, hydrous titanium oxide are industrial by-products, thus can be chemically modified in order to enhance the removal efficiency of heavy metal from the wastewater. Recently several studies have been conducted by Lee et al. (2004) which studied the greensands, which is another by-product produced from the iron foundry industry and is useful for Zn(II) removal. Fly ashes were also investigated for the purpose of adsorbents used for the removal of toxic heavy metals. Gupta et al. (2003) reported the bagasse fly ash, which is solid waste from the sugar industry, for the removal of Cd(II) and Ni(II) from the synthetic solutions having a pH ranging from 6.0 to 6.5. Alinnor (2007) reported using the fly ash coming from the coal-burning in order to be used for the removal of Cu(II) and Pb(II) ions. Sawdust modified with the help of 1,5-disodium hydrogen phosphate was then used for the adsorption of Cr(VI) at pH 2 was reported by Uysal and Ar (2007). Some other types modified sorbents such as iron-based sorbents such as ferrosorp plus (Genc-Fuhrman et al. 2008) and synthetic nanocrystalline akaganeite (Deliyanni et al. 2007) were recently studied for the simultaneous removal of the heavy metals. Ghosh et al. (2003) and Barakat (2005) reported the use of hydrous titanium oxide for the adsorption of Cr (VI) and Cu (II) ions, individually (Fig. 10.1).

10.6.1 Adsorption on Modified Agriculture and Biological Wastes (Bio-Sorption)

Recently, the use of agricultural by-products possesses as a great deal of interest in the field of research as adsorbents for the removal of heavy metals from industrial effluents. The use of agricultural by-products for the process of bioremediation of heavy metal ions is called as bio-sorption technique. This method utilizes the inactive (non-living) microbial biomass in order to bind and concentrate the heavy metals from the water waste streams by the use of physicochemical pathways (which

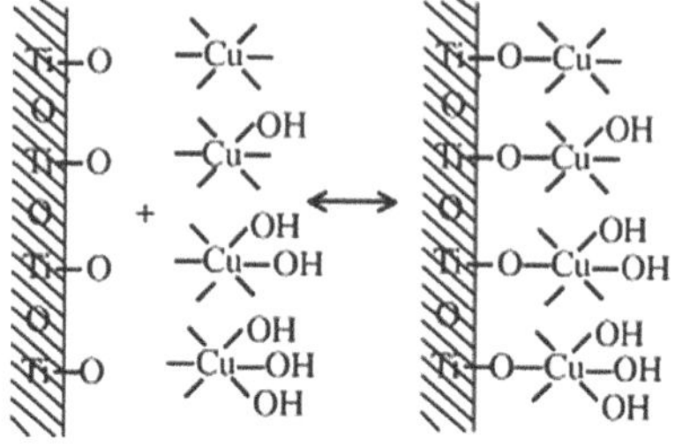

Fig. 10.1 The adsorption mechanism of Cu(II) on hydrous TiO_2. (Barakat 2005)

involves chelation and adsorption mainly) of the uptake mechanism (Igwe et al. 2005). New resources of by-products such as hazelnut shell, rice husk, pecan shells, jackfruit, maize cob or husk can also be used as a heavy metal adsorbent for the uptake after the chemical modification or by converting them into activated carbon by heating. Ajmal et al. (2000) studied the use of orange peel as an absorbent for the Ni(II) removal from contaminated wastewater. They found the maximum extraction of metal occurred at the pH 6.0. The employment of coconut shell charcoal (CSC) when modified with the oxidizing agents and/or chitosan for the removal of Cr(VI) was studied by Babel and Kurniawan (2004).

The removal of Cu(II) and Zn(II) from contaminated wastewater were investigated by using pecan shells-activated carbon (Bansode et al. 2003) and potato peels charcoal (Amana et al. 2008). Bishnoi et al. (2003) studied the removal of Cr (VI) with the help of rice husk-activated carbon from an aqueous solution. They reported the maximum extraction of metal by rice husk at the pH of 2.0. Rice hull, containing cellulose, lignin, carbohydrate, and silica, was also studied for the removal of Cr(VI) from the simulated contaminated solution (Tang et al. 2003). To enhance the property of metal removal, the adsorbents was modified with the ethylenediamine.

The mechanism involved in the uptake of heavy metal ions can also take place by the metabolism-independent metal-binding to the cell walls, and external surfaces were reported by Deliyanni et al. (2007). This mechanism involves the adsorption processes such as the ionic, chemical and physical adsorption. There are a variety of ligands present on the fungal walls are also known to be involved in the metal chelation. These ligands include carboxyl, amine, hydroxyl, phosphate and sulfhydryl groups which play an important role in metal chelation. Metal ions are adsorbed on the surface of the cell by complexation with the negatively charged reaction sites (Table 10.4).

10.6.2 Adsorption on Modified Biopolymers and Hydrogels

Biopolymers are preferable for industrial use due to their capability of lowering the transition metal ion concentrations into subpart per billion concentrations, as well as they are widely available, and are environmentally safe. Another feature present in the biopolymers is that they are composed of a number of different functional

Table 10.4 Adsorption capacities of some agricultural and biological wastes for heavy metals

Adsorbent	Adsorption capacity (mg/g)						References
	Pb^{2+}	Cd^{2+}	Zn^{2+}	Cu^{2+}	Cr^{6+}	Ni^{2+}	
Maize cope and husk	456	493.7	495.9				Igwe et al. (2005)
Orange peel						158	Ajmal et al. (2000)
Coconut shell charcoal					3.65		Babel and Kurniawan (2004)
Pecan shells activated carbon			13.9	31.7			Bansode et al. (2003)
Rice Husk		2.0			0.79		Bishnoi et al. (2003)
Modified rice hull					23.4		Tang et al. (2003)
Spirogyra (green alga)				133			Gupta et al. (2006)
Ecklonia maxima – marine alga	235			90			Fenga and Aldrich (2004)
Ulva lactuca					112.3		El-Sikaily et al. (2007)
Oedogonium species	145						Gupta and Rastogi (2008)
Nostoc species	93.5						Gupta and Rastogi (2008)
Bacillus – bacterial biomass	467	85.3	418	381	39.9		Ahluwalia and Goyal (2006)

groups, such as hydroxyls and amines, which leads to increase in the efficiency of heavy metal ion uptake and capable of maximum loading of chemicals. A new variety of polysaccharide-based-materials were labeled as modified biopolymer adsorbents (which were derived from chitin, chitosan, and starch) used for the removal of heavy metals from the contaminated wastewater. The two main methods for the preparation of biosorbents containing polysaccharides as basic material: (a) crosslinking reactions, where a reaction takes place between the groups of hydroxyl or amino present in the chains with a coupling agent present in order to form a water-insoluble crosslinked networks between them (gels); (b) by immobilization of poly-saccharides on the insoluble substance by using coupling or grafting reactions so as to compose a hybrid or composite materials, these studies were reported by Crini (2005). Pradhan et al. (2005) reported that Chitin being a naturally abundant mucopolysaccharide which is extracted from the crustacean shells, which are a part of waste products left after the processing of seafood by the industries. Chitosan, which is synthesized by deacetylation of chitin, forms an important derivative of chitin synthesis. Chitosan is the partial conversion of crab shell waste, which is a powerful chelating agent and thus interacts efficiently with the transition heavy metal ions present in the wastewater (Pradhan et al. 2005). Recent studies differently modified chitosan beads were suggested for the diffusion of heavy metal ions through the crosslinked chitosan membranes by Lee et al. (2001). Liu et al. (2003) achieved excellent saturation absorption capacity for Cu(II) metal ions with the help of crosslinked chitosan beads which was performed at pH 5 prepared with the help of new hybrid materials that have the capability to adsorb transition metal ions by

Table 10.5 Adsorption capacities of modified biopolymers for heavy metals

Adsorbent	Adsorption capacity (mg/g)					
	Pb^{2+}	Cd^{2+}	Zn^{2+}	Cu^{2+}	Cr^{6+}	As^{5+}
Crosslinked chitosan		150		164		230
Crosslinked starch gel	433			135		
Alumina/chitosan composite				200		

Crini (2005)

immobilizing them by binding them on chitosan present on the surface of non-porous glass beads. The mechanisms employed during the process of sorption are complicated as compared to other are due to the presence of various types of different interactions. Heavy metal complexation with chitosan may undertake two different mechanisms such as (chelation versus ion exchange) also depending on the pH as this parameter may lay an important effect on the protonation of the molecules in question (Crini 2005). Chitosan is characterized by the presence of a high percentage of nitrogen, which is present in the form of amine groups which is responsible for binding of metal ions through the chelation mechanisms. However, chitosan is considered to be a cationic polymer having p*K*a ranges from 6.2 to 7. Therefore, in the case of acidic solutions, it gets protonated and possesses the electrostatic properties. Thus, making it possible to absorb heavy metal ions through the process of anion exchange. Absorbent materials contain the immobilized thiacrown ethers which were prepared by immobilization of the ligands into the sol-gel matrix (Saad et al. 2006) (Table 10.5).

Hydrogels, which are crosslinked hydrophilic polymers, they are capable of expanding their volumes due to their high swelling in water. Accordingly, they are widely used in the purification of wastewater. Different types of hydrogels were remodeled in order to study their adsorption behavior for the contaminants such as heavy metals. Kesenci et al. (2002) synthesized poly(ethylene glycol dimethacrylate-co-acrylamide) hydrogel beads with the affinity for heavy metals such as Pb(II) > Cd(II) > Hg(II) (Fig. 10.2).

10.7 Nanomaterial or Nanocomposite Application in Metal Ion Extraction of Absorption

Nanomaterials are used for the purpose of treating the contaminated wastewater, where they behave as sorbents used for the removal of heavy metal ions. Nanomaterials used for this purpose should fulfil the following criterion:

1. They should be nontoxic in nature.
2. The contaminants adsorbed on the surface of nano adsorbent could be easily removed.

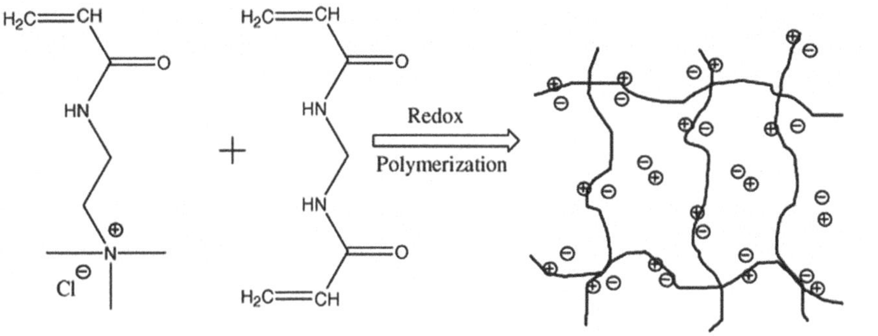

Fig. 10.2 Three-dimensional network formation of cationic hydrogel. (Barakat and Sahiner 2008)

3. The material should possess high sorption capabilities and should be selective to the presence of low concentration of the contaminants.
4. The nanomaterials should have the property to be recycled infinitely.

For adsorption various nanomaterials such as carbon nanotubes, carbon based material composites, graphene, nano metal or metal oxides, and polymeric sorbents are being studied for the purpose of treating the contaminated wastewater and removing the heavy metal ions and the studies have indicated high adsorption capacity of these materials under study.

10.7.1 Carbon Based Nanomaterials

These are widely used for the removal of heavy metal ions, this nanomaterial is preferred over others due to its high sorption capacity as well as its non toxicity. Earlier activated carbon was used as a sorbents. But it is not possible to remove the metal ions from it at the ppb levels. Then with the upcoming technique of nanotechnology, carbon nanotubes, fullerene, and graphene are synthesized and used as nanosorbents.

Carbon nanotubes (CNTs) were discovered by Lijima, they have unique structural, electronic, optoelectronic, semiconductor, mechanical, chemical and physical property, that have been majorly applied in order to remove heavy metal contamination from the wastewater treatment sites. CNTs show high sorption efficiency and are used as nanosorbents for the adsorption of divalent metal ions.

Researchers studied the advantages and disadvantages of the heavy metal ions sorption on the activated carbon, carbon nanotubes, and carbon-encapsulate magnetic nanoparticles, through various sorption studies conducted on Co^{2+} and Cu^{2+}. The studies reveal that carbon (Pyrzynska and Bystrzejewski 2010) nanomaterials showed significantly higher efficiency of sorption when compared to other activated carbon nanomaterials. In order, to increase the sorption abilities of the CNTs, they are modified with the help of oxidation, as well as combing the CNT's with heavy metal ions or metal oxides, and also coupling them with various organic compounds (Afzali et al. 2012).

Whereas, scientists suggested that the conditions of the solution, pH as well as metal ions concentrations, were seen to deter the adsorption characteristics of the carbon nanotubes, and this was also seen to happen through the experimental data as well as the Freundlich adsorption model.

10.7.2 Nanoparticles from Metal or Metal Oxides

Nanoparticles that are formed by metal or metal oxides are used mainly in order to remove the heavy metal ions from the wastewater contamination. Nanosized metals or metal oxides which include nanosized silver nanoparticles, ferric oxides, manganese oxides, titanium oxides, magnesium oxides, copper oxides, cerium oxides, and so on, these nanosized nanoparticles provide high surface area as well as have the specific affinity for metal ions. Metal oxide nanoparticles have the minimal impact on the environmental and have low solubility as well as there is no secondary pollution, thus it have been used as sorbents in order to remove the heavy metal ions (Stafiej and Pyrzynska 2007).

Scientists studied the removal of arsenate with the use of aggregated metal oxide nanoparticle media which is packed in the bed columns. The study was conducted through batch experiments with 16 commercial nanopowders in four water matrices, TiO_2, Fe_2O_3, ZrO_2, and NiO nanopowders were selected by characterization with Freundlich adsorption isotherm parameters, which shows the highest removal of arsenate in all the water matrices under study.

Nanosized metal oxides depict greater efficiency in the removal of heavy metal ions in wastewater, due to their high surface area along with the presence of abundant surface active sites as compared to the bulk materials. But, it is posses a great difficulty in order to separate the adsorbents from the nanosized metal oxides due to their high surface energy and nanosize. Thus, many studies prefer the usage of polymer based nanosorbents (Gupta et al. 2011).

10.7.3 Polymer Supported Nanosorbents

An efficient sorbent having both high capacity and fast rate adsorption should also possess these following important properties which are having a large surface area and presence of functional groups. Mostly used inorganic sorbents do not posses these characteristics at the same time, carbon nanomaterials are known for high surface area, but are not capable of adsorbing a functional group (Mauter and Elimelech 2008).

On the contrary, the organic polymer, polyphenylenediamine, has the property of withholding a large quantity of polyfunctional groups such as amino and imino groups which are effective in adsorbing various heavy metal ions, but due to their small specific area and low rate of adsorption the capability of theses organic polymers is limited.

Thus, there is urgent requirement of new sorbents having both polyfunctional groups and high surface area. Recently, hybrid sorbents have been developed, that has provided us with new opportunities in application of removal of heavy metal ions from the contaminated wastewater.

Polymer-layered silicate nanocomposites exhibit drastic results in the properties that have been studied at very low filler contents.

Hybrid polymers were synthesized by Xu et al. From the ring-opening polymerization of pyromellitic acid dianhydride (PMDA) and phenylaminomethyl trimethoxysilane (PAMTMS). This hybrid polymer was used for the removal of Cu_{2+} and Pb_{2+} ions, adsorption for $Cu_{2+}$and Pb_{2+} ions followed Lagergren second order kinetic model and Langmuir isotherm model, thus depicting the adsorption process might be following the Langmuir monolayer adsorption (Rao et al. 2007).

Nano materials which include the traditional inorganic nano adsorbent materials and the unique polymer composites are also used for the removal of the heavy metal ions in wastewater contamination, because of their novel size- and shape-dependent properties, thus gain excellent metal ions removal efficiency.

10.8 Conclusion

Due to the changing conditions of the environment over the period, the regulations for environmental conservation have become more rigorous. During recent years, a different variety of treatment technologies such as chemical precipitation, adsorption, membrane filtration, electrodialysis, and photocatalysis, have been studied for the removal of heavy metal from contaminated water. From the recent studies, it is evident that lime precipitation proves to be as one of the most effective technique to treat inorganic effluent having a concentration of metal higher than 1000 mg/L; use of new adsorbents and membrane filtration are frequently studied and used for the remediation of the heavy metal-contaminated wastewater. Many techniques have been used for the remediation of contaminated wastewater, it is necessary to select the most effective method for restoration of metal-contaminated sewage based on basic parameters of pH, initial metal concentration, the overall result after the treatment when compared with other technologies along with environmental impact and economics parameter including the capital investment and costs of operation. In the end, the technical applicability along with plant simplicity and cost-effectiveness are major key factors that play an important role in the selection of the suitable treatment system for contaminated wastewater.

References

Afzali D, Jamshidi R, Ghaseminezhad S, Afzali Z (2012) Preconcentration procedure trace amounts of palladium using modified multiwalled carbon nanotubes sorbent prior to flame atomic absorption spectrometry. Arab J Chem 5:461–466

Ahluwalia SS, Goyal D (2006) Microbial and plant derived biomass for removal of heavy metals from wastewater. Bioresour Technol 98(12):2243–2257

Ajmal M, Rao R, Ahmad R, Ahmad J (2000) Adsorption studies on Citrus reticulata (fruit peel of orange) removal and recovery of Ni(II) from electroplating wastewater. J Hazard Mater 79:117–131

Aklil A, Mouflih M, Sebti S (2004) Removal of heavy metal ions from water by using calcined phosphate as a new adsorbent. J Hazard Mater 112(3):183–190

Alinnor J (2007) Adsorption of heavy metal ions from aqueous solution by fly ash. Fuel 86:853–857

Amana T, Kazi AA, Sabri MU, Banoa Q (2008) Potato peels as solid waste for the removal of heavy metal copper(II) from waste water/industrial effluent. Colloids Surf B: Biointerfaces 63:116–121

Aziz HA, Adlan MN, Ariffin KS (2008) Heavy metals (Cd, Pb, Zn, Ni, Cu and Cr(III)) removal from water in Malaysia: post treatment by high quality limestone. Bioresour Technol 99:1578–1583

Babel S, Kurniawan TA (2003a) Low-cost adsorbents for heavy metals uptake from contaminated water: a review. J Hazard Mater B97:219–243

Babel S, Kurniawan TA (2003b) Various treatment technologies to remove arsenic and mercury from contaminated groundwater: an overview. In: Proceedings of the first international symposium on southeast Asian water environment, Bangkok, Thailand, 24–25

Babel S, Kurniawan TA (2004) Cr(VI) removal from synthetic wastewater using coconut shell charcoal and commercial activated carbon modified with oxidizing agents and/or chitosan. Chemosphere 54(7):951–967

Bansode PR, Losso JN, Marshall WE, Rao RM, Portier RJ (2003) Adsorption of metal ions by pecan shell-based granular activated carbons. Bioresour Technol 89:115–119

Barakat MA (2005) Adsorption behavior of copper and cyanide ions at TiO2–solution interface. J Colloid Interface Sci 291:345–352

Barakat MA (2008) Adsorption of heavy metals from aqueous solutions on synthetic zeolite. Res J Environ Sci 2(1):13–22

Barakat MA, Sahiner N (2008) Cationic hydrogels for toxic arsenate removal from aqueous environment. J Environ Manag 88(4):955–961

Barakat MA, Chen YT, Huang CP (2004) Removal of toxic cyanide and Cu(II) ions from water by illuminated TiO2 catalyst. J Appl Catal B Environ 53:13–20

Basaldella EI, Vázquez PG, Iucolano F, Caputo D (2007) Chromium removal from water using LTA zeolites: effect of pH. J Colloid Interface Sci 313:574–578

Bishnoi NR, Bajaj M, Sharma N, Gupta A (2003) Adsorption of Cr(VI) on activated rice husk carbon and activated alumina. Bioresour Technol 91(3):305–307

Bose P, Bose MA, Kumar S (2002) Critical evaluation of treatment strategies involving adsorption and chelation for wastewater containing copper, zinc, and cyanide. Adv Environ Res 7:179–195

Couto MNPFS, Monteiro E, Vasconcelos MTSD (2010) Mesocom trials of bioremediation of contaminated soil of a petroleum refinery: comparison of natural attenuation, biostimulation and bioaugmentation. Environ Sci Pollut Res 17(7):L1339–L1346

Crini G (2005) Recent developments in polysaccharide-based materials used as adsorbents in wastewater treatment. Prog Polym Sci 30:38–70

Deliyanni EA, Peleka EN, Matis KA (2007) Removal of zinc ion from water by sorption onto iron-based nanoadsorbent. J Hazard Mater 141:176–184

Dua M, Singh A, Sethunathan N, Johri A (2002) Biotechnology and bioremediation: successes and limitations. Appl Microbiol Biotechnol 59(2–3):143–152

Eccles H (1999) Treatment of metal-contaminated wastes: why select a biological process? Trends Biotechnol 17:462–465

El-Sikaily A, El Nemr A, Khaled A, Abdelwehab O (2007) Removal of toxic chromium from wastewater using green alga Ulva lactuca and its activated carbon. J Hazard Mater 148:216–228

Fenga D, Aldrich C (2004) Adsorption of heavy metals by biomaterials derived from the marine alga Ecklonia maxima. Hydrometallurgy 73:1–10

Garbarino J (1995) Raising children in a socially toxic environment. Jossey-Bass, San Francisco

Genc-Fuhrman H, Wu P, Zhou Y, Ledin A (2008) Removal of As, Cd, Cr, Cu, Ni and Zn from polluted water using an iron based sorbent. Desalination 226:357–370

Ghosh UC, Dasgupta M, Debnath S, Bhat SC (2003) Studies on management of chromium(VI)-contaminated industrial waste effluent using hydrous titanium oxide (HTO). Water Air Soil Pollut 143:245–256

Gupta VK, Rastogi A (2008) Biosorption of lead(II) from aqueous solutions by non-living algal biomass Oedogonium sp. and Nostoc sp. – a comparative study. Colloids Surf B: Biointerfaces 64:170–178

Gupta VK, Jain CK, Ali I, Sharma M, Saini SK (2003) Removal of cadmium and nickel from wastewater using bagasse fly ash – a sugar industry waste. Water Res 37:4038–4044

Gupta VK, Rastogi A, Saini VK, Jain N (2006) Biosorption of copper(II) from aqueous solutions by Spirogyra species. J Colloid Interface Sci 296:59–63

Gupta VK, Agarwal S, Saleh TA (2011) Synthesis and characterization of alumina-coated carbon nanotubes and their application for lead removal. J Hazard Mater 185:17–23

Igwe JC, Ogunewe DN, Abia AA (2005) Competitive adsorption of Zn(II), Cd(II) and Pb(II) ions from aqueous and non-aqueous solution by maize cob and husk. Afr J Biotechnol 4(10):1113–1116

Kajitvichyanukul P, Ananpattarachaia J, Pongpom S (2005) Sol–gel preparation and TiO2 thin film for reduction of chromium (VI) in process. Sci Technol Mater 6:352–358

Kesenci K, Say R, Denizli A (2002) Removal of heavy metal ions from water by using poly (ethyleneglycol dimethacrylate-co-acrylamide) beads. Eur Polym J 38:1443–1448

Kurniawan TA, Babel S (2003) A research study on Cr(VI) removal from contaminated wastewater using low-cost adsorbents and commercial activated carbon. In: Second International Conference on Energy Technology towards a Clean Environment (RCETE), vol. 2. Phuket, Thailand, 12–14 February, pp. 1110–1117

Kurniawan TA, Chan GYS, Lo WH (2006) Physico-chemical treatment techniques for wastewater laden with heavy metals. Chem Eng J 118:83–98

Kurniawan TA, Chan GYS, Lo WH, Babel S (2005) Comparisons of low-cost adsorbents for treating wastewaters laden with heavy metals. Sci Total Environ 366(2–3):409–426

Lee ST, Mi FL, Shen YJ, Shyu SS (2001) Equilibrium and kinetic studies of copper(II) ion uptake by chitosan–tripolyphosphate chelating resin. Polymer 42:1879–1892

Lee TY, Park JW, Lee JH (2004) Waste green sands as a reactive media for the removal of zinc from water. Chemosphere 56:571–581

Lenntech K (2004) Water treatment and air purification. Rotter Dam Seweg, Netherlands

Leung WC, Wong MF, Chua H, Lo W, Leung CK (2000) Removal and recovery of heavy metals by bacteria isolated from activated sludge treating industrial effluents and municipal wastewater. Water Sci Technol 41(12):233–240

Liu XD, Tokura S, Nishi N, Sakairi N (2003) A novel method for immobilization of chitosan onto non-porous glass beads through a 1,3-thiazolidine linker. Polymer 44:1021–1026. Anal. Chim. Acta 555, 146–156

Mauter MS, Elimelech M (2008) Environmental applications of carbon-based nanomaterials. Environ Sci Technol 42:5843–5859

Mirbagherp SA, Hosseini SN (2004) Pilot plant investigation on petrochemical wastewater treatment for the removal of copper and chromium with the objective of reuse. Desalination 171:85–93

Mohammadi T, Mohebb A, Sadrzadeh M, Razmi A (2005) Modeling of metal ion removal from wastewater by electrodialysis. Sep Purif Technol 41:73–82

Mouflih M, Aklil A, Sebti S (2005) Removal of lead from aqueous solutions by activated phosphate. J Hazard Mater 119(1–3):183–188

Nah IW, Hwang KY, Jeon C, Choi HB (2006) Removal of Pb ion from water by magnetically modified zeolite. Miner Eng 19(14):1452–1455

Pan BC, Zhang QR, Zhang WM, Pan BJ, Du W, Lv L, Zhang QJ, Xu ZW, Zhang QX (2007) Highly effective removal of heavy metals by polymer-based zirconium phosphate: a case study of lead ion. J Colloid Interface Sci 310(1):99–105

Pedersen AJ (2003) Characterization and electrolytic treatment of wood combustion fly ash for the removal of cadmium. Biomass Bioenergy 25(4):447–458

Pradhan S, Shyam S, Shukla K, Dorris KL (2005) Removal of nickel from aqueous solutions using crab shells. J Hazard Mater B125:201–204

Pyrzynska K, Bystrzejewski M (2010) Comparative study of heavy metal ions sorption onto activated carbon, carbon nanotubes, and carbon-encapsulated magnetic nanoparticles. Colloids Surf A Physicochem Eng Asp 362:102–109

Ríos CA, Williams CD, Roberts CL (2008) Removal of heavy metals from acid mine drainage (AMD) using coal fly ash, natural clinker and synthetic zeolites. J Hazard Mater 156(1–3):23–35

Rao GP, Lu C, Su F (2007) Sorption of divalent metal ions from aqueous solution by carbon nanotubes: a review. Sep Purif Technol 58:224–231

Rockne K, Reddy K (2003) Bioremediation of contaminated sites. University of Illinois, Chicago

Saad B, Chong CC, Ali AM, Bari MF, Rahman IA, Mohamad N, Saleh MI (2006) Selective removal of heavy metal ions using sol–gel immobilized and SPE-coated thiacrown ethers. Anal Chim Acta 555:146

Shanahan P (2004) Bioremediation, waste containment and remediation technology, Spring 2004. Massachsetts Institute of Technology/MIT Open Course Ware

Skubal LR, Meshkov NK, Rajh T (2002) Thurnauer. J Photochem Photobiol A Chem 148:393

Şölener M, Tunali S, Safa Özcan A, Özcan A, Gedikbey T (2008) Adsorption characteristics of lead (II) ions onto the clay/poly(methoxyethyl)acrylamide (PMEA) composite from aqueous solutions. Desalination 223(1–3):308–322

Stafiej A, Pyrzynska K (2007) Adsorption of heavy metal ions with carbon nanotubes. Sep Purif Technol 58:49–52

Tang P, Lee CK, Low KS, Zainal Z (2003) Sorption of Cr(VI) and Cu(II) in aqueous solution by ethylenediamine modified rice hull. Environ Technol 24:1243–1251

Uysal M, Ar I (2007) Removal of Cr(VI) from industrial wastewaters by adsorption. Part I: determination of optimum conditions. J Hazard Mater 149:482–491

Vengris T, Binkien R, Sveikauskait A (2001) Nickel, copper and zinc removal from waste water by a modified clay sorbent. Appl Clay Sci 18(3–4):183–190

Wu T, Crapper M (2009) Simulation of biopile processes using a hydraulics approach. J Hazard Mater 171(1–3):1103–1111

Chapter 11
Removal and Recovery of Heavy Metal Ions Using Natural Adsorbents

Amjad Mumtaz Khan and Sajad Ahmad Ganai

Abstract Now a day's heavy metal pollution has become a serious environmental problem. The presence of heavy metal ions is a major problem due to their toxicity to many life forms on this planet. Therefore the removal of heavy metals from the environment is of special concern due to their persistence. During these days natural adsorbents are most frequently studied and widely applied for the metal contaminated water. Adsorption processes are being widely used by various researchers for the removal of heavy metals from the waste streams. The need for the safe and economical methods for the elimination of heavy metals from contaminated waters has developed the interest of researchers towards the production of low cost adsorbents. Therefore there is an urgent need that all possible sources of agro-based inexpensive adsorbents should be explored and their role for the removal of heavy metals should be studied in detail.

Keywords Adsorbents · Adsorption · Agro-based · Heavy metals · Toxicity · Environment

11.1 Introduction

Most of the chromatographic methods employed for the separation and analysis of components form mixtures are based on adsorption. We know that the particles on the surface of a solid or a liquid experience a strong inward pull because of unbalanced attractive interaction with other particles which surround them only one side and not all sides. At the same time when a new solid surface is created by breaking a solid, some interatomic bonds are broken and some valencies of the surface atoms are left unsatisfied. As a result the surface of a solid has a tendency to attract and hold molecules of a gas, a liquid, a dissolved solute or other particles with

A. M. Khan (✉)
Department of Chemistry, Aligarh Muslim University, Aligarh, India

S. A. Ganai
Department of Chemistry, NIT Sri Nagar, Kashmir, India

M. Oves et al. (eds.), *Modern Age Waste Water Problems*,
https://doi.org/10.1007/978-3-030-08283-3_11

which it comes in contact. By holding them on the surface, the surface energy of the solid is decreased and consequently a more stable state is acquired by the solid. The molecules or other particles held by the solid, however, remain only at the surface of the solid and don't penetrate into the bulk of the solid. This phenomenon of adhesion of other substances on the surface of a solid without any penetration into the bulk phase in known as adsorption. The solid on the surface of which adsorption takes place is called an adsorbent and the substance which gets adsorbed is called adsorbate. Activated charcoal, silica gel, alumina etc. are some of the well-known adsorbents. Since adsorption is a surface phenomenon, the greater the surface area of the adsorbent, the greater is adsorption. The following simple experiments illustrate the phenomenon of adsorption of a gas and a dissolved solute respectively.

(i) Take some ammonia gas in a gas tube over mercury and note the level of mercury. Now introduce a piece of activated charcoal in the tube. The level of mercury starts rising. This shows the disappearance of ammonia due to its adsorption on the surface of charcoal.
(ii) Dissolve a small quantity of some organic dye in water. Add some animal charcoal to it and filter, the filtrate is found to be colorless because the dye gets adsorbed by charcoal.

The term adsorption should be clearly distinguished from absorption. Adsorption involves the concentration of a substance at the surface while absorption involves the penetration of substance into the bulk of a solid or a liquid. Adsorption is a fast process while absorption is a slow process. This is because absorption involves the penetration into the interior of mater.

Sometimes a substance may undergo adsorption as well as absorption such process is termed as sorption. For example, if ammonia gas is passed through water containing some charcoal ammonia first dissolves in water (i.e. absorption takes place) and is then adsorbed by charcoal.

11.2 Types of Adsorption

11.2.1 Physical Adsorption

This type of adsorption is due to van der waals' forces between the gases and solids. These are weak forces but in case of polar gases these are relatively stronger. This type of adsorption is not permanent and can be decreased by decreasing pressure or increasing temperature.

11.2.2 Chemical Adsorption

In some solids there exist some unsatisfied valencies on the surface. There is a possibility of adsorbate forming chemical bond with these free valencies of adsorbent. These types of adsorption are called chemical adsorption or chemisorption. Chemisorption is much stronger than physical adsorption. Chemisorption may or may not decrease by raising temperature or reducing pressure.

The extent to which a gas is adsorbed on a solid adsorbent depends upon the following factors:

(i) Nature of the solid and of the gas.
(ii) Temperature
(iii) Pressure

11.2.3 Nature of the Gas and the Solid

Adsorption of gases on solids generally involves physical adsorption. Therefore, it is non-specific in nature and as such every gas is adsorbed to a small or large extent on every solid surface. However, it has been found that gases such as NH_3, HCl & SO_2 which are non-soluble and more easily liquefiable are adsorbed to a large extent then gases like N_2 & H_2. The reason for greater adsorption of easily liquefiable gasses is that van dar Waals' forces are more predominant in their case.

As far as solid adsorbents are concerned, adsorbents having large surface area such as charcoal & silica gel acts as a good adsorbent.

11.2.4 Temperature

The adsorption of a gas is always accompanied by evolution of heat. Therefore, according to La Chatelier's Principle, extent of adsorption decreases with the rise in the temperature.

However, chemisorptions of a gas may be endothermic in nature far a small initial range of temperature.

Therefore, chemisorption shows an initial increase in extent of adsorption with increase in temperature and then a regular decrease takes place.

The effect of temperature on the amount of gas adsorbed, $^{\mathbf{x}}/\mathbf{m}$, where **x** is the amount of gas adsorbed and **m** is the weight of the adsorbent is shown is in Figs. 11.1 and 11.2 for physical and chemical adsorption respectively. There curves are known as adsorption isobars.

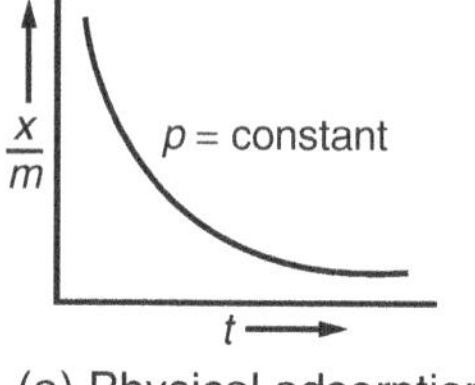

Fig. 11.1 Physisorption

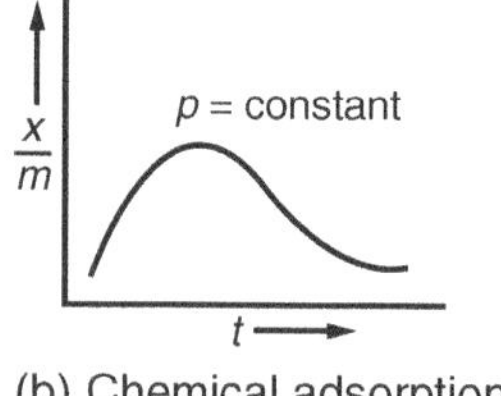

Fig. 11.2 Chemisorption

11.2.5 Pressure

The adsorption of a gas by an adsorbent depends upon the pressure of the gas. Initially the amount of gas adsorbed from a given amount of adsorbent increases rapidly with increase in pressure. However, as the pressure becomes high and almost the enter surface of the adsorbent gets saturated with the gas, the effect of pressure becomes very small. Ultimately a stage is reached when no more gas is adsorbed even if the pressure is increased. This stage known as saturation stage and the pressure applied is known as saturation pressure.

The relationship between the magnitude of the adsorption and pressure can be expressed mathematically by an empirical equation called Freundlich Isotherm which may be written as:

$$x/m = Kp^{1/n}$$

Where x is the amount gas adsorbed, m is the weight of the adsorbent, p is the pressure & k and n are constants which depend upon the nature of gas & adsorbent.

The effect of pressure on the amount of gas adsorbed, x/m, is shown graphically in Fig. 11.3. This curve is known as adsorption isotherm.

We know that water is the main source of energy and important for life. But due to some ignorance it has got polluted to a large extent, mainly due to the heavy metal ions. The contamination of water due to heavy metals is a challenging problem due to their toxic effects. These metals like Zn, Hg, Cu, Pb, Cd etc. are responsible for the damage of liver and nerves and can block functional group of vital enzymes and bones (Awwad et al. 2013). These metal ions can be present in water naturally form anthropogenic sources or from leaching of ore deposits, which may include solid

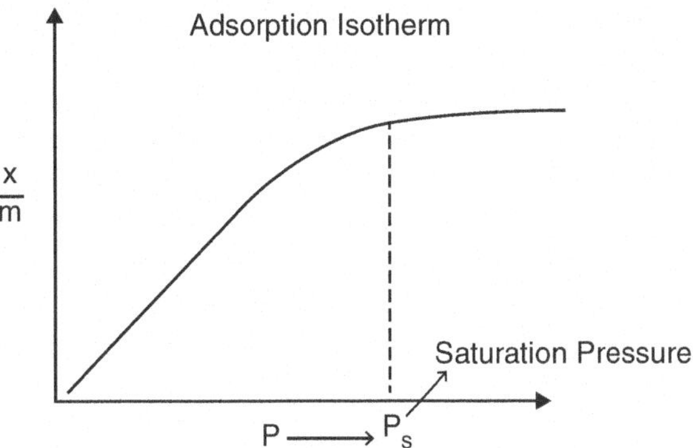

Fig. 11.3 Adsorption isotherm

waste disposal and industrial effluents. The metal ions are being added to water sources at a much higher concentration, hence leading to health hazards and environmental degradation.

- **Mercury**: it is not present naturally in living organisms. It is a toxic substance. The major natural sources of Hg are degassing of earth's crust, emissions from volcanoes and evaporation from natural bodies of water. The Hg is commonly used in industrial processes and in different products. It is mostly present in a relatively uncreative form as a gaseous element. The methylated forms of Hg are bioaccumulated over a million fold and concentrated in living beings especially fish these forms of Hg are highly toxic causing neurotoxicological disorders. Monomethyl Hg causes damage to central nervous system, while foetal and postnatal exposures have given rise to abortion, congenital malformations and development changes in young children.
- **Cadmium**: The cadmium derives it toxicological properties from its chemical similarity to Zn. As an impurity, Cd is present is several products like phosphate fertilizers, detergents and refined petroleum products. As reported average daily intake of cadmium for humans is 0.15 μg from air and 1 μg from water. Cadmium effects kidney if exposed for a long time. Its high exposure may cause obstructive pulmonary disease and lung cancer. In humans and animals it is reported that cadmium. Causes bone defects. It can also cause increased blood pressure and myocardial disease in animal.
- **Lead**: children are more sensitive towards lead then the adults. The Pb exposure can be through drinking water, food, air, soil & dust from old paint. The high levels of Pb may results in toxic effects in humans which in turn cause problems in the synthesis of hemoglobin, effects kidneys, reproductive system, joints and damage to nervous system.
- **Selenium**: it is needed in small amounts by humans & animals. But if it level increases, it can damage to nervous system & causes fatigue and irritability, it accumulates in living tissues. Its high content in fish and other animals can cause health problems in human beings over a lifetime of overexposure. There may be damage to nervous system, circulating tissues & liver tissues and may be hair and fingernail loss.

- **Nickel**: it is needed in small amounts to produce red blood cells, but in excess quantity, it becomes toxic and results in decrease in body weights. It damages heart and liver. It can also cause skin irritation.
- **Copper**: if copper is present in high doses anemia may occur. It damages kidney liver and stomach if its level gets increased. During Wilson's disease, it effects greatly.
- **Chromium**: The low level Cr can irritate skin and can produce ulcer. Its chronic exposure can damage liver and kidney. It also damages circulatory and nervous tissues. It is accumulated in aquatic animals and can cause toxicity to eating fish. Chromium occurs in several oxidation states from (+2) to (+6) in environment. The most abundantly occurring forms of chromium are +4 and + 6, which are being toxic to plants, animals and humans. Chromium occurs naturally by burning coal, petroleum, fertilizers, oil mill drilling and metal plating tenures. Fertilizers and sewages are the common sources of release of chromium in environment anthropogenically. Most of the industries using chromium play important role in environmental pollution and causing the adverse effects on the biological and ecological species with chromium. Agricultural and industrial practices on chromium increase the level of toxicity in surrounding of living being. Due to the presence of oxygen in the environment, the forms of chromium are oxidized to very harmful and poisonous substances, which are highly soluble in water. Chromate manufacturing process causes series pollution to farm land. Chromium also causes toxicity in plants. Consumption of toxic plant effects the biological factors. Phytotoxicity due to chromium is the common feature in reduction of root growth, inhibition of seeds, germination, suppresses biomass and chlorosis of leaf. It is reported that hexavalent chromium is considered group 1 in bone cancer causing.
- **Iron**: Iron is important for growth and survival of human beings. Many people exposed to iron through drinking water that was collected from ground. Fishes are effected by iron contamination, causing disturbances is respiration. Rice, the main food item, is also effected by iron toxicity. Iron stops the rice production and causes zinc deficiency. When iron fails to bind with protein, they form harmful free radical. This harmful red radical which effects the level of iron in human cells, damages digestive track, they also penetrate into the cells of vital organs like liver, brain & heart. Over intake of iron can increase the risk of these free radicals to cause further DNA damage. As reported these free radicals are also responsible for mutations and transformations.
- **Aluminum**: We are exposed to aluminum by water, food and drugs containing aluminum. Symptoms of high level of aluminum are vomiting, ulcer, skin rashes and bone pain. Aluminum effects brain and causes loss of memory and imbalance in human posture.
- **Arsenic**: It is toxic in nature and cancer causing. It is found in different forms, mainly in the form of oxide and sulphide. It is also found in salt form with sodium, copper and iron. Arsenate and its compounds are dangerous to human beings and also to the surroundings. Arsenic causes cell respiration malfunctioning. Inorganic methylated arsenic compounds are transformed

biologically into harmful forms due to bacteria, fungi, humans and algae to provide mono and dimethyl arsenic acids. This biological process of transformation converts inorganic arsenic species enzymatically into methylated arsenic which shows the severe exposure of arsenic. Arsenic minerals are the causes of ground water contamination. So higher amounts of arsenic exposure to environment are due to fertilizers and certain pesticides and animal feeding operation. Arsenide and arsenate are cancer causing to lungs, liver, skin & bladder. Arsenic exposure to humans is mainly by air, food and water.

For the removal of these metal ions from water various methods like ion-exchange, reverse osmosis, chemical precipitation, chemical oxidation or reduction, electrodialysis and ultrafiltration are used. But these techniques have their own inherent limitations such as sensitive operating conditions, less efficiency and production of secondary sludge and further the disposal is a costly affair (Fu and Wang 2011; Ahluwalia and Goyal 2005). On the other hand, adsorption technique has gained high attention due to its advantages. These days' natural absorbents have received much attention for the removal of heavy metal ions. Untreated plant wastes e.g. peanut hull pellets (Johnson et al. 2002), Papaya stem (Saeed et al. 2005a), rise husk ash and neem bark (Bhattacharya et al. 2006) Saltbush (Atriplex canescens) leaves have received a wide attention of adsorption studies.

The various adsorbents used for this purpose are:

11.3 Agricultural Waste Adsorbents

These are different forms of inexpensive and non-living plant material such as black gram husk (Saeed et al. 2005b), eggs hell (Park et al. 2007) Sugar-beet pectin gels (Mata et al. 2009) and citrus peels (Schiewer and Patil 2008) are considered as potential adsorbents. Annaduri et al. (2003) performed the adsorption of heavy metal ions like Cu^{2+}, Zn^{2+}, Co^{2+}, Ni^{2+} & Pb^{2+} onto acid and alkali treated banana and orange peels. It has been reported that residues of banana & orange peels are cellulose-based wastes. Hence, can be processed and converted to be adsorbents as they have large surface areas, high swelling capacities, and good mechanical strength, are convenient to use, and have great potential to adsorb harmful contaminants like heavy metals. Based on regeneration studies, it was reported that the peels could be used for two regenerations for removal and recovery of heavy metal ions.

The HCl treated carrot residues can remove the heavy metals like Cr^{3+}, Cu^{2+}, Zn^{2+} from waste water. The adsorption of metal ions onto carrot residues was possible due to cation exchange properties of these residues which was attributed to the presence of carboxylic and phenolic functional groups, which exist in either the cellulosic matrix or in the materials associated with cellulose, such as hemicelluloses and lignin (Nasernejad et al. 2005).

11.3.1 Biomass Adsorbents

Many biomass source adsorbents have seen widely investigated as potential biosorbents for heavy metals. Algae, a renewable natural biomass which proliferates ubiquitously and abundantly in the littoral zones of the world, have attracted the attention of many investigators as organisms to be tested and used as new adsorbents to adsorb metal ions. The biosorption of Cu^{2+} and Zn^{2+} by dried marine green macroalga (c.linum) investigated by Ajjabi and chouba (2009).

11.3.2 Byproduct Adsorbents

- **Sawdust**: It is obtained from wood industry as a byproduct. It has been found that sawdust contains several organic compounds (e.g. cellulose, lignin and hemicelluloses) with polyphenolic groups that could bind heavy metal ions through different mechanisms. Sawdust was used by Sciban et al. (2006) for the removal of heavy metals.
- **Lignin**: Lignin is also used as an adsorbent for the metal ions. The high adsorption capacity of lignin is in part due to polyhydric phenols and other functional groups on the surface. Ion-exchange may also play a role in the adsorption of metals by lignin
- **Rice Husk and Rice Husk Ash**: Rice husk is insoluble in water. It has high mechanical strength and has good chemical stability. It has a granular structure. It is a good adsorbent material for the removal of heavy metals from waste water. Rice husk has been extensively reviewed by Chuah et al. (2005) for the removal of heavy metal. It has been used for the removal of heavy metals either in untreated form or modified form by different modification methods. The common chemical treatment methods of rice husk are sodium carbonate and hydrochloric acid (Kumar and Bandyopadhyay 2006), epichlordhydrin & sodium hydroxide (Bhattacharya et al. 2006) and tartaric acid. Rick husk's pretreatment can remove hemicelluloses and lignin, reduce cellulose crystallinity and increase the surface area or porosity. It was reported by Kumar and Bandyopadhyay (2006) that rice husk treated with sodium carbonate, sodium hydroxide and epichlorihydrin enhanced the adsorption capacity of heavy metals. Rice husk ash is used for the removal of Zn^{2+} (Bhattacharya et al. 2006). Rice bran was evaluated for its potential as an adsorbent for Cd^{2+}, Cu^{2+}, Pb^{2+} & Zn^{2+} (Montanher et al. 2005). Rice bran adsorbent is able to adsorb metal ions from aqueous solutions.
- **Fly Ash**: Since the industrial revolution, a wide scale coal burning for power generation started. Although that were many millions of tons of ash that have been generated. It was estimated that the current annual production of coal ash (worldwide) is around 600 million tons, with fly ash forming about 500 million tons at 75–80% of the total ash produced. Thus the amount of fly ash as coal waste released by thermal power plants and factories has been increasing throughout the world.

The utilization of fly ash for the removal of heavy metals from industrial waste waters was reported by Gangoli et al. (1975).

As a law-cost adsorbent, fly ash has been widely used for use removal of heavy metals. As early as 1975, the use of fly ash for the removal of heavy metals was reported. Bayat (2002a) investigated the removal of heavy metal ions (Bayat 2002b), using lignite-based fly ash and activated carbon and discovered that fly ash was effective as activated carbon.

A series of investigations was conducted on the adsorption of heavy metals, using bagasse fly ash adsorbent bagasse fly ash from sugar industries was used for the removal of zinc and copper form aqueous solution (Gupta and Ali 2000; Gupta and Sharma 2003).

After the adsorption process, fly ash can be regenerated using certain reagents. Batabyal et al. (1995) reported that saturated fly ash can be regenerated using 20% of aqueous H_2O_2 solution.

11.4 Conclusion

Overall adsorption processes are most conviniest method for removal of heavy metals from the solvent and waste water. Biomaterials based biosorbents are safe and economical to eliminate heavy metals from contaminated waters. The gain the interest of researchers towards the production of low cost adsorbents for more application and sustainable development of ecosystem. Therefore there is an urgent need that all possible sources of agro-based inexpensive adsorbents should be explored and their role for the removal of heavy metals.

References

Ahluwalia SS, Goyal D (2005) Removal of heavy metals by waste tea leaves from aqueous solution. Eng Life Sci 5(2):158–162

Ajjabi LC, Chouba L (2009) Biosorption of Cu^{2+} and Zn^{2+} from aqueous solutions by dried marine green macroalga charctomorpha linum. J Environ Manag 90(11):3485–3489

Annaduroi G, Juang RS, Lee DJ (2003) Adsorption of heavy metals from water using banana & orange peels. Water Sci Technol 47(1):185–190

Awwad NS, El-Zahhar AA, Founda AM, Ibrahim HA (2013) Removal of heavy metal ions from ground water and surface water samples using carbons derived from date pits. J Environ Chem Eng 1(3):416–423

Batabyal D, Sahu A, Chaudhuri SK (1995) Kinetics and mechanism of removal of 2,4- dimethyl phenol from aqueous solutions with coal fly ash. Sep Technol 5(4):179–186

Bayat B (2002a) Combined removal of Zinc (II) and cadmium(II) from aqueous solutions by adsoption onto high-calcium Turkish fly ash. Water Air Soil Pollut 136(1–4):69–92

Bayat B (2002b) Comparative study of adsorption properties of Turkish fly ashes: I. the case of nickel (II), Copper (II) and Zinc (II). J Hazard Mater 95(3):251–273

Bhattacharya AK, Mandal SN, Das SK (2006) Adsorption of Zn (II) from aqueous solution by using different adsorbents. Chem Eng J 123(1–2):43–51
Chuoh TG, Jumosiah A, Azmi I, Katayan S, Thomas Choong SY (2005) Rice husk as a potential low-cost biosorbent for heavy metal and dye removal: an overview. Desalination 175(3):305–316
Fu F, Wang Q (2011) Removal of heavy metal ions from wastewaters: a review. J Environ Manag 92(3):407–418
Gangoli N, Markey D, Thodas G (1975) Removal of heavy metal ions, from aqueous solutions with fly ash. In: Proceedings of the national conference on complete water use, Chicago, IL, USA, pp 270–275
Gupta VK, Ali I (2000) Utilization of bagasses fly ash (a sugar industry waste) for the removal of copper and zinc from waste water. Sep Purif Technol 18(2):131–140
Gupta VK, Sharma S (2003) Removal of Zinc from aqueous solutions using message fly ash a low cost adsorbent. Ind Eng Chem Res 42(25):6619–6624
Johnson PD, Watson MA, Brown J, Jefcoat IA (2002) Peanut hull pellets as a single use sorbent for the capture of cu (II) form wastewaer. Waste Manag 22(5):471–780
Kumar U, Bandyopadhyay M (2006) Sorption of cadmium from aqueous solution using pretreated rice husk. Bioresour Technol 97(1):104–109
Mata YN, Blazquez ML, Ballester A, Gonzalez F, Munoz JA (2009) Sugar-beet pulp pectin gels as biosorbent for heavy metals: preparation and determination of biosorption and desorption characteristics. Chem Eng J 150(2–3):289–301
Montanher SF, Oliveira EA, Rollemberg MC (2005) Removal of heavy metal ions from aqueous solutions by sorption onto rice bran. J Hazard Mater 117(2–3):207–211
Nasernejad B, Zadeh TE, Pour BB, Bygi ME, Zamani A (2005) Comparison for biosorption modeling of heavy metals from waste water by carrot residues. Process Biochem 40(3–4):1319–1322
Park HJ, Jeong SW, Yang JK, Kim BG, Lee SM (2007) Removal of heavy metals using waste egg shells. J Environ Sci 19(12):1436–1441
Saeed A, Akhtar MW, Iqbal M (2005a) Removal and recovery of heavy metals from aqueous solution using papaya wood as a new biosorbent. Sep Purif Technol 45(1):25–31
Saeed A, Iqbal M, Akhtar MW (2005b) Removal and recovery of lead (II) from single and multimetal (Cd, Cu, Ni, Zn) solutions by crop milling waste (black gram husk). J Hazard Mater 117(1):65–736
Schiewer S, Patil SB (2008) Modeling the effect of PH on biosorption of heavy metals by citrus peals. J Hazard Mater 157(1):8–17
Sciban M, Klasnja M, Skrbic B (2006) Modified hardwood sawdust as adsorbent of heavy metal ions from water. Wood Sci Technol 40(3):217–227

Chapter 12
Wastewater Treatments Plants and Their Technological Advances

Ngangbam Sarat Singh, Ranju Sharma, and Talat Parween

Abstract The biggest challenge of the twenty-first century is the provision of pure and affordable water to meet up human needs. Water supply is struggling to keep up the pace with the ever-increasing demand worldwide, which is being aggravated by various factors like population explosion, rapid urbanization, water quality deterioration and global climatic change. Continued depletion of freshwater resources, has shifted the focus more towards the recovery, reuse, and recycling of water, which mandates advanced wastewater treatment. Numerous innovations have been observed in the field of water treatment in the last few years, which have resulted in the blooming of better alternatives to conventional wastewater treatment systems. Some of those technologies are nanotechnology, Membrane Filtration bioreactors, an advanced oxidation process (AOP) and microbial fuel cells which have promising prospects and broad applications. These new treatment technologies have been proven to remove a wide range of challenging contaminants from wastewater successfully. These revolutionary techniques in the field of wastewater treatment have been discussed in this chapter.

Keywords Climate · Urbanisation · Bioreactors · Nanotechnology

12.1 Introduction

Water is the primary requisite for the survival of water, without water life cannot exist. Though 71% of the earth is covered with water, only 29% is in the form of fresh water (Sonune and Ghate 2004). In other words, it is a scarce resource which

N. S. Singh
University of Delhi to Department of Zoology, Dr. SRK Government Arts College, Yanam, Puducherry (UT), India

R. Sharma
Indian Institute of Technology, New Delhi, India

T. Parween (✉)
Department of Bioscience, Jamia Millia Islamia, New Delhi, India

M. Oves et al. (eds.), *Modern Age Waste Water Problems*,
https://doi.org/10.1007/978-3-030-08283-3_12

we need to use judiciously. Water supply is struggling to keep up the pace with the ever-increasing demand worldwide, which is being aggravated by various factors like population explosion, rapid urbanization, water quality deterioration and global climatic change (Gehrke et al. 2015). Hence, the biggest challenge of the twenty-first century is the provision of pure and affordable water to meet up human needs. Continued depletion of freshwater resources has shifted the focus more towards the recovery, reuse, and recycling of water, which mandates wastewater treatment. Around 70–80% of the water supply for domestic and industrial use comes out as wastewater. Wastewater consists of suspended and dissolved organic solids, which are biodegradable (Rupani et al. 2010). It contains sewage and hospital waste which carry pathogens. Wastewater may comprise of toxic chemicals as well as heavy metals are known to have environmental and health impacts. It consists agricultural runoff which includes fertilizers, pesticides and other carcinogens magnifying the risk of cancers in humans and also have environmental implications (Islam and Tanaka 2004). Majority of the cases results in the let out of untreated wastewater which either contaminates the groundwater or surface water. Therefore, there is a need to for innovative technologies in wastewater treatment for recycling and reusing wastewater.

Wastewater treatment is an essential aspect of the water industry which safeguards public health, natural environment (Yüksel 2010). Wastewater treatment is the conversion or removal of highly complex and biodegradable organic solids into relatively inorganic or stable organic solids. It involves a reduction in the concentration of pollutants present in wastewater in compliance with proper maintenance and operation of the treatment plant. It consists of three stages of treatment primary, secondary and advanced treatment (Oller et al. 2011). Primary and secondary treatment eliminates most of the Biochemical Oxygen Demand and total suspended solids present in the wastewater (Sonune and Ghate 2004). But, in the majority of cases, this treatment stands inadequate in treating the polluted water or in providing reusable water for domestic and industrial uses. Therefore, additional advanced treatment steps have been appended to the treatment plants to achieve better removal of organic, nutrients and toxic materials. So, it can be said that advanced wastewater treatments have turned into a common focal point as people, industries, and countries are trying to endeavor for approaches to garner one of the essential resources for life available and usable. Hence, advanced wastewater treatment technology, combined with various wastewater reduction and recycling initiatives, renderaspiration of decelerating or possibly stopping the unavoidable reduction in the amount of usable water. Numerous innovations have been observed in the field of water treatment in the last few years, which have resulted in the blooming of better alternatives to conventional wastewater treatment systems.

Some new technologies are being used and introduced for wastewater treatment globally to reclaim the resources. Some of those technologies like nanotechnology and Membrane Filtration bioreactors, an advanced oxidation process (AOP) and microbial fuel cells with promising prospects and wide applications have been discussed in this chapter. These new technologies of waste water treatment have been proven to efficaciously remove various noxious waste.

12.2 Primary Treatment

Wastewater contains large solid objects and grits which may result in mechanical damage to the waste water treatment equipment. Primary treatment does primary screening which consists of physical removal of large objects like clothes, shoes, plastics, sticks, rags, and anything that can harm or clog the equipment, small pipes, etc. (Parcher 1998). Moden-day treatment plants also use excellent screening system along with the coarse screen system. Coarse screen typically has openings of larger than 6 mm while fine screens have the opening 6–0.1 mm. Size of the coarse solid particles is reduced for their easy removal during the downstream processes (Metcalf and Eddy 1991). These coarse particles (6–19 mm) are grinded and shredded into small pieces using devices like comminutors and grinders so that these can remain in the wastewater.

From the screening stage wastewater moves into the grit chamber, the water is kept stagnant for a small duration. The selected duration (detention time) for this step is long enough to allow the settling of grit which includes heavy materials like gravels, pebbles, sand, etc. which have higher specific gravity as compared to other organic matter. Detention time is calculated as the volume of the tank divided by the flow rate. Grit is removed in order to prevent any type of wearing of the equipment and abrasions. The collection of these heavier particles simplifies their disposal problem as these are easy to dispose of in the landfills after washing.

After the grit chamber, the sewage enters the primary settling tank or primary clarifier, where the flow rate is diminished adequately to allow the settling of suspended solids under the effect of gravity (Fig. 12.1). Primary settling tank's sizing is done very cautiously as it determines the success of the procedure. Duration of detention of wastewater is kept typically for 2–3 h which helps in removing 55–

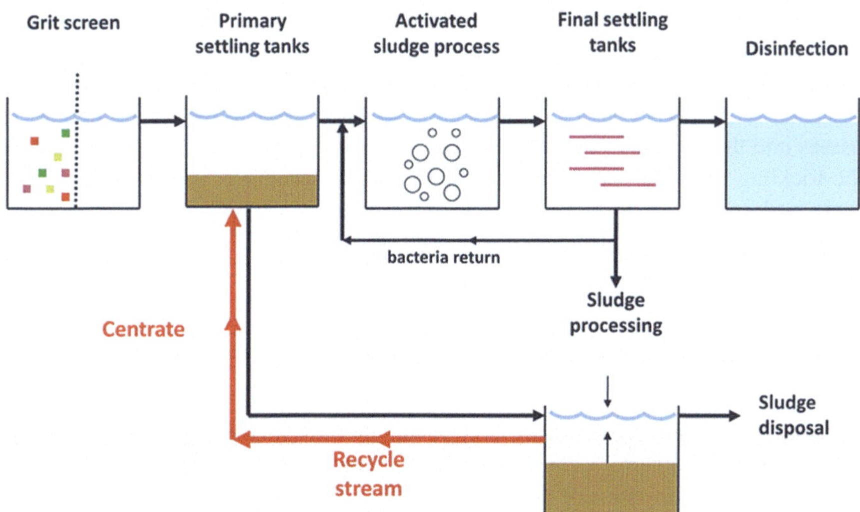

Fig. 12.1 Schematic diagram of a typical wastewater treatment plant. (Zhou et al. 2011)

65% of suspended solids and 30–40% of BOD (Gray 1999) The shape of the primary settling tank may be square or rectangle. The solid that settles at the base of the settling tank is known as raw sludge or primary sludge and is removed for further processing. The scum or grease that floats over the wastewater is also removed. If the plant carrying these processes is a primary treatment plant, at this stage the effluents if treated with chlorine in order to disinfect the water from the bacteria as well as for odor control and finally released.

12.3 Secondary Treatment

Secondary treatment reduces 90% of BOD and 90% of suspended solids. It primarily aims at removing BOD by the breakdown of organic compounds with the help of aerobic microorganisms (Gupta et al. 2012). Some of the traditionally used processes are trickling filters, activated sludge method and oxidation ponds, where the first two approaches follow primary treatment while the third one does require primary treatment.

12.3.1 Trickling Filters

Trickling filters have a circular bed which may be of a 20–30 m diameter and 2–3 m deep, which consist of gravels, pebbles, peat moss, and other broad materials over which a rotating arm distributes wastewater (Fig. 12.2). The coarse boulders have small spaces between them which allow the movement of air resulting in aerobic conditions. The bed is covered with biofilm (microbial slime) that imbibe and degrade the waste present in the wastewater aerobically. The biofilm majorly contains many bacterial species, algae, fungi, protozoa and many microfauna. This biofilm may form into several millimeters thick layer. This slime periodically gets collected at the base of the trickling filter together with the wastewater. From there it passes into the secondary settling chamber and is removed. The small rocks used in the trickling filters are being replaced by the plastic media which are lighter than rocks and provide more surface area for microbial growth and much deeper beds. These were first used in 1893 and were used successfully since then.

12.3.2 Rotating Biological Contactors

Rotating biological contactors are another variation of the trickling filter. Trickling filter depends on micro-organisms grown on rocks, slag, plastic, etc. RBC is a variation to this, consisting of closely spaced, circular, parallel plastic discs typically 3–4 m in diameter and 3 mm in thickness fixed to a rotating shaft (Fig. 12.3). The plastics used in this process are PVC, polyethylene and expanded polystyrene. The

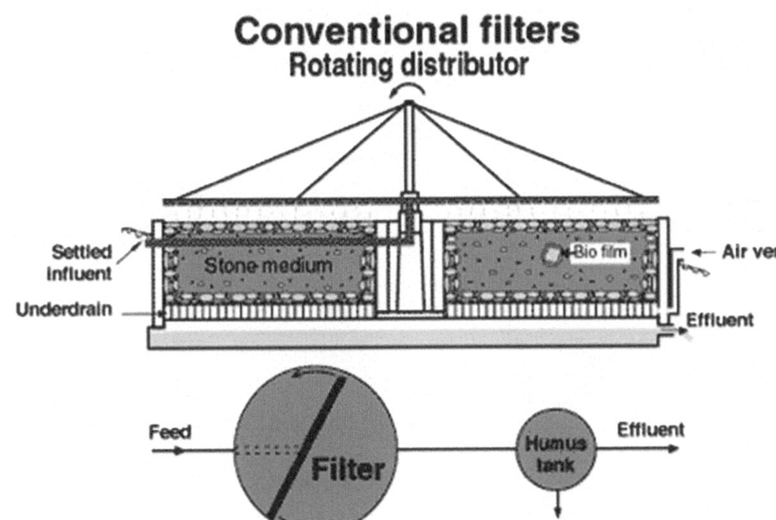

Fig. 12.2 Diagram of a convention al trickling filter. (http://www.unep.or.jp/ietc/Publications/TechPublications/TechPub)

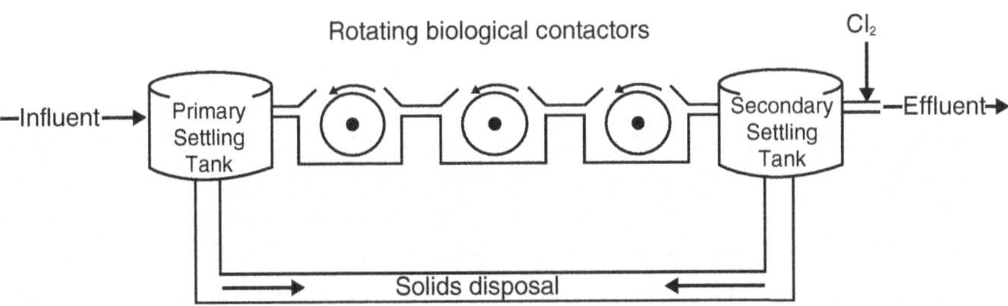

Fig. 12.3 Schematic diagram of Rotating Biological Contractors. (https://water.me.vccs.edu/courses/ENV295WWII/Lesson5_print.htm)

40% area of the discs is immersed in wastewater. The biofilm attached to the surface of the discs moves in and out of the wastewater with the discs. The oxygen is supplied to the microorganisms when they are out of the water, and they degrade the waste while in the water. The performance can be enhanced by adding a number of discs in the series in order to attain the treatment requirements. RBC was installed for the first time in the world Germany in 1959.RBCs are easy to use and easy to handle.

12.3.3 Activated Sludge Method

Activated sludge method is more efficient and cost-effective than trickling filters. As shown in the (Fig. 12.4) the key unit in this method is the aeration tank. After the

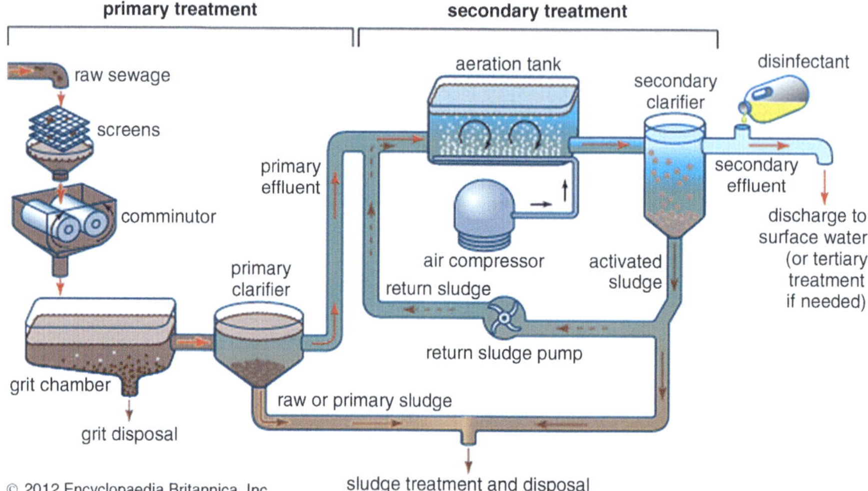

Fig. 12.4 Comprehensive diagram showing primary and secondary treatment using activated sludge method. (www.britannica.com)

primary treatment, the effluent from the primary sedimentation tank enters the aeration tank. Aeration tank named so because of the presence of air. As the wastewater enters the aeration tank activated sludge (microbial mass from the secondary settling tank) is added into it as inoculums. Air is pumped into the tank, and the wastewater mix is agitated thoroughly for 6–8 h to maintain aerobic conditions. After that wastewater (now known as mixed liquor) is transferred into the secondary settling chamber or clarifier where bacterial flocs are separated from the water by subsidence. Some amount of this solid is taken for inoculums, and the remaining sludge is further processed and disposed off.

Activated sludge method provides maximum contact between the microorganisms and the wastewater that too in less space as compared to Trickling filters and gives similar results. These are economical, show few problems and reduce BOD significantly. The only negative aspect being their high operation cost.

So, these methods are the most conventional methods of Secondary treatment from the beginning. But there have been a lot of advancements in this field during the past few decades. Some of them are discussed here.

12.4 Advanced Wastewater Treatment

Wastewater treatment plants provide three-stage treatment to the wastewater primary, secondary or tertiary treatment on the basis of purification requirement. Majority of suspended solids and BOD is removed during primary and secondary treatment. Although, many times it doesn't prove to be sufficient to provide recycled water for domestic or industrial supply. Hence, few additional treatment stages are required for the treatment of wastewater in treatment plants to further enhance the water quality by removing any further presence of organic solids, nutrients, and toxic contaminants. Thus, advanced wastewater treatment is said to be designed to provide a higher quality effluent than normally achieved by secondary treatment processes. It includes various categories like tertiary treatment, physicochemical treatment, and combined biological-physical treatment. Some of the conventional methods of advanced treatments include reverse osmosis (RO), distillation, ion exchange, etc. Advanced wastewater treatment involves various stages like nutrient removal, removal of additional organic and suspended solids, removal of toxic substances like heavy metals and hazardous wastes.

But in the recent years, a number of new technologies are being used and introduced for wastewater treatment globally to reclaim the diminishing resource. Some of the most promising and widely applicable processes have been discussed here.

12.5 Membrane Bioreactor Technology

Membrane bioreactor technology is being extensively used for advanced treatment to produce water for reuse by the industries. MBRs are a promising combination of activated sludge treatment and membrane filtration for biomass retention (Melin et al. 2006). With MBRs, complete biological treatment and the retention of pathogens including viruses has become possible. Treatment with membrane bioreactor produces a highly clarified effluent that can be more easily disinfected. It is a pressure-dependent process that relies on the pore size of the membrane to filter components present in the wastewater (Melin et al. 2006).

Low-pressure membrane filtration techniques are utilized for the separation of effluent from the activated sludge, i.e., either ultrafiltration or microfiltration. Thekey membrane configurations involved in MBR technology side stream membranes or submerged membranes (Fig. 12.5). Submerged MBRs is the more often applied configuration in wastewater treatment. Submerged MBRs involve the use of either hollow fiber membranes or plate membrane design. But this technique requires sufficient pre-treatment in order to avert membrane clogging by hairs, fibers, etc. In many membrane filtration processes the flow rate declines during filtration, the major reason behind this is membrane fouling. Thus, the Control of membrane fouling is the major issue in MBR operation. Major factors influencing the membrane fouling are membrane type, hydrodynamic conditions, module configuration and also the presence of compounds having a high molecular weight which are being produced by microbial metabolism or have been introduced into the sludge bulking process (e.g., poly-electrolytes).

There has been a drastic change in the number of MBR installations around the world, since the 1990s, when this technology just arrived (Melin et al. 2006). Comparison with other technologies used for water recycling reveals that MBRs not only produce lower residual concentrations but do so more robustly than the alternatives (Jefferson et al. 2001 & Jefferson et al. 2000).

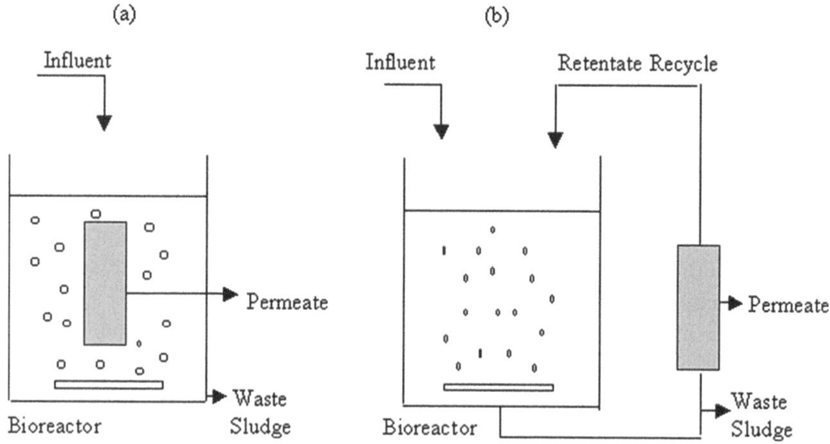

Fig. 12.5 Design of (a) Submerged MBR and (d) Side stream MBR. (Melin et al. 2006)

12.6 Advanced Oxidation Processes (AOPs)

Sometimes a single oxidation system is not sufficient for the degradation of total organic pollutants present in wastewater. Advanced Oxidation Processes (AOPs) involves usage of multiple oxidation processes simultaneously and entails escalated production of the highly reactive species hydroxyl radical (Yoon et al. 2001; Comninellis et al. 2008; Chong et al. 2010). The abovementioned processes include techniques like ultraviolet (UV) photolysis, Fenton's reagent oxidation, sonolysis and posses the capability to carry out organic pollutant's degradation under ambient conditions. The key benefit of this process is the production of CO_2 as a result of oxidation of organic contaminants. A number of AOPs are available such as combined ozone and peroxide (Mahamuni and Pandit 2006), chemical oxidation processes using ozone (Jyoti and Pandit 2001), UV enhanced oxidation like photo-Fenton or UV/Fenton (Ghaly et al. 2001; Gernjak et al. 2003) UV/ozone, UV/hydrogen peroxide (Esplugas et al. 2007).

12.7 Nanotechnology

Many conventional techniques like adsorption, extraction and chemical oxidation are generally effective while being much expensive. Thus, the reduction of toxic contaminants like heavy metals to the below permissible limit at a much reasonable price is quite necessary. Nanotechnology can play a crucial role in this regard (US EPA 2018). As with the advent of nanotechnology and embodiment of living microorganisms into biological microelectronic devices, the water treatment process has revolutionized. The top feature of nanotechnology is its easy merger with other technologies which helps to improvise, advocate and explain the existing concepts. It proposes a revolutionary approach to evolve and utilize these methods in novel ways. Nanotechnology concepts are being investigated for higher performing membranes with fewer fouling characteristics and improved hydraulic conductivity. A number of new researches are being conducted for producing fabrication of membranes from nanomaterials for decomposition of toxic compounds during the treatment (Dastjerdi and Montazer 2010). It will also provide effective segregation of metals, bimetallic nanoparticles, mixed oxides, zeolites and carbon compounds, etc. from the wastewater resources. With improved membranes and configurations, more efficient pumping and energy recovery systems will be possible.

Many conventional techniques like adsorption, extraction, and chemical oxidation are generally effective while being much expensive. Thus, the reduction of toxic contaminants to the below permissible limit at a much reasonable price is quite important. Nanotechnology can play an crucial role in this regard (US EPA 2018). Due to their unique active surface area, nanomaterials can be used over a number of applications like bioactive nanoparticles, nanosorbents, catalytic membranes, and metal nanoparticles such as silver, iron, titanium oxides, etc. (Xu et al. 2012; Ghasemzadeh et al. 2014).

Nanoparticles can be used in the treatment of wastewater in a number of ways as discussed below.

12.7.1 Nanocatalysts

Nanocatalysts have also been used enormously water treatment as it poses a unique property of having a large surface area with shape dependent properties that result in increased surface catalytic activity. It increases the reactivity as well as degradation of pollutants. The generally used catalytic nanoparticles are zero-valence metals, semiconductor materials, and bimetallic nanoparticles for the degradation of various environmental contaminants like organochlorine pesticides, azo dyes, PCBs (polychlorinated biphenyls), halogenated herbicides, halogenated aliphatic and nitroaromatics (Bhattacharya et al. 2013). This catalytic activity has proved very effective on a laboratory scale for various pollutants.

12.7.2 Nanosorbents

Nanosorbents have high and peculiar sorption capacity and also have widespread application in wastewater treatment. U. S and Asia are the biggest users of commercialized nanosorbents. (Bhattacharya et al. 2013). Magnetic nanosorbentshave proved quite interesting especially for the removal of organic contaminants. Magnetic separation by nanosorbents have revolutionized this field, to prepare them magnetic nanoparticles are coated with specific ligands furnishing them specific affinity (Roco et al. 2010). To evade any unwarranted toxicity, various methods like cleaning agents, magnetic forces, ion exchangers, etc. are utilized for the removal of nanosorbents from the site of treatment. Regenerated nanosorbents are inevitably an economical technique and encouraged more for commercialization.

12.7.3 Nanostructured Catalytic Membranes (NCMS)

Nanostructured catalytic membranes are extensively utilised for the treatment of contaminated water. It has many advantages such as high optimization capability, catalytic sites uniformity, small contact time of catalyst, sequential reaction capability and ease in industrial scale up. Nanofiltration membranes already have wide application in the removal of micro-pollutants and dissolved salts, water softening and wastewater treatment (Bhattacharya et al. 2013). These membranes behave like a physical barrier and can capture micro-organisms and particles larger than their pore size while rejecting the substances selectively. It is expected that nanotechnology will further upgrade membrane technology and shall also lessen the unduly high price of desalination technology for fetching fresh water from saline water.

Nanostructured TiO_2 films and membranes under UV and visible-light irradiation may perform several functions like inactivation of microorganisms, decomposition of organic pollutants, physical separation of water contaminants and anti-biofouling action (Chorawalaa and Mehta 2015). The N-doped "nut-like" ZnO nanostructured material that forms multifunctional membrane is very useful in eliminating water pollutants by escalating photodegradation activity under visible light. It has also shown good antibacterial activity and has aided in producing clean water with constant good flow rate benefiting the water treatment field.

Multiple studies have researched about metallic nanoparticles' immobilization in the membrane-like polyvinylidene fluoride (PVDF), cellulose acetate, chitosan, polysulfone. For attaining efficient degradation of toxic contaminants which provide many advantages like organic partitioning, high reactivity, lack of agglomeration, prevention of nanoparticles and reduction of surface passivation (Prachi et al. 2013).

12.7.4 Membrane Filtration Technology

Advancements in nanotechnology have resulted in the advent of multiple new catalytic membranes which are known to be highly selective, permeable and resistant to fouling (Bhattacharya et al. 2013). Nanofiltration is a liquid separation membrane technology positioned between ultrafiltration and reverse osmosis systems. While Removal efficiency of ROlies in the range of particles of 0.0001-micron diameter, nanofiltration (NF) removes particles of 0.001-micron diameter. NF is a selective membrane process that rejects particles having 1-nanometer approx diameter (Bhattacharya et al. 2013). NF membranes can eliminate selected salts and organic compounds at a lower pressure as compared to RO systems because of their large pore sizes. It also possesses the ability to remove toxic contaminants like viruses, bacteria, and organic pigments without generating any harmful side product (Chorawalaa and Mehta 2015).

NF is used in the removal of various organic pollutants like pesticides from water both surface and ground (Plakas and Karabelas 2012). It is also a promising alternative to lime softening or zeolite softening technologies that's why also called membrane softening. It also incurs lower energy cost than RO; energy costs are lower than that.NF is best suited for the treatment of the well water or water from rivers or lakes (Bhattacharya et al. 2013).

12.7.5 Nanofibers

Nanofiber technology in amalgamation with the biotic removal of toxic xenobiotics is an advanced technique for industrial wastewater treatment. Here, the formation of microbial biofilm is significantly supported by nanofiber structures and thus results in escalated biodegradation (Prachi et al. 2013).

Nanofiber carriers have been analyzed on multiple parameters like the cleansing ability of toxic compounds, carrier layer's stability, desirable microorganism's transport in growth rate and nanofibers' disintegration (Dhand et al. 2015). Each biomass carrier is required to the necessary parameters like physical and chemical stability, microorganism colonization ability, surface morphology, etc.

The properties like high porosity, large specific surface, and small pore size make these nanofiber carriers extraordinary. These are also highly efficient in heavy metal removal from wastewater for, eg. ElectrospunPolyacrylonitrilenanofiber mats have been proved phenomenal for the removal of heavy metal).

Nanofibers are also durable, chemical resistant and easily moldable. The principal advantage of nanofiber materials is the fast colonization of microorganisms on the nanofiber surface because of their surface morphology, comparability with microorganism dimensions and biocompatibility (Sharma and Sharma 2012).

A principal benefit of this technology is that there is a possibility of a buildup of bacterial biofilm on the surface as well as inside the carrier, where bacteria are safer against harmful effects of the surrounding environment. Moreover, it provides better oxygen and substrate to the microorganisms. Nanofiber layer allows excellent adhesiveness to the bacteria because of its large specific surface and hence simplifies microbial immobilization, providing special help during initial stages of colonization and during harsh conditions).

Following an extended colonization period, the microbial biomass starts growingspontaneouslyeven without nanofibers resulting in a more efficient wastewater treatment system. One of the example of the use of Fe-Grown Carbon Nanofibers reported for the remediation of Arsenic (V) in wastewater (Adeleye et al. 2016).

12.8 Microbial Fuel Cells

A microbial fuel cell is an example of a rapidly developing microbial electrochemical technology (Logan and Rabaey 2012). This technology is a quantum leap innovation where organic matter present in wastewater can be used for the extraction of electrical energy by the utilization of electron transfer for capturing the energy that microorganisms produced (Pant et al. 2010). Microorganisms are cultivated on an electrode as a biofilm, electrical current is established via proton exchange membrane which separates electron donor from the electron acceptor (Liu and Logan 2004). This technology is still in growing phase, and considerable advancements in various aspects like the efficiency of the method, etc. are necessary before its widespread use in electrical energy generation right from the organic matter present in wastewater.

12.9 Conclusion

The recent trends in wastewater treatment technologies discussed in this chapter open a new horizon in wastewater treatment. These technologies like MBR, nanotechnology, etc. give better results in cleaning wastewater, while Microbial fuel cells along with the wastewater problem are tackling another problem of energy scarcity by producing energy from the waste. However, further research is required for new development of simple, cost-effective and better technologies.

References

Adeleye AS, Conway JR, Garner K, Huang Y, Su Y, Keller AA (2016) Engineered nanomaterials for water treatment and remediation: costs, benefits, and applicability. Chem Eng J 286:640–662

Bhattacharya S, Saha I, Mukhopadhya A, Chattopadhyay D, Ghosh UC, Chatterjee D (2013) Role of nanotechnology in water treatment and purification: potential applications and implications. Int J Chem Sci Tech 3(3):59–64, ISSN 2249-8532

Chong MN, Jin B, Chow CW, Saint C (2010) Recent developments in photocatalytic water treatment technology: a review. Water Res 44(10):2997–3027

Chorawalaa KK, Mehta MJ (2015) Applications of nanotechnology in wastewater treatment. Int J Innov Emerg Res Eng 2(1):21–26

Comninellis C, Kapalka A, Malato S, Parsons SA, Poulios I, Mantzavinos D (2008) Advanced oxidation processes for water treatment: advances and trends for R & D. J Chem Technol Biotechnol 83(6):769–776

Dastjerdi R, Montazer M (2010) A review on the application of inorganic nano-structured materials in the modification of textiles: focus on anti-microbial properties. Colloids Surf B: Biointerfaces 79(1):5–18

Dhand C, Dwivedi N, Loh XJ, Ying ANJ, Verma NK, Beuerman RW et al (2015) Methods and strategies for the synthesis of diverse nanoparticles and their applications: a comprehensive overview. RSC Adv 5(127):105003–105037

Esplugas S, Bila DM, Krause LGT, Dezotti M (2007) Ozonation and advanced oxidation technologies to remove endocrine disrupting chemicals (EDCs) and pharmaceuticals and personal care products (PPCPs) in water effluents. J Hazard Mater 149(3):631–642

FACT SHEET, Emerging contaminants – nanomaterials, solid waste and emergency response (5106P), EPA 505-F-09-011, United States Environmental Protection Agency September 2009

Gehrke I, Geiser A, Somborn-Schulz A (2015) Innovations in nanotechnology for water treatment. Nanotechnol Sci Appl 8:1

Gernjak W, Krutzler T, Glaser A, Malato S, Caceres J, Bauer R, Fernández-Alba AR (2003) Photo-Fenton treatment of water containing natural phenolic pollutants. Chemosphere 50(1):71–78

Ghaly MY, Ha¨rtel G, Mayer R, Haseneder R (2001) Photochemical oxidation of p-chlorophenol by UV/H2O2 and photo-Fenton process. A comparative study. Waste Manag 21:41–47

Ghasemzadeh G, Momenpour M, Omidi F, Hosseini MR, Ahani M, Barzegari A (2014) Applications of nanomaterials in water treatment and environmental remediation. Front Environ Sci Eng 8(4):471–482

Gray NF (1999) Water technology. John Wiley & Sons, New York, pp 473–474

Gupta VK, Ali I, Saleh TA, Nayak A, Agarwal S (2012) Chemical treatment technologies for wastewater recycling—an overview. RSC Adv 2(16):6380–6388

http://www.unep.or.jp/ietc/Publications/TechPublications/TechPub

https://water.me.vccs.edu/courses/ENV295WWII/Lesson5_print.htm

Islam MS, Tanaka M (2004) Impacts of pollution on coastal and marine ecosystems including coastal and marine fisheries and approach for management: a review and synthesis. Mar Pollut Bull 48(7–8):624–649

Jefferson B, Laine A, Parsons S, Stephenson T, Judd S (2000) Technologies for domestic wastewater recycling. Urban Water 1(4):285–292

Jefferson B, Laine AL, Stephenson T, Judd SJ (2001) Advanced biological unit processes for domestic water recycling. Water Sci Technol 43(10):211–218

Jyoti KK, Pandit AB (2001) Water disinfection by acoustic and hydrodynamic cavitation. Biochem Eng J 7(3):201–212

Liu H, Logan BE (2004) Electricity generation using an air-cathode single chamber microbial fuel cell in the presence and absence of a proton exchange membrane. Environ Sci Technol 38 (14):4040–4046

Logan BE, Rabaey K (2012) Conversion of wastes into bioelectricity and chemicals by using microbial electrochemical technologies. Science 337(6095):686–690

Mahamuni NN, Pandit AB (2006) Effect of additives on ultrasonic degradation of phenol. Ultrasonics 13(2):165–174

Melin T, Jefferson B, Bixio D, Thoeye C, De Wilde W, De Koning J, van der Graaf J, Wintgens T (2006) Membrane bioreactor technology for wastewater treatment and reuse. Desalination 187 (1–3):271–282

Metcalf and Eddy (1991) Wastewater engineering treatment, disposal, reuse. McGrawHill Book Company N, New Delhi

Oller I, Malato S, Sánchez-Pérez J (2011) Combination of advanced oxidation processes and biological treatments for wastewater decontamination—a review. Sci Total Environ 409 (20):4141–4166

Parcher MJ (1998) Wastewater collection system maintenance. Technomic Publishing Company Inc, Basel

Pant D, Van Bogaert G, Diels L, Vanbroekhoven K (2010) A review of the substrates used in microbial fuel cells (MFCs) for sustainable energy production. Bioresour Technol 101(6):1533–1543

Plakas KV, Karabelas AJ (2012) Removal of pesticides from water by NF and RO membranes—a review. Desalination 287:255–265

Prachi, Gautam P, Madathil D, Brijesh Nair AN Nanotechnology in waste water treatment: a review. Int J Chem Tech Res CODEN(USA): IJCRGG ISSN: 0974-4290 5(5):2303–2308, July–Sept 2013, 2304–2306

Roco MC, Mirkin CA, Hersam MC (2010) Nanotechnology research directions for societal needs in 2020. Retrospective and outlook. National Science Foundation (Sponsor)

Rupani PF, Singh RP, Ibrahim MH, Esa N (2010) Review of current palm oil mill effluent (POME) treatment methods: vermicomposting as a sustainable practice. World Appl Sci J 11(1):70–81

Sharma V, Sharma A (2012) Nanotechnology: an emerging future trend in wastewater treatment with its innovative products and processes. Nanotechnology 1(2)

Sonune A, Ghate R (2004) Developments in wastewater treatment methods. Desalination 167:55–63

U.S. Environmental Protection Agency. National recommended water quality criteria. http://www.epa.gov/ost/criteria/wqctable/. Accessed 15 July 2018

Xu P, Zeng GM, Huang DL, Feng CL, Hu S, Zhao MH, Lai C, Wei Z, Huang C, Xie GX, Liu ZF (2012) Use of iron oxide nanomaterials in wastewater treatment: a review. Sci Total Environ 424:1–10

Yoon J, Lee Y, Kim S (2001) Investigation of the reaction pathway of OH radicals produced by Fenton oxidation in the conditions of wastewater treatment. Water Sci Tech J Int Assoc Water Pollut Res 44(5):15–21

Yüksel I (2010) Hydropower for sustainable water and energy development. Renew Sust Energ Rev 14(1):462–469

Zhou W, Li Y, Min M, Hu B, Chen P, Ruan R (2011) Local bioprospecting for high-lipid producing microalgal strains to be grown on concentrated municipal wastewater for biofuel production. Bioresour Technol 102(13):6909–6919

Chapter 13
Bioremediation of Oil-Spills from ShoreLine Environment

Ranju Sharma, Ngangbam Sarat Singh, Neha Dhingra, and Talat Parween

Abstract In the present day society, we are in great need of petroleum hydrocarbons for our energy need. In spite of recent advance technologies, crude oil accidental spill occurs at constant rate during its extraction, transportation, storage, refining, and distribution. Marine shorelines are essential ecological and human resources that serve as a home of a variety of wildlife habitat. Marine oil spill causes extensive damage to coastal marine environments. Unlike, higher organisms that are adversely affected by oil spill, specific microorganisms are capable of degrading these hydrocarbons into the non-toxic compound and mineralize them. They play an essential role in the bioremediation of oil spill and reduce the overall impact of the oil spill disaster. Microbial bioremediation of petroleum hydrocarbons is a useful approach and have been used practically in recent years. In this chapter, we studied various factors responsible for the oil spill disaster, its ill effect and ways to overcome these effects. We also explored the importance of bioremediation technique over the traditional methods.

Keywords Oil-spill · Bioremediation · Bio-augmentation · Biodegradation

R. Sharma (✉)
Indian Institute of Technology, New Delhi, India

N. S. Singh
University of Delhi to Department of Zoology, Dr. SRK Government Arts College, Yanam, Puducherry (UT), India

N. Dhingra
Department of Zoology, University of Delhi, New Delhi, India

T. Parween
Department of Bioscience, Jamia Millia Islamia, New Delhi, India

M. Oves et al. (eds.), *Modern Age Waste Water Problems*,
https://doi.org/10.1007/978-3-030-08283-3_13

13.1 Introduction

Globally, inter-regional trade of oil proliferates in the past few decades. Transporting liquid petroleum-based products and oils from production sources to consumption location involve high risk including the risk of accidental spillage of oil. Oil spills are a form of pollution in which release of liquid petroleum based hydrocarbon occur accidentally in land or marine ecosystem. Since, the transportation of oil occurs mainly through marine, therefore, oil spillage is one of the major concerns of the marine ecosystem. Despite the leakage of crude oil from tankers, wells and drilling rings, the primary factor of oil spillage in the marine ecosystem are automobiles, boats, industrial plant, and machinery. This oil finally reaches the ocean and harms the marine ecosystem (Chang et al. 2014).

Factors that influence the oil spill pollution in the shoreline environment are the type of spillage, the quantity of oil, the effect of tidal waves, weather condition, biological and ecological attributes of the area and sensitivity of crucial species to oil spill pollution. The oil spill has broad-ranging impact on environments which include both immediate as well as extended-term effect. It affects birds, wild life, marine ecosystem and human. Many cases of the oil spill have been recorded in the history. They have both commercial and environmental impacts. Here we discuss some of the major oil spill disaster listed by extensive damage they have done.

13.2 Deep Water Horizon Spill

Also known **BP –oil spill,** it is one of the most significant oil spills in the history. It occurred on **20th April 2010** in the Gulf of Mexico on the Transocean operated and drilling for BP-owned Macondo prospect oil field (Fig. 13.1). Total estimated discharge of about 4.9 million barrel was reported. Current studies showed that the well site is still leaking (Pallardy 2017).

- *Marine Consequence:* It causes extensive damage to wildlife habitat, fishing, and tourism industries. Marine life was continued to die six times faster than the average rate. It causes cardiotoxicity to wildlife habitat. Water contains 40 times more PAH then present before the oil spillage. PAH combine with oil spill creates various carcinogen and chemicals that cause a different health risk to human and marine life. It decreases the dissolved oxygen content as it contains 40% methane by weight which suffocates marine life and creates a dead zone (Pallardy 2017).

Footprints of oil spillage found under the shell of tiny blue crab larvae. In coastal marshes, insect population gets affected significantly. Pelican eggs were found to contain petroleum compounds. Fish with oozing sores and lesions were observed on the Al Jazeera shore. By July 2010, damage to the ocean floor endangered the pancake batfish. By, 2013 number of dolphin stranded with oil increases four fold. NWF (National Wildlife Federation) reports about endangered Kemp's ridley sea

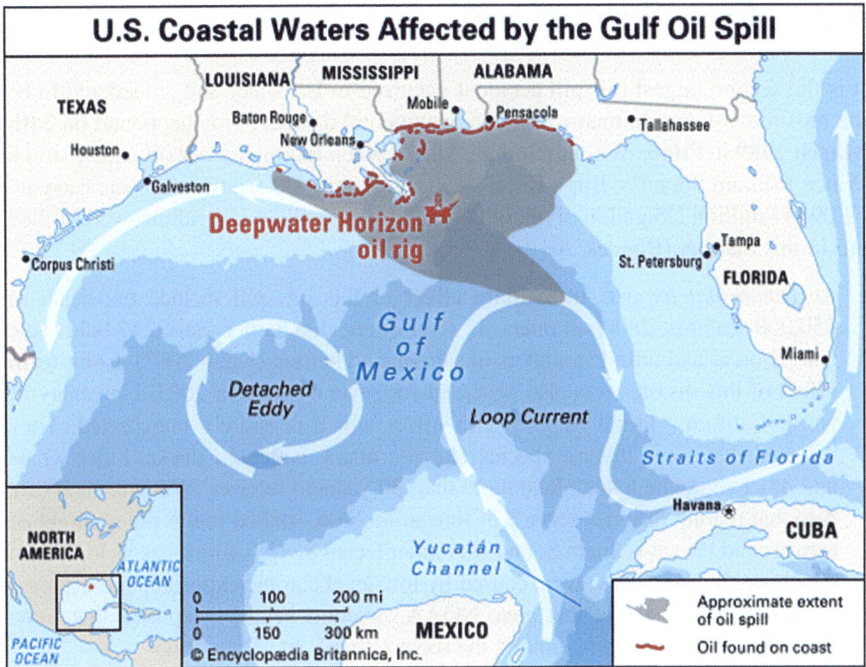

Fig. 13.1 Area affected by the Deep Horizon oil spill of the Gulf of Mexico. (Pallardy 2017)

turtles (https://www.nwf.org/). Long-term effect of oil spill remains in the food chain for a generation due to clumping of oil on the sea floor, known as an underwater rain of oily particle. On 12th April, 360 babies or stillborn within the spill area had abnormal or "underdeveloped lungs" (Pallardy 2017).

- *Health Consequence:* By June 2010, 143 oil spill exposure cases have been reported to the Louisiana Department of health and hospital. Out of these 143 cases, 108 were involved in clean-up efforts while 35 were resident of that area. This is the public health crisis from a chemical poisoning in the history of Louisiana (Gee 2010). Environmental scientist – Wilma Subra observed the presence of chemicals in the blood due to exposure to oil spillage. Survey in 2012, showed different types of human health effect including eye, nose and throat irritation, respiratory problems, blood in urine, vomit, rectal bleeding, nausea, skin irritation, burning and lesion, short-term memory loss and confusion, nose bleeding and early start of menstruation among girls (Blumenfeld 2011).
- *Economic Effects:* According to David Abramson, Director of Columbia University of National Centre for Disaster Preparedness, this oil spill has a significant effect on the economy sectors such as offshore drilling, fishing, and tourism. They lost coastal economy up to the cost of 22.7 billion through 2013. Economy loss occurs due to tourism and fishing industry become 153 dollar and 247 dollar respectively. Due to this economic loss, BP has dropped from the second position to fourth most significant among the four oil companies by 2013 (Schleifstein 2013).

13.2.1 Exxon Valdez Oil Spill

It is the second largest oil spill accident occurred in US water and considered to be the most devastating human-caused environmental disaster. This happened on **24th March 1989** in Prince William Sound, Alaska when Exxon Valdez oil tanker struck Prince William Sound's, Bligh Reef. According to the US reports ship was carrying 53.09451 million US gallon of oil, out of which 10.8 million US gallons were spilled out in this disaster (Raunek 2017).

- *Environmental impact:* **Immediate effect** of this oil spill include the death of 250,000 seabirds, 2800 sea otters, 12 river otters, 300 harbor seals, 247 bald eagle and 22 orcas and infinite numbers of salmon and herring (Sarah 2003). **Long-term effect** of this disaster was also observed for several years. In 2003, University of North California found its long-term effect which is entirely unexpected. They reported that some diverse species like sea otters, harlequin ducks, killer whale and other shoreline habitat take more than 30 years to recover (Williamson 2003). National marine fisheries service in Juneau in 2006, studied that 6 miles shoreline area around Prince William Sound was still affected by spill which has its long-term ecological impact. These were caused by low-level chronic exposure, and decrease the wildlife diversity of that area. NOAA, Scientist in 2014, reported that some amount of oil is remaining among the rocks (Patrick and Dan 1991).
- *Economic impact:* It has its both long term and short-term economic effects. Loss of recreational sports, fisheries, reduces tourism. This spillage adversely affects the economy of Cordova, Alaska. After 20 years, people still have an adverse health effect of oil spillage (Patrick and Dan 1991).

13.2.2 Gulf War Oil Spill

It occurs due to gulf war in 1991. Bagdad reports said that American airplane strikes two oil tankers which caused oil spillage. Sea Island in Kuwait is the primary source of oil. **On 26th January 1991**, three US fighter-bombers (F-117) destroyed Kuwaiti oil refinery (Bultmann 2001)

- *Environmental and economic effects:* Approximately 11,000,000 barrels of oil was spilled during this disaster. About 101 miles of the area were affected. A study sponsored by UNESCO, Bahrain, Iran, Iraq, Kuwait, Oman, Qatar, Saudi Arabia, the United Arab Emirates and the United States showed that this disaster has its long-term effects (Bultmann 2001). Oil remnants are still found after 9 years later in marshland and mud tidal. During the quantitative survey in 2002 and 2003 by Dr. Jacqueline Michel, he reported that due to the absence of the clean-up process spill oil remained over 800 km of the shore. The spilled oil penetrate 30–40 cm to the tidal flats (Michel 2010). In 2001, Dr. Hans-Jörg Barth, a German geographer stated in his research report that several coastal areas still had significant oil impact. After 10 years, 50% of coastline still had the most massive effect (Barth 2011).

- **Ixtoc I:** On **3rd June 1979**, an accidental explosion was occurred due to hydrostatic pressure build-up during the semi-submersible drilling in the bay of Campeche of Gulf of Mexico. Due to the build-up of pressure, oil caught fire, and Sedco 135 collapsed into the sea. Initially, 30,000 barrels of oil spilled per day. Later it gets reduced to 20,000 barrels by July 1979, due to pumping of mud (Milne 2008).
- *Environmental and Economic effects:* Due to this accidental disaster, the various offshore region, as well as the coastal zone near the Gulf of Mexico, gets polluted. About 6000 metric tonnes of oil observed on sea beaches. This accidental spill also had their adverse effect on marine life. Littoral crab and mollusks fauna were profoundly affected. After 9 months, crab's population like ghost crab, *Ocypode quadrata*, were entirely eliminated over a wide range of area. (Jernelöv and Lindén 1981).

This also affects the breeding and growth rate of several sea animals lived in estuaries and coastal lagoons lining the bay. Kemp's Ridley turtle at the Rancho Nuevo beach which has a most concerns in Mexico gets critically endangered. Shrimps, fish, and mollusks are representing the most important commercial fishing species for Texas economy. Due to the oil spill effect on these species, fishing at the contaminated area at the north and south of the well were banned entirely. Due to this, the fishing industry showed an economic reduction. This oil spill continues to affect the more abundant species of marine habitat for a generation (Teal and Howarth 1984).

It also reduces the tourism economy due to the presence of thick oil layer on the sea beaches which cover the whole beauty and fun of those beaches. It also dramatically affects the livelihoods of local people (Teal and Howarth 1984).

13.2.3 Aegean Captain Oil Spill

On the stormy evening of **19th July 1979** Atlantic express, a Greek crude oil tanker collided with Atlantic express, another oil tanker at the off coast of Trinidad and Tobago. 90 million gallons of oil were spilled out during the explosion. This is the first massive spill out without any environmental damage (http://www.popularmechanics.com/science/energy/g1765/biggest-oil-spills-in-history/).

13.2.4 Fergana Valley Oil Spill

Fergana valley oil spill occurs on **2nd March 1992** in Mingbulak oil field located in Fergana valley, Uzbekistan. It is also known as a Mingbulak oil spill. Near about 88 gallons of oil were spilled out from well and it is one of the largest inland oil spill disaster ever reported (http://www.popularmechanics.com/science/energy/g1765/biggest-oil-spills-in-history/).

13.2.5 Nowruz Field Oil Spill

This oil spill disaster held on **10th February 1983** in Nowruz oil field situated in the Persian Gulf, Iran. This oil spill disaster occurs during the Iran-Iraq war. An oil tanker hit the platform attacked by Iraqi helicopter and spill caught fire. It has various ill effects on marine life. (10 biggest oil spill in history, http://www.popularmechanics.com/science/energy/g1765/biggest-oil-spills-in-history/).

13.2.6 ABT Summer Oil Spill

On **28th May 1983**, an oil tanker named ABT summer exploded and caught fire while routeing to Rotterdam. At the time of this disaster, it was 900 miles off the coast of Angola. About 80 gallons of oil was spilled out. The ship continued to burn for 3 days. Out of 32 crews, five were burned, and four were reported as missing. Since the incident occurred in an offshore location little damage was caused to the environment (http://www.popularmechanics.com/science/energy/g1765/biggest-oil-spills-in-history/).

13.2.7 Castillo de Bellver Oil Spill

On **6th August 1983**, Spanish oil tanker known as Castillo de Bellver route from the Persian Gulf to Spain caught fire while it crossed Saldanha bay off the coast of South Africa which is approximately 70 miles away at the northwest of Cape Town, South Africa. The ship carried about 79 million gallons of crude oil, of which around 78.9 gallons were spilled out during this disaster (Moldan 1997).

- *Environmental impact:* Although it cause little ecological damage, because this disaster occurs in ecologically and economically sensitive area which is rich in flora and fauna, many seabirds were found whose feather become glued with oil. Within the 24 h of this incidence, black rain occurred as an immediate effect due to the presence of oil droplet in the air (Joye 1983).

13.2.8 *Amoco Cadiz Oil Spill*

On **16th March 1978** Amoco Cadiz ship which enrooted from the Persian Gulf to Rotterdam, Netherlands. Due to heavy stormy weather, the ship was broken down into three pieces and sank into the deep sea. About 4000 tons of fuel oil was spilled out due to this disaster. About 76 beaches were affected by the accident due to the penetration of oil in the soil. It became one of the most studied disasters in the history (Visser 2010).

- *Environmental damage:* It causes one of the most significant losses of marine life ever reported. Millions of dead mollusks, sea urchin and other bottom-dwelling organisms were found after 2 weeks following the incident. Around 20,000 dead birds were recovered; 9000 tons of oysters died, echinoderm and crustaceans become entirely disappeared. In the disaster core area, fishes become suffered from skin ulceration, tumor and also had a strong taste of petroleum while eating those fishes. This is the largest ever recorded oil spill and was the first oil spill in which estuarine tidal river was studied. It cost about 250 million US dollar to the fishery and tourism industries (Hartog and Jacobs 1980).

13.3 Oil Spill Pollution Act

In 1990, US Congress passes an oil pollution act to require various oil storage facilities like Riveted and bolted tanks, Field welded storage tanks, open top tank, fixed roof tank, floating roof tank, etc. and prevent oil spill pollution. This act was again revised by EPA in 1994 and renamed as "Worst Case-Scenario" of oil discharge. This Act was prepared to prevent oil spill disaster and develop an immediate response to the oil spill. It includes:

- The establishment of new requirements for oil spill response and made amendments in the U.S. Federal Water Pollution Control Act.
- The requirement of coast guard to monitor oil tank vessels and make new oil spill clean-up specifications for companies (https://www.environmentalpollutioncenters.org/oil-spill/).

13.3.1 *Effects of Oil Spill*

Oil spill at the shoreline environment has various direct and indirect effects. They may affect wildlife as well as a human in multiple ways (https://www.environmentalpollutioncenters.org/oil-spill/).

13.3.1.1 Direct Effect of the Oil Spill

- **Direct contact with the skin-** Some organic compounds penetrate into our body and causes irritation when entering through direct skin absorption.
- **Through inhalation-** Volatile compounds were easily evaporated in the environment and enter our body through inhalation.
- **Through ingestion-** enter directly through contaminated water and food.

13.3.1.2 The Indirect Effect of the Oil Spill

- Disrupting professional activities in the polluted area.
- Decrease property rate in the polluted area.
- Overall economic costs.

13.3.1.3 Behaviour of Oil Spill in the Environment

Behavior of oil spill change is different environment depending on the place and condition it occurs (https://www.environmentalpollutioncenters.org/oil-spill/).

- ***Behavior of the oil spill in the marine environment:*** In marine environment, oil accumulate on the water surface and makes a film. It usually degrades faster in the water environment as water is the best medium for dispersion, emulsification, and microbial degradation.
- ***Behavior of the oil spill inthe land environment:*** On land, oil may penetrate deep into the soil through water capillary and eventually reaches the groundwater.
- ***Behavior of the oil spill in the underground environment:*** This will occur due to leaching into underground water and cause groundwater pollution.

13.4 Traditional Method Used for Clean-up Oil Spill in Marine Shorelines and Freshwater Environments

Traditionally there were three main methods, i.e., physical method, chemical method and natural method for oil spill clean-up from marine and shoreline environment (NOAA 1992; NOAA and API 1994; U.S. EPA 1999; Doerffer 1992).

13.4.1 Physical Methods

This method was used for clean-up of physical contaminant and recovery of free oil in marine and shoreline environment. Most Commonly used physical methods are:

- ***Booming and Skimming:*** To control the movement and recover the floating oil.
- ***Mechanical removal:*** Collection and extraction of oil from sediment only if a limited amount of oil spillage occurs.
- ***Washing:*** This is done by applying low-pressure cold water and high-pressure hot water flushing. The oils get adhered along the shoreline to the water edge and are available for collection.
- ***Wiping with adsorbent material:*** In this method, hydrophobic material is used for wiping up of oil from the water surface.
- ***Sedimentation relocation and tilling:*** In this process, the oil contaminated sediment from one place to another is transferred to enhance the natural clean-up process. This will facilitate the dispersion of oil into the water column of sediment and increase the interaction between oil and mineral particles.
- ***In situ burning:*** This includes the combustion of oil present on vegetation, wood logs, and debris. But this method causes air pollution due to burning.

13.4.2 Chemical Methods

In this cleanup method, surfactant is used which act as dispersant and emulsifiers (Lessar and Demarco 2000). In the United States, this method is not extensively used due to long-term toxic environmental effect of these chemicals. Most commonly used chemicals are:

- ***The surfactant used as dispersing agent:*** These chemicals are used to remove the oil from the water surface before it reaches the shoreline environment through the water column. This will decrease the toxicity effect of oil and increase the biodegradation rate of the oil spill by increasing the surface area (Fig. 13.2). They increase the degradation rate of the oil spill by 50%. *Example:* **Corexit 9527A** and **Corexit 9500 A** were used during the Exxon Valdez Oil Spill Clean up the process. ***Petroleum distillates, Ethanol 2-butoxy, Propylene Glycol, Organic Sulfonic Acid Salt,*** **are the active ingredient of this two dispersant.**

- ***The surfactant used as de-emulsifier:*** These chemicals are used to break the oil in water emulsion which results in increasing natural dispersion. They are generally Epoxy resins, Polyethyleneimines, Polyamines, Di-epoxides, Polyols based chemicals.
- ***The surfactant used as Solidifier:*** These chemicals results in the polymerization of oil due to which the spreading of oil gets minimized. These surfactants were generally used for physical recovery operations.
- ***The surfactant used for Surface film formation:*** These chemicals help in the creation of oil film on the water surface which helps in easy removal of surface oil through the washing process of oil spill clean-up.

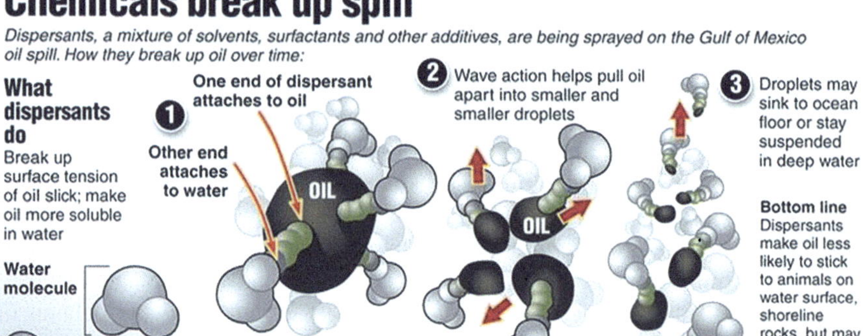

Fig. 13.2 Mechanism of dispersant to degrade oil spill. (http://www.mcclatchydc.com/news/nation-world/national/article24582025.html)

13.4.3 Natural Methods

This refers to the natural means for oil spill degradation process. These methods are cost effective and the environment-friendly. These methods are effective where natural removal rate is fast. Application of these methods needs regular monitoring so that we can observe the performance of natural attenuation. It includes:

- ***Evaporation:*** It is the most commonly used oil spill clean-up method. It generally used to remove lighter weight oil component. We can eliminate 50% of the spilled oil through this process within 12 h following the oil spill.
- ***Photo-oxidation:*** Breakdown of complex oil compound into simpler ones through sunlight in the presence of oxygen.
- ***Biodegradation:*** Degradation of oil through biological means such as microbes. This will remove the non-volatile component of petroleum and are a relatively slow process. It takes months or years to degrade a significant amount of oil.

13.5 Bioremediation Technology for Oil-Spills Clean-up

Bioremediation is a technique in which we use soil microbes for the remediation of toxic chemicals. Bacteria may use these chemicals as a source of energy and mineralize them into the more straightforward end product. It is the best technique to degrade the petroleum-based toxic component into nontoxic form and finally eliminate them from the environment.

Degradation of hydrocarbon begins with the conversion of polycyclic aromatic hydrocarbon (PAH) into alcohol. Further, these alcohols oxidized into aldehyde and then mineralize into water, carbon dioxide and biomass (Venosa et al. 1997). This is considered to be one of the most promising techniques for the oil-spill removal from the environment (Hoff 1993; Swannell and Head 1994). Bioremediation operation applied after EXXOn Valdez oil spill gave excellent success (Atlas and Bartha 1992; Bragg et al. 1994; Prince 1993; Pritchard and Costa 1991).

- Bioremediation technique for oil spill clean-up process can be applied in two ways (Venosa et al. 1997).
- ***Bio-stimulation*** by adding nutrients which enhance the growth of existing oil-degrading microorganism.
- ***Bio-augmentation*** by adding oil-degrading microorganism which degrades the toxic compound.
- **Bio-stimulation:** It is a process in which nutrients are added to the oil spill site to enhance the growth of existing natural oil-degrading microbial population for the clean-up process of the marine oil spill (Venosa et al. 1997). When an oil spill incidence occur, it leads to increase in the amount of carbon content in the soil which stimulates the growth of oil-degrading microbes already existing in the soil. However, these oil-degrading microorganisms have limited growth and remediation capability due to available nitrogen and phosphorus. By the addition of supplemental nutrient in an appropriate concentration, existing microbes get maximum growth rate and hence increased rate of oil degradation. Addition of nitrogen at oil spilled area gave maximum growth rate and thus achieved maximum degradation (Boufadel et al. 1999). This is an effective way for the bioremediation of oil from the affected area. The concentration of nutrient plays a significant role in the bio-stimulation process. Appropriate concentration of nutrient as well as environmental conditions such as wave movement, the influence of tidal and difference in the densities of various nutrients. When the supplement nutrient applied in water, they dissolved in water and move along with it. They move downward during rising tides and seaward during falling tides. Therefore, for effective bio-stimulation and maximum bioremediation of oil, nutrient should be applied during low tide at the high tide line so that they get a maximum time of contact with the oil (Boufadel et al. 1999).
- **Bio-augmentation:** It is a process in which we add bio-degrader such as microbes to the polluted site when the existing microbes are unable to degrade them. There are various types of oil spill degrading microorganisms such as bacteria, fungus, and yeast which are capable of degrading hydrocarbon and used as a source of energy (Zobell 1964).

Bio-augmentation is not very often used because millions of hydrocarbon-degrading microorganisms are naturally present in the environment. It is also very less effective against oil spill clean-up process because sometimes inoculation of non-native microbes may have competition with existing bacteria for their survival. Generally, native bacteria within sediments (Button et al. 1992; Lee and Levy 1987; Prince 1993; Venosa et al. 1997), seawater (Atlas 1993; Pierce et al. 1975), and sea ice

(Delille et al. 1997) naturally increase in numbers. Various researchers also show that the addition of non- native oil-degrading microbes did not significantly increase the rate of biodegradation (Fayad et al. 1992; Venosa et al. 1996). Another reason behind the less frequent use of this technique is that there is a criticism about whether it is safe to inoculate new microbial species to the environment which is non-native (Venosa et al. 1996).

13.6 Factor Affecting the Bioremediation

Oil spill disaster is an environmental phenomenon so, physical, chemical and ecological factor, as well as conditions of the contaminated area, plays a major role in the bioremediation of oil. Physical factor such as temperature, water energy and surface area of the oil spill and chemical factors like pH, amount of nutrients and oxygen content and soil composition affect the rate and efficiency of bioremediation process (Zhu et al. 2001).

13.6.1 Physical Factors

Since bioremediation is a microbial process, so **temperature** play an essential role in the growth rate of microbes as well as the degradation rate of oil (Nedwell 1999). At the lower temperature, the viscosity of oil gets increases which change the solubility and toxicity of oil. Coulon in 2006, studied that population of oil-degrading area increases with increase in temperature, i.e., increases two, three and four orders of magnitude at 4 °C, 12 °C and 20 °C respectively (Coulon et al. 2006).

Oil-degrading microbes generally resides at the interface of water and oil. Larger the **surface area** of spilled oil, higher the number of bacteria (Venosa et al. 1996). Addition of chemical dispersant increases the microbial activity by increasing the interface area between oil and water. A dispersing agent such as powder peat and bio-surfactant are generally used for this purpose (Lee et al. 1985, 1999; Swannell and Daniel 1999). Studies on clean up procedure of the Exxon Valdez oil spill demonstrated the significance of clay-oil flocculation processes on the natural cleansing of oil residues from impacted shoreline sediment (Bragg and Owens 1994).

Water energy also plays an essential role in the bioremediation process as well as the efficiency of the oil spill disaster to harm the marine environment. Rough water will disperse and dilute supplement nutrients while on the other hand they also spread the oil to more vast areas (Zhu et al. 2001).

13.6.2 Chemical Factors

Rate of microbial degradation is very low under the anoxic conditions (Atlas and Bartha 1992; Lee and Levy 1991). Sediment tilling and raking increase the penetration depth of oxygen and supplement nutrient thus increase the bioremediation (Sendstad et al. 1984; Sergy et al. 1998).

13.7 Advantages of Bioremediation

- ***Bioremediation is a cost-effective technique.***

Exxon Valdez oil spill clean-up process for 120 km shoreline is quite less as compared to 1 day physical washing of shoreline area (Atlas 1995).

- ***It saves manpower as well as time.***

Financial saving of this technique has many benefits over traditional clean-up process. It does not require the constant need of the worker.

- ***It is environment-friendly technique.***

It does not require toxic chemicals to degrade oil into simpler ones. It does not interrupt the natural habitat, unlike physical and chemical method.

In the today's world, where we have a limited number of resources, bioremediation is an environmental benefit technique.

13.8 Disadvantages of Bioremediation

- ***It is a slow process:*** On the basis of the degree of that that oil spill possesses, bioremediation is a slow process as it takes a few months to several years for complete remediation of the contaminated area.
- ***It is affected by climatic conditions:*** Effectiveness of bioremediation process depends upon the climatic conditions of the contaminated area. If there are more waves and windy climate, then nutrient will not completely dissolve in water and causes quick loss and dilution of nutrients. In the case of bio-augmentation, there may be a competitive association (Venosa et al. 1996).
- ***Regular monitoring:*** It is challenging to conduct and monitor bioremediation test in fields because there are various factors which we do not control in the fields like environmental and climatic conditions (Venosa et al. 1996).

13.9 Exxon Valdez and Bioremediation

It is one of the most devastating incidences in the history occur on 24th March 1989 in Prince William Sound. Due to the remoteness of the oil spill area, there are many complications in the clean-up process. Only the boat and helicopters were accessed at that time (EPA 2006). Firstly, they burn the oil but due to bad weather, this method of oil spill clean-up was failed. After this, they adopt a mechanical approach to remove the oil with skimmer and boom. Again, due to very dense oil, they were unsuccessful as the oil easily clogged the skimmers. Following this, chemical dispersant was used but due to bad weather dispersing agent was not completely mixed with water and again they had an unsuccessful attempt for oil spill clean-up (EPA 2006).

After all the attempts, bioremediation technique was used and got successful results. It is well known that there were indigenous hydrocarbons degrading microorganisms present in Prince William Sound. After the oil spill disaster, the microbial population of these oil-degrading microbes were increases 10,000 times. Supplement nutrients were added to increase the growth rate and degradation efficacy of microbes, and within 10–14 days of application, noticeable results were found. With this success, EPA declares bioremediation is one of the best techniques for the marine oil spill clean-up process (Pritchard et al. 1992).

13.10 Future Prospectives

Although bioremediation is greener and economically efficient technique, future developments are still required to overcome its disadvantages. We still needed to construct a database having information regarding nature and type of oil, available methodologies of bioremediation like kind of microorganism available and its frequency of degradation against oil, environmental and climatic conditions of that particular area. We are in still need of further development of bioremediation microbes which are oil-degrading.

13.11 Conclusion

The degree of the hazardous impact of oil spill or explosion depends upon the volume of oil spilled, environmental factor at the time of incidence and chemical nature of oil. Bioremediation is an emerging tool for oil spill explosion in the twenty-first century. Due to its coat effectiveness and environment-friendly nature, bioremediation is a toolbox for oil spill treatment strategy. Although bioremediation is a greener and sustainable approach, continuous research is needed to overcome the disadvantages and makes it a feasible technique.

References

10 biggest oil spill in history. http://www.popularmechanics.com/science/energy/g1765/biggest-oil-spills-in-history/. Accessed 13 Dec 2017

Atlas RM (1993) Bacteria and bioremediation of oil spills. Oceanus 36:71–73

Atlas RM (1995) Bioremediation of petroleum pollutants. Int Biodeterior Biodegrad:317–327

Atlas RM, Bartha R (1992) Hydrocarbon biodegradation and oil spill bioremediation. Adv Microb Ecol 12:287–338

Barth HJ (2011) The coastal ecosystems 10 years after the 1991 Gulf War

Blumenfeld E (2011) Exposing the human side of BP's oil spill. http://www.aljazeera.com/indepth/interactive/2011/05/2011512141926468292.html. Accessed 7 Dec 2017

Boufadel MC, Suidan MT, Rauch CH, Ahn C-H, Venosa AD (1999) Nutrient transport in beaches subjected to freshwater input and tides. In: Proceedings of the International oil spill conference. American Petroleum Institute, Washington, DC

Bragg JR, Owens EH (1994) Clay-oil flocculation as a natural cleansing process after oil spills: Part 1: studies of shoreline sediments and residues from past spills. In: Proceedings of the 17th Arctic and Marine Oil Spill Program (AMOP) Technical Seminar, Vancouver, British Columbia, pp 1–24

Bragg JR, Prince RC, Harner EJ, Atlas RM (1994) Effectiveness of bioremediation for the Exxon Valdez oil spill. Nature 368:413–418

Bultmann PR (2001) Environmental warfare: 1991 Persian Gulf War. State University of New York, Oneonta, p 13820

Button DK, Robertson BR, McIntosh D, Juttner F (1992) Interactions between marine bacteria and dissolved phase and beached hydrocarbons after the Exxon Valdez oil spill. Appl Environ Microbiol 58:243–251

Chang SE, Stone J, Demes K, Piscitelli M (2014) Consequences of oil spills: a review and framework for informing planning. Ecol Soc 19:26

Coulon F, McKew BA, Osborn AM, McGenity TJ, Timmis KN (2006) Effects of temperature and biostimulation on oil-degrading microbial communities in temperate estuarine waters. Environ Microbiol 9:177–186

Delille D, Basseres A, Dessommes A (1997) Seasonal variation of bacteria in sea ice contaminated by diesel fuel and dispersed fuel oil. Microb Ecol 33:97–105

Doerffer JW (1992) Oil spill response in the marine environment. Pergamon Press, Oxford

EPA (2006.) http://www.epa.gov/oilspill/oiilprofs.htm. Accessed 30 Dec 2017

Fayad NM, Edora RL, El-Mubarak AH, Polancos AB (1992) Effectiveness of a bioremediation product in degrading the oil spilled in the 1991 Arabian Gulf war. Bull Environ Contam Toxicol 49:787–796

Gee RE (2010) Louisiana DHH releases oil spill-related exposure information. Louisiana Department of Health & Hospitals (DHH). http://new.dhh.louisiana.gov/index.cfm/newsroom/detail/124. Accessed 7 Dec 2017

Geographie/phygeo/downloads/barthcoast.pdf. Accessed 10 Dec 2017

Hartog C, Jacobs RPWM (1980) Effects of the Amoco Cadiz oil spill on an Eelgrass Community at Roscoff (France) with special reference to the mobile benthic fauna. Helgol Mar Res 33:182–191

Hoff R (1993) Bioremediation: an overview of its development and use for oil spill cleanup. Mar Pollut Bull 26:476–481

Jernelöv A, Lindén O (1981) The Caribbean: Ixtoc I: a case study of the World's largest oil spill. Ambio Allen Press R Swed Acad Sci 10: 299–306

Joye BS (1983) Marine oil spills. http://joyeresearchgroup.uga.edu/public-outreach/marine-oil-spills/other-spills/castillo-de-bellver. Accessed 15 Dec 2017

Lee K, Levy EM (1987) Enhanced biodegradation of a light crude oil in sandy beaches. In: Proceedings of the International oil spill conference. American Petroleum Institute, Washington, DC, pp 411–416

Lee K, Levy EM (1991) Bioremediation: Waxy crude oils stranded on low-energy shorelines. In: Proceedings of the International oil spill conference. American Petroleum Institute, Washington, DC, pp 541–547

Lee K, Cobanli SE, Gauthier J, St.-Pierre S, Tremblay GH, Wohlgeschaffen GD (1999) Evaluating the addition of fine particles to enhance oil degradation. In: Proceedings of the International oil spill conference. American Petroleum Institute, Washington, DC

Lee K, Wong CS, Cretney WJ, Whitney FA, Parsons TR, Lalli CM, Wu J (1985) Microbial response to crude oil and Corexit 9527: seafluxes enclosure study. Microb Ecol 11:337–351

Lessar RR, Demarco G (2000) The significance of oil spill dispersants. Spill Sci Technol Bull 6:59–68

Michel J (2010) Lessons learned from Gulf War oil spill. The World (radio program). Public Radio International. https://www.pri.org/node/11782/popout. Accessed 10 Dec 2017

Milne CJ (2008) Sedco 135 series. http://www.oilcity.co.uk/home/article.asp?pageid=645. Accessed 13 Dec 2017

Moldan A (1997) Response to the *Apollo Sea* oil spill, South Africa. In: International oil spill conference proceedings, vol. 1, pp 777–781

National wilf life Federation. https://www.nwf.org/. Accessed 7 Dec 2017

Nedwell DB (1999) Effect of low temperature on microbial growth: lowered affinity for substrates limits growth at low temperature. FEMS 30:101–111

NOAA (1992) Shoreline countermeasure manual. National Oceanic & Atmospheric Administration, Seattle, Washington

NOAA and API (1994) Options for minimizing environmental impacts of freshwater spill response. National Oceanic & Atmospheric Administration and American Petroleum Institute

Oil spill Effect. Environmental pollution centers. https://www.environmentalpollutioncenters.org/oil-spill/. Accessed 20 Dec 2017

Oil Spill. Preliminary report. http://www.uni-regensburg.de/Fakultaeten/phil_Fak_III/

Pallardy R (2017) Deepwater Horizon oil spill of 2010 Environmental Disaster, Gulf of Mexico. https://www.britannica.com/event/Deepwater-Horizon-oil-spill-of-2010. Accessed 5 Dec 2017

Patrick D, Dan O (1991) Sad is too mild a word. Press coverage of the Exxon Valdez oil spill. J Commun 41:42–57

Pierce RH, Cundell AM, Traxler RW (1975) Persistence and biodegradation of spilled residual fuel oil on an estuarine beach. Appl Microbiol 29:646–652

Prince RC (1993) Petroleum spill bioremediation in marine environments. Crit Rev Microbiol 19:217–242

Pritchard PH, Costa CF (1991) EPA's Alaska oil spill bioremediation project. Environ Sci Technol 25:372–379

Pritchard PH, Mueller JG, Rogers JC, Kremer FV, Glaser JA (1992) Oil spill bioremediation: experiences, lessons and results from the Exxon Valdez oil spill in Alaska. Biodegradation:315–335

Raunek (2017) The complete story of the Exxon Valdez oil spill. https://www.marineinsight.com/maritime-history/the-complete-story-of-the-exxon-valdez-oil-spill/. Accessed 9 Dec 2017

Sarah G (2003) Environmental effects of Exxon Valdez spill still being felt. https://www.scientificamerican.com/article/environmental-effects-of/. Accessed 8 Dec 2017

Schleifstein M (2013) BP deepwater horizon spill: scientists say seafood safe, but health effects being measured. http://www.nola.com/news/gulf-oil-spill/index.ssf/2013/01/bp_ deepwater_ horizon_spill_sci.html. Accessed 7 Dec 2017

Sendstad E, Sveum P, Endal LJ, Brattbakk Y, Ronning O (1984) Studies on a seven-year-old seashore crude oil spill on Spitsbergen. In: Proceedings of the 7th Arctic and Marine Oil Spill Program (AMOP) technical seminar, Environment Canada, pp 60–74

Sergy G, Guénette C, Owens E, Prince RC, Lee K (1998) The Svalbard shoreline experimental oil spill field trials. In: Proceedings of the 21st Arctic and Marine Oilspill Program (AMOP) Technical seminar. June 10–12, 1998, Edmonton, Alberta. pp 873–889

Swannell RPJ, Daniel F (1999) Effect of dispersants on oil biodegradation under simulated marine conditions. In: Proceedings of the International oil spill conference. American Petroleum Institute, Washington, DC

Swannell RPJ, Head IM (1994) Bioremediation comes of age. Nature 368:396–397

Teal JM, Howarth RW (1984) Oil spill studies: a review of ecological effects. Environ Manag 8:27–43

U.S. EPA (1999) Understanding oil spills and oil spill response, EPA 540-K-99-007. Office of Emergency and Remedial Response, U.S. Environmental Protection Agency

Venosa AD, Haines JR, Eberhart BL (1997) Screening of bacterial products for their crude oil biodegradation effectiveness. Methods in biotechnology, 2. In: Sheehan D (ed) Bioremediation protocols. Humana Press, Totowa, pp 47–58

Venosa AD, Suidan MT, Wrenn BA, Strohmeier KL, Haines JR, Eberhart BL, King D, Holder EL (1996) Bioremediation of an experimental oil spill on the shoreline of Delaware Bay. Environ Sci Technol 30:1764–1775

Visser A (2010) Amoco Cadiz. International Super Tankers. http://www.aukevisser.nl/supertankers/part-1/id532.htm. Accessed 20 Dec 2017

Williamson D (2003) Exxon Valdez oil spill effects lasting far longer than expected, scientists say. http://www.unc.edu/news/archives/dec03/peters121803.html. Accessed 8 Dec 2017

Zhu X, Venosa AD, Suidan MT, Lee K (2001) Guidelines for the bioremediation of marine shorelines and freshwater wetlands. U.S. EPA Office of Research and Development, National Risk Management Research Laboratory, Land Remediation and Pollution Control Division, 26 W. Martin Luther King Drive, Cincinnati, OH 45268

Zobell CE (1964) The occurrence, effects, and fate of oil polluting the sea. Adv Water Pollut Res 3:85–119

Chapter 14
Rainwater Harvesting and Current Advancements

Neha Dhingra, Ngangbam Sarat Singh, Ranju Sharma, and Talat Parween

Abstract Our natural resources such as water, soil, forests, etc. are limited. As the cities are burdened with new residential areas, industries, food supplies with increasing population, the demand for these resources have been elevated as per they are already being exploited beyond their limits. Water is the basis of our life as crucially involved in simple chores of drinking, cooking to high scale industrial set ups. But with increasing urbanization and industrialization the water bodies have been exhausted to their utmost limit. Which is why there is a need to develop the alternate water sources. For this, the concept of rainwater harvesting comes up as a robust approach for sustaining the human needs. Rainwater harvesting is not a new policy; only it has been revitalized with the modern scientific approaches since there is numerous evidence of ancient civilizations of conserving and harvesting rainwater. The common method involves rainwater harvesting is a collection of rainwater in different structures and then making use of it in daily life, conserving it for future applications or it recharges the groundwater bodies. The implementation of varying rainwater harvesting techniques and methods varies from place to place depending upon their specific climatic conditions, land topography, hydrogeological conditions, etc. Also, it strengthens the relationship between the humans and the environment, making them aware of the need for conserving nature and natural resources along with sustainable development.

Keywords Rain water · Water harvesting · Runoff coefficient · Rain water harvesting system · IRCSA

N. Dhingra
Department of Zoology, University of Delhi, New Delhi, India

N. S. Singh
University of Delhi to Department of Zoology, Dr. SRK Government Arts College, Yanam, Puducherry (UT), India

R. Sharma
Indian Institute of Technology, New Delhi, India

T. Parween (✉)
Department of Bioscience, Jamia Millia Islamia, New Delhi, India

M. Oves et al. (eds.), *Modern Age Waste Water Problems*,
https://doi.org/10.1007/978-3-030-08283-3_14

14.1 Introduction

Water is the basis of our origin of life. It forms an indispensable part of our life, it is of utmost importance for our survival. Water, is one of the most important natural resource and it is depleting at a very rapid pace. The earliest mankind used the riverbeds as their water sources which can also be seen in the present day case. The accessibility to clean drinking water as well as sanitation conveniences are the most basic fundamental rights to a human being There is rapid population increase in most of our urban cities, thus creating a heavy demand for the timely supply of sufficient water to meet societal requirements (Kahinda et al. 2007). In order to fulfill the needs of freshwater, there are two alternatives; those are – to find the alternate water resources using conventional approaches; or to carry out the judicious use of the limited amount of currently available water resources (Meter et al. 2014). Due to the depleting water levels, people have now become increasingly aware of the need of water and urgent need to save it. The main reason behind the depletion of water levels was the pollution in the environment. As a result of increasing pollution levels the underground table of water as well as the layer of soil is deteriorating at a fast pace. Due to the depletion of water table, the tube wells and taps are dry in summer season. The human population starve for the water, which are situated in the arid areas having scanty to no rainfall. Thus forcing them to turn towards the contaminated sources of water such as stagnant water, contaminated ponds and lakes. This also leads to the increase in health issues faced by them and also gives way to epidemics. Due to the water crisis people not only facing the issues related to water scarcity but also facing health related issues after consuming the polluted water from the contaminated sources. Thus this situation has led to the urgent need to device different methods of conserving water and utilizing it properly.

With the help of recent technological developments, there is a significant portion of the global population that is suffering from lack in the basic amenity of having safe water to drink and for sanitation practices as well. Having various substitute technologies that are available in order to supplement the freshwater resources, rainwater harvesting and its application is a most expedient solution, whereas many other environmental problems can be negated by using this technology on a conventionally large-scale. Rainwater harvesting is a conventional water management practice which is increasingly accepted as a substantial dogma for ensuring food security and improving water shortage issues (Minkley 2012; Meter et al. 2014). This practice involves the collection of rainwater with large structures like rooftops, land surfaces or rock catchments and ultimately storing it in simple jars, pots as well as in some highly complex engineered assemblies. Conservation of the rainwater is not a new practice but this practice was also practiced by the ancient people about 4000 years ago. This technology proves to be is a central water source in many areas which are lacking in having substantial rainfall as well as lacks in any integrated water supply system. Also, plays an important role in the areas which are deficient in fresh surface water or groundwater, the application of suitable rainwater harvesting technology can prove to be a significant water resource in all the problematic areas.

14.2 The Increasing Water Crisis in World

There has been a drastic increase in the last century in the world population. The current population estimation from the United Nations includes that, in a medium-fertility scenario, global population is going to reach about 8.9 billion in 2050(Ray K et al. Rainwater Harvesting and Utilization:– Blue Drop Series 2).

Along with these demographic changes, the demographic shifts including migration of a large population from rural to urban areas for work opportunities, are creating extra pressure on the available water resources in these urban areas. This rapidly growing population with increasing industrialization, urbanization, and agricultural pressure is leading us towards a global water crisis (Singh et al. 2014). According to reports, two-thirds of the global population possibly go through shortage of water in the next decade (Tamaddun et al. 2018). Asia contributes a larger part of about 65% of the population and Africa 28% people, who are living without safe water (Ray et al. Rainwater Harvesting and Utilization:– Blue Drop Series 2). The pressure on resources can be easily estimated from this, that Asia, Africa, and Latin America consist of about three times the size of the urban population of the rest of the world. Due to this population increase, the natural resources (such as water, soil, forests and food stocks) are already being exploited in excess, and substantial efforts need to be done to fulfill the needs of the increasing population in the near future (Kahinda et al. 2007; Minkley 2012).

Among the various issues depleting water tables are causing problems (Singh et al. 2014). Besides the contamination of drinking water from rivers, lakes and reservoirs with heavy metal pollution is another alarming issue throughout the world. With more number of people are becoming reliant on limited fresh water supplies, the available water resources are getting more polluted. Thus, Water resourcing is emerging as a significant international primacy (Tamaddun et al. 2018).

14.3 Rainwater Harvesting

The rainwater collection and storage comes in practice thousands of years ago to when farming of crops started and an irrigation system was required (Jatiya and Hosokawa 2004; Jones and Hunt 2010). In areas with hotter climates, the scarcity of water leads to a collection of rainfall water (Mbilinyi et al. 2005). Earliest civilizations like the Indus Valley were more advanced in using huge vats that were cut into the rock to collect water when there was torrential rainfall (Qureshi et al. 2009). An alternative technique which has in use for hundreds of years in India was to build the water harvesting rooftops at the top of the houses. This technology is still in use and has widespread across the world. Beside that the construction of rainwater harvesting aqueduct in seventeenth century's Malta Island and the Sunken Palace of Istanbul were used to collect rainwater but later were stopped due to the continuous migration of people lead to clogging of systems which might harbor disease-causing organisms

(Mays et al. 2013). Rainwater harvesting systems were also in practices in ancient Roman times, where there is evidence of roof catchment systems from Roman cities. In the Roman period, the whole towns were planned to conserve rainwater as their primary source for drinking (Yannopoulos et al. 2016). In ancient Iraq about 5000 years ago, the civilizations were using storing tanks for runoff from hillsides to be used for both domestic and agricultural purposes (Mbilinyi et al. 2005). Also, the rainwater used to be collected in cisterns in the middle and late bronze age (2200–1200 B.C). Which was conserved in short rainy season and would be enough for at least one dry season? In old Palestine, they were the main source of drinking water. A 1921 census from Jerusalem counted 7000 cisterns collecting runoff water systems (Dreiblatt 1982). These systems were also in use in earliest middle east countries like Jordan where many of these systems are still functional today, where these tanks were about of a capacity ranging from 200 to 2000 m^3 (Jones and Hunt 2010). In Asia, about 2000 years ago, people in Thailand were using small-scale rainwater collection systems from the roofs or via simple gutters into traditional jars and pots (Guangfei Lo 2015). In this way where the old civilizations were practicing these rainwater harvesting techniques, the modern world people need to realize and revive the traditional tool, merge them with modern advancements of science and technologies to achieve the required need of water for our present and future generations.

14.4 Rainwater Harvesting Technology

Rainwater harvesting technology works on the collection as well as the storage of rainwater, where the collected water can be used in various other such as drinking, agricultural and can also be used for maintaining the groundwater levels. In addition to it also prevent the loss of water through evaporation and seepage (Meter et al. 2014). As according to the hydrological cycle rain is a primary source of water. All the other water sources like rivers, lakes, and groundwater are all secondary sources of water. Rain is the primary source which feeds into all other secondary sources. There are many cities which do get a lot of annual rainfall but still, they are in water shortage, which is only due to our failure to realize the importance of rainwater. By using the various water harvesting techniques, we can make the maximum use of rainwater to obtain self-sufficient, without being dependent on remote water sources. Rainwater harvesting system is very important in the present, based on the current scenario it is to be implemented immediately in order to capture the rain failing over a region and is stored to its maximum holding capacity, which can be used directly for the household purpose and various other purposes (Traboulsi and Traboulsi 2015). Many of the urban cities experiencing the great floods every year in rainy seasons and on another hand acute water shortage. The one and only solution to this water crisis is rainwater harvesting, where surface water is insufficient to encounter our demand and utilization of groundwater resource has resulted in a decline in water levels, only rainwater harvesting methods application can save us from the future serious consequences (Tamaddun et al. 2018).

14.5 Analyzing the Rainwater Harvesting Potential of an Area

Rainwater endowment of an area is the the total amount of water that is received by that area in the form of rainfall. Water harvesting potential is the amount of water which can be harvested efficiently. There are various factors which affect the rainwater harvesting potential of an area, such as: -

- **Rainfall quantity**: Rainfall is the most detrimental factor to decide the potential rainwater supply for a given catchment, for this a reliable past rainfall data is required, preferably for a period of at least 10 years. We can also use the rainfall data of an area with the similar climatic condition as a reference (Traboulsi and Traboulsi 2015; Stephens 1994).
- **Rainfall Pattern**: This is the most variable and unpredictable factor for describing the rainwater harvesting potential of an area. As lesser the availability of rain or a long dry period suggests more need for rainwater collection in a region (Rahman 2017). For such conditions, it is more feasible to use rainwater to recharge groundwater bodies or underground water table rather than for surface storage.
- **Catchment Area features**: – The catchment area determine the storage conditions for the harvesting facility. There are calculations required in order to decide the production of catchment systems; it involves the use of **runoff coefficient** (*it takes the fact that all the rainfall falling on a catchment cannot be collected, the loss will be in the form of evaporation and retention on the surface itself. In this the size and texture of the catchment area also act a contributing factor, where the roofing material texture may slow down the flow; evaporation; and affect the collection process*) to access the losses due to leakage, permeation, surface wetting, and evaporation, which ultimately reduce the amount of runoff (Farreny et al. 2011).
- **Runoff coefficient** for any catchment is the ratio of the volume of water that runs off a surface to the volume of rainfall that falls on the surface(http://www.fao.org/docrep/u3160e/u3160e05.htm)

  Water harvesting potential = Rainfall (mm) × Area of catchment x Runoff coefficient

  OR

  Water harvesting potential = Rainfall (mm) × Collection efficiency

The **collection efficiency** includes the fact that all the total rainwater falling over an area cannot be harvested at its full potential, because of various factors like evaporation, leakage, etc. For calculating the collection efficiency, the runoff coefficient of an area is used.

For e.g., a building with a rooftop catchment area of 200 m^2, receives an average annual rainfall of about0.3 m. Area of catchment = 200 m^2.

Height of rainfall = 0.3 m.

Total rainfall volume = **Catchment Area** × **Height of rainfall** = 200 m^2.× 0.3 m = 60 mm^3 = 60,000 L.

Let's assume that only 50% of the total rainfall is efficiently harvested,
Total volume of water harvested = 60,000 L × 0.5 = 30,000 L.

- **Storage of rainwater: -** The harvested rainwater can be stored for direct use or can also discharge into the groundwater bodies. The choice to store or recharge ground water is determined by the rainfall pattern of that specific region. For an area of a short dry span, water tanks can serve as the storing body and can be used for domestic purposes. Whereas for those areas where the total annual rainfall duration is small, it is more reasonable to recharge groundwater aquifers rather than use the rainwater for storage. Ultimately the selection should be made carefully in order to ensure maximum harvesting of the rainwater.

14.6 Sites for Harvesting Rainwater

Rainwater harvesting can be done on following surfaces: Rooftops: buildings with water-resistant roofs provide the catchment area for collecting water. Also, utilizing the landscapes, open fields, parks, storm water drains, roads and pavements and other open areas for effectively harvesting the runoff is also a part of rainwater harvesting (Mbilinyi et al. 2005). The ultimate advantage is that besides reducing the loads from water bodies such as lakes, rivers, tanks, and ponds, it increases their water level. This harvested rainwater also replenishes the groundwater table (Meter et al. 2014). There are many simple and cost-effective means for harvesting rainwater systems depending upon the rainfall a particular area is getting annually (Table 14.1).

14.7 Types of Rainwater Harvesting Systems

A typical rainwater harvesting system involves these basic elements: the collection system, the conveyance system, and the storage system. The collection system may be of simpler types in the household system and bigger systems are installed in the

Table 14.1 Average annual rainfall in different regions

Region	Annual rainfall (mm)
Desert	**0–100**
Semi-desert	**100–250**
Arid	**250–500**
Semi-arid	**500–750**
Semi-humid	**900–1500**
Wet tropics	**Over 2000**

Source: Van Hatum and Worm (2006)

industrial sector having a large catchment area thatserves for the purpose of reservoir from where the water is either used for recharging the groundwater bodies or it is pumped into different water treatment plants. Rainwater harvesting systems have various categories depending on factors such as the size as well as nature of the catchment areas and also depends that the systems are present in the urban or in the rural settings. Such as:-

- **Simple Roof Water Collection Systems:-**The cisterns are the main components of a simple roof water collection system. There are cost-effective systems in which the reservoirs are made with Iron-cement, etc. Here, the harvested rainwater may be subjected to filtration or disinfection.
- **Larger Systems for Industries, Institutions, Airports, and Other Facilities: -** In the case of larger facilities the overall setup becomes more complicated. The harvesting can be done on multiple levels such as on roofs and grounds, and the storage can be built in the form of underground reservoirs followed by treatment and then use for non-potable applications.
- **High-rise Buildings in Residential Areas: -**In such cases, the roofs can serve as the catchment facility and the collected roof water can be retained in distinct cisterns on the rooftops for other uses.
- **Land Surface Catchments**: - As compared to the rooftop catchment techniques, ground surface techniques are the broader service provider. As the flow of water retained by small reservoirs (either on surface or underground) which is a low-cost, effective idea can reduce water shortage in dry periods (Roman et al. 2017). This is mainly used for agricultural purposes as a larger part of water gets seeps into the ground.
- **Collection of Storm Water**: - The stormwater can be collected in reservoirs, and these reservoirs need to be subjected to various pollution controlling policies as they are at high risk of contaminations (Roman et al. 2017).

14.8 Development of Rainwater Harvesting System

The process of rainwater harvesting consists of collecting rainwater, guiding it to a suitable location, followed by filtering it if to be directly used and storing it for further use (Mbilinyi et al. 2005). For storage purposes tanks, jars, ponds or lakes are the potential candidates. Otherwise, it could be used for groundwater recharge (Sharda et al. 2006). The most common domestic rainwater harvesting encompasses the rooftop catching of and storage in tanks/reservoirs/groundwater aquifers. It could also serve as conservation of rooftop rainwater in urban areas and utilize it in recharge of groundwater. Which needs an outlet pipe connecting the rooftop to a storage body in the form of existing well/tube well/bore well or a specially designed well.

Rainwater harvesting involves structure development from a small rooftop to large areas such as that of institutions and industries. Among all the most critical step to keep the system working well is to maintain the hygiene and sanitation, where all the catchments should be clean, separating all kinds of solid, liquid and other waste from harvested water. As there is large number of methods available for rainwater harvesting and each of them is particular to a specific site.

14.9 Components of a Rainwater Harvesting System

All rainwater-harvesting systems comprise six necessary components irrespective of the size of the system.

- **Catchment area/roof**: - It comprises the surface upon which the rain falls; responsible for catching the rainwater, where the roof has to be correctly inclined towards the direction of storage and recharge. As the catchment area consisting of rooftop area where the rainwater is supposed to be collected, so it should be cleaned appropriately (Farreny et al. 2011; Traboulsi and Traboulsi 2015).
- **Gutters and downspouts**: - This includes the transport channels from the catchment surface to storage; these have to be carefully designed **depending on site, rainfall characteristics and roof characteristics** (Farreny et al. 2011**).** These connections are to be constructed, so they carry the collected water to the recharge location.
- **Leaf screens and roof washers**: - These are systems which are in control for the removal of impurities and debris. To avoid the entry of leaves and other debris into the storage system, the transport system should have continuous leaf screen.
- **Cisterns or storage tanks**: - It includes the large jars, tanks, etc. where the collected rain-water is firmly stored for usage or then used for recharging the groundwater through open well channels, bore wells or percolation pits, etc. (Roman et al. 2017). The storage tanks for collecting rainwater may either be located above or below the ground, where they are a part of building or present as a separate structure at some distance from the building (Mbilinyi et al. 2005). The tank should be made up of inert material to avoid stored water quality degradation.
- **Conveying**: - It involves the transport system for the treated rainwater, either by gravity or pump. As the water pressure for a gravity system depends on the difference in elevation between the storage tank and the faucet. As the water gains, 1 psi of pressure for every 2.31 feet of rising or lift and hence designs of the structure should be made accordingly.
- **Water treatment**: - in order to improve the quality of rainwater it is subjected to several types of filters for the removal of solids and organic material (Rahman 2017). These types of treatment (Table 14.2) includes filtration, disinfection, and buffering for controlling the pH. For drinking and cooking purposes, filtration followed by some disinfection is necessary like by using ultraviolet light or chemical disinfection (Rani et al. 2012).

Table 14.2 Types of treatment techniques

Method	Location	Result
Screening		
Strainers and leaf screens	Gutters and leaders	Prevent leaves and other debris from entering the tank
Settling		
Sedimentation	Within tank	Settles particulate matter
Filtering		
In-line/multi cartridge	After pump	Sieves sediment
Activated carbon[a]	At tap	Removes chlorine
Reverse osmosis	At tap	Removes contaminants
Mixed media	Separate tank	Traps particulate matter
Slow sand	Separate tank	Traps particulate matter
Disinfecting		
Boiling/distilling	Before use	Kills microorganisms
Chemical treatments (chlorine or iodine)	Within tank or pump	Kills microorganisms (liquid, tablet or granule)
Ultraviolet light	Should be located after the activated carbon filter before the trap	Kills microorganisms
Ozonation	Before tap	Kills microorganisms

Source: Ray. K et al. Rainwater Harvesting and Utilization: - Blue Drop Series 2
[a]Should only be used after chlorine or iodine has been used as a disinfectant. Ultraviolet light and ozone systems should be located after the activated carbon filter but before the tap

In summary, the rainwater harvesting systems production differs with the size and texture or the material used for the construction of the catchment area, Where roofing material is smoother and impermeable and thus provides an increase water quality and also in greater quantity (Farreny et al. 2011) (Fig. 14.1).

14.10 Benefits of Rainwater Harvesting

The resources which are provided to us by Mother Nature are limited, so we should use them judiciously as they will remain available to our future generations as well. During the last centuries we have exhausted all our water resources thus now we are facing their extreme shortage. So, it is about time that we should lay more emphasis on how much we should use and we can use our reminant resources more economically in (Vieira et al. 2014). Earth consists of about 70% of water mostly in the form of the ocean. We have rivers, lakes, glaciers, and groundwater as reservoirs which provide us with serviceable water, which gets restocked on a regular basis when it rains. This comprises of total 1% of water, i.e., drinkable and limited, and with our ever-growing population, this will not be enough.

- The ecological benefits of rainwater harvesting are colossal: where a simple usage of water in toilet flushing accounts for about 35% of the average household's

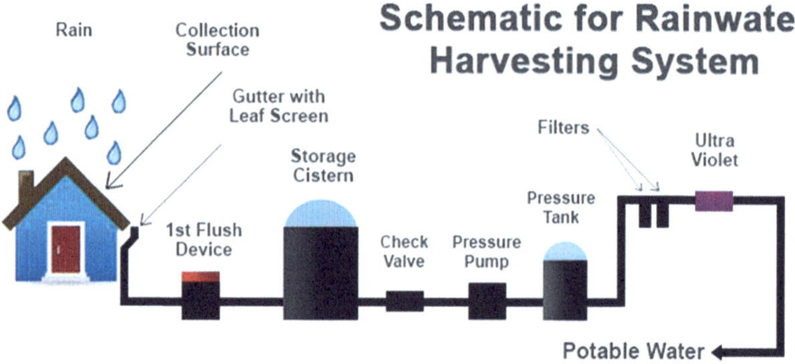

Fig. 14.1 Components of rainwater harvesting system. (Source:- Google Images)

water usage. A further addition to this in the form of washing clothes or watering the garden and the possible savings begin to grow. As for the industrial purposes, these figures are much higher. With rainwater harvesting, we can reduce wastage of fresh water to a significant level (Singh et al. 2014).

- The foremost ecological advantage obtainable from the use of rainwater harvesting is to reduce the pressure on natural sources and also this harvested water can be used for drinking, bathing and cooking purposes. Besides that it also reduces the load on drainage systems, prevents soil erosion reducing the calamities like a flood (Farreny et al. 2011), and that water can be put through recycling purposes.
- Rainwater harvesting also comprises financial benefits it is not necessary to pay the bill for the collected rainwater. But a large number of population is adapting the metered water provision, as to save a large sum of money by avoiding or limiting the use of water. By using the collected rainwater for domestic purposes, about 40–50% reduction can be made on water bills.
- Rainwater harvesting systems are fairly easy to maintain because they are not dependent on any installation of any expensive technology and monitoring. In most of the cases, the buildings are provided with catchment area, capable of collecting a significant amount of water (Mbilinyi et al. 2005).

14.11 Rainwater Harvesting as a Sustainable Water Approach

- **Restoration of Hydrological Cycle**: - The hydrological cycle takes place in nature at different levels, which depends upon climates, geography, and the present biological factors (Kawachi et al. 2005; Elhag and Bahrawi 2014). The ultimate goal of rainwater is to recharge the water bodies or groundwater table (Sharda et al. 2006), but due to the construction of concrete surfaces and asphalt structures, their impermeable surfaces have disrupted the natural hydrological cycle, and reduce the amount of rainwater permeating underground. Which finally leads to the overflow in drainage system as water is unable to reach underground, causing huge problems like floods. Also, the activities like deforestation, pollution causing excessive heating and contamination of water bodies ultimately leading to drying of water bodies in cities. For the betterment of the situation, sustainable development with the restoration of the urban hydrological cycle is required (Elhag and Bahrawi 2014). For this conserving or restoring the natural water in the form of rainwater and facilitating its permeation to groundwater sources is required which finally leads to maintaining natural groundcover and greenery cycle (Rahman 2017) (Fig. 14.2).

- **Development of water sources in Remote Water Sources:-**There is a large number of cities which fulfill their water needs from farther distances, which is not sustainable. Practices such as building dams lead to flooding the houses, fields and wooded areas present at the downhill areas. Also, these dams remove the base sediments which reduces the quality of water collected.
- **Decentralized "Life-Points," Versus the Conventional "Life-Line" Approach:-** Due to increased urbanization, the dependence of large cities on remote water resource has been dramatically increased, and even if rainfall in these upstream water sources gets compromised, the function of these cities gets affected (Roman et al. 2017). Similarly, the centralized site of water supply pipeline (or "life-line") is susceptible to large-scale natural disaster. Therefore, a number of widely distributed water resource "life-points" within a city should be present to draw rainwater and groundwater, providing the city with greater flexibility in the face of water shortages and natural calamities. Hence, a shift from centralized "life-line" to decentralized "life-points" should be encouraged to ensure the crucial water supply at the time of crisis
- **Cycle capacity: -** In order to restore the hydrological cycle one should consider the perspective of sustainable development. With this, the concept cycle capacity refers to the time taken by the nature to recuperate the hydrological cycle (Meter et al. 2014). This concept should be utilized wisely in terms of exploiting the natural water source of an area considering the amount of time required by the water body to revive itself, in short overexploitation can be avoided with this. As rainwater seeps underground and over a period of time it becomes shallow stratum ground water and then becomes deep stratum ground water. To carry out the sustainable use of groundwater, an over-extraction within a short period of time should be avoided (Rahman 2017).
- **Water demand management:-**As the established cities continue to expand regarding population and so in terms of demands of water supply also. The cost of these water utilities in cities recovered through water rates. But when water is abundant in the resource area, the concept of conservation is usually ignored. Which leads to conflict at the time of drought. Due to the lack of promotion of water conservation and rainwater harvesting such a crisis occur more frequently.

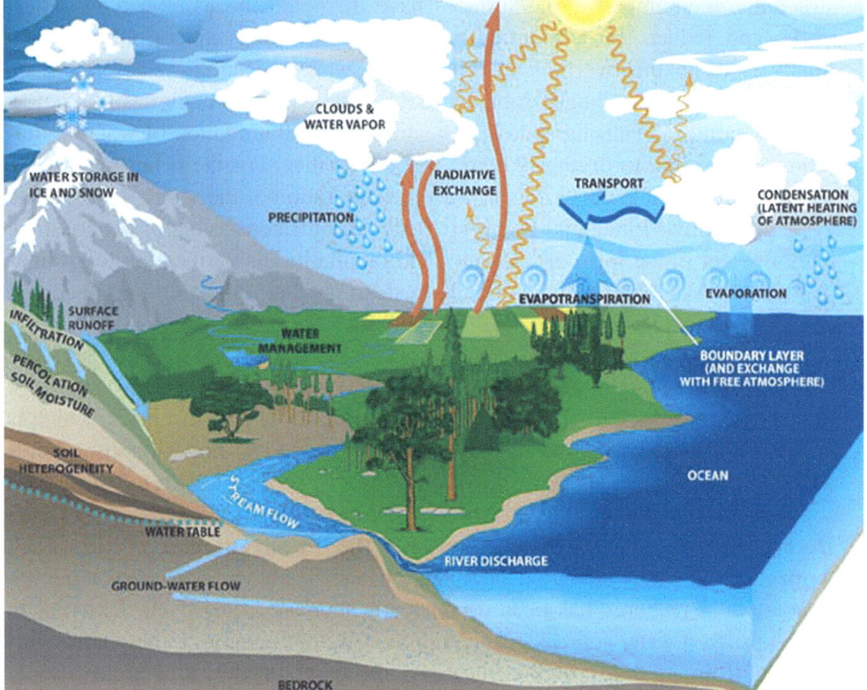

Fig. 14.2 Different stages of hydrologic cycle. (NASA Earth Science: wtere cycle weblink or article refernce)

For this sustainable use of water management programs are needed to be introduced, where the urban water supply requires a large change such that water supply should be in controlled demand (Meter et al. 2014). The application of demand-side management inspires the local population to accept water conservation approaches, including harvesting rainwater as well (Jatiya and Hosokawa 2004).

14.12 Future Perspective in Rainwater Harvesting

With the growth of population and the drastic changes in our climate, we need to realize that water is not an unlimited source, and soon it going to be depleted seeing how much we are exploiting our precious water sources (Kawachi et al. 2005). The idea of **Rainwater harvesting** comes as a boon to our societies, and now it's being revived in many forms for small houses to large-scale industries and buildings with the strong guidelines for sustainable use of water (Rahman 2017).

In a wider view, increased use of rainwater harvesting systems can lead to a more reasonable supply of the water, as it encourages the practice of a natural resource in the form of readily available rainwater instead of groundwater sources. Also, the rainwater harvesting system as a part of an integrated strategy helps in the management of our water needs along with maximizing the availability of water (Jatiya and Hosokawa 2004; Meter et al. 2014). Besides that, it may help us in improving our irrigation efficiency in agricultural field reducing the non-beneficial evaporation from soil or supply sources (Jatiya and Hosokawa 2004; Kawachi et al. 2005; Ward et al. 2012; Roman et al. 2017). However, further research is required to create rainwater harvesting model systems according to regional hydrological conditions (Meter et al. 2014). Considering the fact of coevolutionary dynamics, the human-water system interactions can help us in understanding the possibility of rainwater to affect the social and hydrological landscapes of water-stressed areas throughout the world.

- In the near future, there will be more sustainable and proficient systems or methods for collecting rainwater and linking it with the main sources to help reduce the pressure on them in areas with a dense population. Globally, as our need for water is continuously escalating with the exhausting water sources, sustaining a constant water supply is indeed an important task (Rahman 2017; Tamaddun et al. 2018). To revive these water resources, the cost-effective community-level water conservation and storage concepts provided by rainwater harvesting systems can play an immense role in meeting our water needs along with replenishing our water resources (Kawachi et al. 2005; Meter et al. 2014). There are various reasons to promote the rainwater harvesting as listed:-
- It provides a systematic approach as along with rainwater application, water conservation and wastewater management can also be achieved. For this strict laws and regulations should be implemented into municipal ordinances and regulations (Jones and Hunt 2010). It is necessary that every house and different industries should participate and incorporate the harvesting structure used in the rain water harvesting system. Also, the concerned governments should take the edge to promote the notion of rainwater harvesting as resource independent, economically efficient and natural hydrological cycle conserving idea (Kahinda et al. 2007; Elhag and Bahrawi 2014).
- It is vital to encourage the development of the effectual and reasonably priced for water conservation purposes. Together with this, there is an urgent need for have rain specialists to understand the usage of such an advanced system and devices (Jones and Hunt 2010).
- To promote rainwater harvesting and utilization, a comprehensive range networking system should be present connecting government machinery with citizens, architects, and technician and system developers (Kahinda et al. 2007; Jones and Hunt 2010; Traboulsi and Traboulsi 2015; Rahman 2017).

14.13 IRCSA: – International Rainwater Catchment Systems Association

In order to serve the purpose of rainwater harvesting on the global platform, an international organization was established in August 1989 by the fourth International Rainwater Cistern Systems Conference in Manila. It gets legally launched as **IRCSA** (International Rainwater Catchment Systems Association) in August 1991 at the fifth conference in Taiwan (http://ircsa.in). It intends to encourage and develop the rainwater catchment systems technology with appropriate planning, management, research technology, and educating worldwide. For this, it also found an international forum for scientists, engineers, educators, and administrators concerned in this field for setting up such international guidelines for this technology and also, provide new studies or techniques, publicize new information; and also collaborate with international awareness programmes for conserving our natural resources (Mbilinyi et al. 2005).

References

American Rainwater Catchment Systems Association. http://www.arcsa.org/. Accessed 20 Dec 2017

Center for Science and Environment. http://www.cseindia.org/. Accessed 12 Dec 2017

Dreiblatt D (1982) United Nations, ground water in the Eastern Mediterranean and Western Asia. Nat Res Forum 6:369–370

Elhag M, Bahrawi J (2014) Potential rainwater harvesting improvement using advanced remote sensing applications. Sci World J 2014:1–8

Farreny R, Morales-Pinzón T, Guisasola A, Tayà C, Rieradevall J, Gabarrell X (2011) Roof selection for rainwater harvesting: quantity and quality assessments in Spain. Water Res 45:3245–3254

Guangfei Lo A (2015) Rainwater harvesting: global overview. Rainwater Harvesting Agric Water Supply:213–233. https://unhabitat.org/books/blue-drop-series-on-rainwater-harvesting-and-utilisation-book-3project-managers-and-implemetation-agency/

International Rainwater Catchment Systems Association. http://ircsa.in/. Accessed 10 Dec 2017

Jatiya V, Hosokawa T (2004) Significance of rainwater harvesting in mega-diversity water resource country: India. J Rainwater Catchment Syst 10:21–28

Jones M, Hunt W (2010) Performance of rainwater harvesting systems in the southeastern United States. Resour Conserv Recycl 54:623–629

Kahinda J, Rockström J, Taigbenu A, Dimes J (2007) Rainwater harvesting to enhance water productivity of rainfed agriculture in the semi-arid Zimbabwe. Phys Chem Earth Parts A/B/C 32:1068–1073

Kawachi T, Aoyama S, Yangyuoru M, Unami K, Matoh T, Acquah D, Quarshie S (2005) An irrigation tank for harvesting rainwater in semi-arid Savannah areas: design and construction practices in Ghana/West Africa. J Rainwater Catchment Syst 11:17–24

Mays L, Antoniou G, Angelakis A (2013) History of water cisterns: legacies and lessons. Water 5:1916–1940

Mbilinyi B, Tumbo S, Mahoo H, Senkondo E, Hatibu N (2005) Indigenous knowledge as decision support tool in rainwater harvesting. Phys Chem Earth Parts A/B/C 30:792–798

Meter K, Basu N, Tate E, Wyckoff J (2014) Monsoon harvests: the living legacies of rainwater harvesting systems in South India. Environ Sci Technol 48:4217–4225

Minkley G (2012) Rainwater harvesting, homestead food farming, social change and communities of interests in the eastern cape. S Afr Irrig Drain 61:106–118

Qureshi A, McCornick P, Sarwar A, Sharma B (2009) Challenges and prospects of sustainable groundwater Management in the Indus Basin. Pak Water Resour Manag 24:1551–1569

Rahman A (2017) Recent advances in modelling and implementation of rainwater harvesting systems towards sustainable development. Water 9:959

Rainwater Harvesting and Utilisation.. An environmentally sound approach for sustainable urban water management: an introductory guide for decision-makers. http://www.unep.or.jp. Accessed 2 Jan 2018

Rani D (2012) Rainwater harvesting practices: a key concept of energy-water linkage for sustainable development. Scientific Research and Essays 7:538–543

Ray K, Dzikus A, Singh K, Mathur PS, Prasad D, Sharma DK., Un Habitat, Rainwater harvesting and utilization: - Blue drop series, Book 2

Roman D, Braga A, Shetty N, Culligan P (2017) Design and modeling of an adaptively controlled rainwater harvesting system. Water 9:974

Sharda V, Kurothe R, Sena D, Pande V, Tiwari S (2006) Estimation of groundwater recharge from water storage structures in a semi-arid climate of India. J Hydrol 329:224–243

Singh S, Samaddar A, Srivastava R, Pandey H (2014) Ground water recharge in urban areas – experience of rain water harvesting. J Geol Soc India 83:295–302

Stephens D (1994) A perspective on diffuse natural recharge mechanisms in areas of low precipitation. Soil Sci Soc Am J 58:40

Tamaddun K, Kalra A, Ahmad S (2018) Potential of rooftop rainwater harvesting to meet outdoor water demand in arid regions. J Arid Land 10:68–83

Texas Water Development Board. https://www.twdb.texas.gov/innovativewater/rainwater/. Accessed 15 Dec 2017

The Renewable Energy Hub. https://www.renewableenergyhub.co.uk/. Accessed on 7 Dec 2017

Traboulsi H, Traboulsi M (2015) Rooftop level rainwater harvesting system. Appl Water Sci 7:769–775

United Nation Environment Program. https://www.unenvironment.org/. Accessed 7 Dec 2017

Van Hattum T, worm J (2006) Rainwater harvesting for domestic use. Agromisa Foundation/ Digigrafi, Wageningen, Netherlands

Vieira A, Beal C, Ghisi E, Stewart R (2014) Energy intensity of rainwater harvesting systems: a review. Renew Sust Energ Rev 34:225–242

Ward S, Memon F, Butler D (2012) Performance of a large building rainwater harvesting system. Water Res 46:5127–5134

Yannopoulos S, Antoniou G, Kaiafa-Saropoulou M, Angelakis A (2016) Historical development of rainwater harvesting and use in Hellas: a preliminary review. Water Sci Technol Water Supply 17:1022–1034

Chapter 15
Adsorptive Removal and Recovery of Heavy Metal Ions from Aqueous Solution/Effluents Using Conventional and Non-conventional Materials

Ashitha Gopinath, Kadirvelu Krishna, and Chinnannan Karthik

Abstract The development of industries over the past few decades has turned the world into a modernized era. The ultimate pros of this industrialization are benefited by humans on one side whereas they are also severely affected by the cons on the other side. The intended release of industrial wastewater into the water bodies often contaminates it heavily with toxic substances particularly heavy metals, as they are non-biodegradable and persistent in the environment. Remediation of heavy metals by various physical and chemical approaches is not advisable as it is uneconomical and generates a large number of secondary wastes. Hence, utilization of low cost conventional and non-conventional adsorbents offers natural and eco-friendly statuary for metal removal. Hence, it is considered an efficient and alternative tool for metal remediation. Based on the facts stated above, the present chapter described the sources and environmental significance of heavy metals as well as remediation strategies using low-cost adsorbents.

Keywords Wastewater · Heavy metals · Remediation · Adsorbent · Removal

15.1 Introduction

Contamination of the biosphere is a major environmental risk, which has been increased dramatically since the beginning of rapid industrialization and modern lifestyle. A wide variety of organic (hydrocarbons, volatile organic compounds and solvents) and inorganic (heavy metals) pollutants are continuously released into the environment by various industrial activities, long time usage of fertilizer and pesticides in agriculture, mining, burning of fossil fuels and sewage sludge amendments (Abdelatey et al. 2011). Among the pollutants, heavy metals play a crucial role on environment and living organisms due to its toxic nature. The term heavy metal

A. Gopinath · K. Krishna · C. Karthik (✉)
DRDO – BU – CLS, Bharathiar University Campus, Coimbatore, Tamil Nadu, India

M. Oves et al. (eds.), *Modern Age Waste Water Problems*,
https://doi.org/10.1007/978-3-030-08283-3_15

denotes to any metallic element with high density (higher than 5 g cm^{-3}) and toxic/poisonous to the environment at low concentrations. Heavy metals also called as trace elements, because of their presence in the environment at a trace level (mg/kg) or in ultra-trace level (μg/kg). Moreover, they are a heterogeneous group of elements, which greatly differ in their properties and biological functions (Mukesh et al. 2008). Heavy metals enter into the environment through natural (rocks, metalliferous minerals, erosion, volcanic activity, wind-blown dust, rain, snow, and hail) and various anthropogenic resources (metallurgy, energy production, microelectronics, mining, sewage sludge, waste disposal and agriculture). Heavy metals are deliberately released into the environment from anthropogenic resources (Wuana and Okieimen 2011). Moreover, the use of wastewater to irrigate agricultural land is another major environmental risk as it enters the edible as well as nonedible plants via soil posing serious health issues (Kumar Sharma et al. 2007, 2009). Due to these anthropogenic resources, heavy metals enter into the food chain. Heavy metals have been detected in various living forms like microorganisms, plant, animal and human alike through the food chain migration.

15.2 Heavy Metal Contamination and Toxicity

Out of naturally occurring 90 elements, 21 nonmetals, 16 light metals and the remaining 53 are considered as heavy metals. Majority of heavy metals are transition elements, with incompletely filled d-orbitals. These d-orbitals provide heavy metal cations with the ability to form complex compounds, which may or may not be redox-active. Among them, the most common toxic heavy metals are Lead (Pb), Cadmium (Cd), Copper (Cu), Chromium (Cr), Mercury (Hg), Zinc (Zn), Aluminum (Al) and Manganese (Mn) (Duruibe et al. 2007). Based on the biological properties, heavy metals are classified into four major groups such as essential, non-essential, less toxic and highly toxic heavy metals. Some of the heavy metals such as Co, Cr, Cu, Mn, Fe, and Zn are classified as essential metals based on their biological functions (Bruins et al. 2000). A lesser quantity of these metals are common in our environment and essential for the metabolic process, good health, growth, disease resistance, vigor production, and reproduction. On the other hand, elevated concentration can be a potential carcinogen to the environment (Gogoasa et al. 2006; Aelion et al. 2008). Heavy metals like Ba, Al, Li, and Zr, are classified as non-essential metals. These non-essential metals are not needed by living organisms for their metabolic process and growth. Similarly, metal ions like Sn and Al classified as less toxic heavy metals based on their toxicity nature. However, heavy metals like Hg, Pb, and Cd are classified as high toxic heavy metals. These metals are well-known toxicants even at lower concentrations, which affect cellular organelles and components mostly, involved in detoxification, DNA or cell damage repair and metabolism (Lim and Schoenung 2010; Mukesh et al. 2008). Because of its mutagenic and carcinogenic nature, heavy metals are considered as a "Priority pollutant"

(USEPA 1997). Toxicity and biological significance of the primary heavy metals have been discussed elaborately in Table 15.1. This imposes massive pressure on the society for the removal and recovery of it from the contaminated site. Henceforth, it is essential to establish a method suitable and practicable for the removal of heavy metals on-site, which can be performed by the common man as well.

In the last few decades, several heavy metal removal strategies such as ion exchange, reverse osmosis, electrodialysis, flocculation and chemical precipitation have been extensively explored for their effectiveness in the removal of metals in industrial effluents. However, these strategies become uneconomical and unsuitable for a large quantity of heavy metal removal from industrial wastewater. Hence, the development of an alternative, low cost, and eco-friendly strategies are highly desirable for heavy metal removal. In this scenario, utilization of various conventional and non-conventional adsorbents could be cost-effective and more efficient approach over the traditional methods. Thus, the current chapter has been discussed in detail about the efficiency of various conventional and non-conventional adsorbents in heavy metal removal from aqueous solution.

15.3 Conventional Technologies for Heavy Metal Removal

Given the above context, a vast array of techniques has been witnessed by the scientist's for the removal of heavy metals from aqueous solution. Selection of a method depends on many factors such as cost, time and easiness to operate, safety, efficiency, long term and short-term environmental effects. This section presents an overview of different techniques adopted for the elimination of heavy metals from wastewater. Heavy metals can be specifically removed with the aid of ion exchangers that employs acidic and basic resins (Sofinska-Chmiel and Kolodynska 2017). Each heavy metal requires the use of unique resins implying that multi-components cannot be removed in a single process or single resin. This makes the treatment process impractical for high-end applications. Chemical precipitation arises to be a well-known method due to its safety, ease of operation and it does not demand the use of any sophisticated types of equipment as well. The method employs a precipitating agent such as lime to convert the dissolved metal ions into an insoluble metal hydroxide (Marchioretto et al. 2005). However, it suffers from many limitations including slow settling rate, huge sludge generation and the expense incurred during its disposal. A similar method which generates an excessive amount of sludge is coagulation-flocculation since it uses a large amount of coagulating agents such as alum and ferric salts to increase the particle size (floccules) for fast settling (Ferhat et al. 2016). Henceforth, the above mentioned two methods are not acceptable for large-scale remediation purposes. The reverse process of coagulation involves flotation or collection of heavy metals from suspension when it is in the floating stage rather than by settling of particles (Salmani et al. 2013). This reduces the retention time but utilizes a huge amount of collectors to remove low

Table 15.1 Toxicity effects of major heavy metals on living forms

Heavy metals	Source	Effects on human	Effects of plant	Effects on microorganisms	Reference
Ar	Industrial manufacturing, wood preservatives, cosmetics, agrochemicals, glass and glass wares	Respiratory, nervous and cardiovascular disorders, skin cancer, dermatitis, conjunctivitis, brain, and kidney damage	Reduce the metabolic activity, inhibition of growth, and photosynthesis rate and induced the oxidative stress	Inhibit the cell division and growth, enzymatic activities	Bissen and Frimmel (2003)
Cd	Agro-chemicals, pigments, and plastics, mining, refining, welding	Bone & kidney diseases, gastroenteric distress, and pain, lung and prostate cancer, lymphocytosis, coughing, emphysema, and vomiting	Reduce the nutrient uptake, alter the stomatal function and photosynthesis, inhibition of seed germination rate and growth behavior	Long time exposure induces DNA and protein damage, inhibit the transcription and cell division	Nagajyoti et al. (2010), Fashola et al. (2016)
Cr	Dyeing, tanning, electroplating, textile, steel fabrication and paint	Lung cancer, liver diseases, nausea, skin irritation, itching of the respiratory track, inflamed mucosa, reproductive toxicity, bronchopneumonia, chronic bronchitis, diarrhea	Inhibition of seed germination rate, growth, and photosynthetic rate, affect the antioxidant expression and induce lipid peroxidation and proline	Growth inhibition, elongated lag phase, modulation in physiological and biochemical processes	Karthik et al. (2016), Cervantes et al. (2001)
Cu	Industrial machinery, electrical equipment, mining, painting, combustion of fossil fuels and wastes, wood production	Liver and kidney disorders, modulating the metabolic activities, nose, mouth, and eyes irritation, diarrhea, vomiting, stomach cramps, headaches, dizziness, and nausea,	Growth retardation, imbalanced antioxidant expression, chlorosis and reduce the water and nutrient transport and genotoxicity	Reduce the size of microbial biomass, enzymatic activities and disrupt cellular function	Nagajyoti et al. (2010), Salem et al. (2011)
Co	Smelting process, phosphate fertilizers, fossil fuels, vehicular exhaust and other industrial activities	Lungs disorder including asthma, pneumonia and wheezing, vomiting, skin rashes, allergy and hearing problems	Decrease plant growth, biochemical functions and change the antioxidant system expression	Induce cell surface modifications, imbalanced metabolic process	Gopal et al. (2003)

Mn	Mining and mineral processing, emissions from ferroaa lloy, steel and iron production as well as combustion of fossil fuels	Neurodegenerative disorder, behavioral changes, lungs irritation, loss of sex drive and sperm damage.	At higher concentration decrease growth rate, induce chlorosis, necrosis, modulating both enzymatic and non-enzymatic antioxidant expression,	At elevated concentration inhibit the growth and metthe abolic process of the cells	Zhao et al. (2012), Moroni et al. (2003)
Pb	Batteries manufacturing, mining, paint and pigment industries, automobile exhaust of leaded petrol and soldiering	Modulation in various enzymatic function, high blood pressure, neurons damage, reduced fertility, anorexia, hyperactivity and damage in the renal system	Affect seed germination, growth, antioxidant system, photosynthesis, and chlorosis	Cell membrane disruption, denaturation of DNA and protein and inhibits transcription	Wuana and Okieimen (2011)
Hg	Coal burning, mining, paint, paper, electrical and electronic industries, methylmercury fungicides and batteries	Psychological changes, neurological damage, Minamoto disease, fertility problems, gastrointestinal irritation and loss of immunity	Retardation of growth, inhibition of seed germination, growth and photosynthesis rate, modulate the antioxidant system and affect nutrient uptake	Inhibit cell cycle, reduce the biomass and disrupt cell membrane biochemical processes	Fashola et al. (2016), Wuana and Okieimen (2011)
Ni	Electroplating, power plants, and trash incinerators,	Allergy, lung and nasal cancer, kidney disease, nausea, cardiovascular diseases, asthma	Decrease the growth, biomass, chlorophyll content, reduce enzymatic activities induce reactive oxygen species and genotoxicity	Affect the morphological and physiological behavior, inhibit enzyme activities and oxidative stress	Chibuike and Obiora (2014), Fashola et al. (2016), Malik (2004)
Zn	Mining, oil refinery, steel production, fertilizer, coal burning, burning of wastes and fertilizers.	Anemia, kidney and liver damage, stomach cramps, skin irritation, vomiting, nausea, and depression	Affect the reproductive percentage, induce reactive oxygen species, physiology, morphology, and genotoxicity, reduced the germination, photosynthesis and chlorophyll content.	Reduce the growth rate and biomass, induce cell death (at higher concentration) and inhibit the cell cycle.	Chibuike and Obiora (2014), Gumpu et al. (2015)

concentration of heavy metals efficiently. Membrane filtration, the process that underlies the size of the particles is another option to separate heavy metals from contaminated water. Specific particles of varying size could be retained depending on the nature of the filtration membrane, which is being categorized as ultrafiltration (UF), nanofiltration (NF) and reverse osmosis (RO) (Vinodhini and Sudha 2017; Mehdipour et al. 2015; Mohsen-Nia et al. 2007). UF retains macromolecules but rejects low molecular weight compounds whereas NF process lies in between that of UF and RO. RO involves pressure dependent rejection of water and retention of heavy metals, which makes it more suitable for the different membrane processes. Although membrane filtration can be used at large scale, its long time efficiency is doubtful, as fouling is a common problem encountered infiltration processes that necessitates frequent cleaning and replacement of membrane. This leads to operational as well as maintenance costs, which makes the process non-economical. A more recent technique, which has been practiced, is an electro-based treatment that includes electrolysis, electroprecipitation, electrodialysis (Gherasim et al. 2014; Yasri and Gunasekaran 2017) and electrocoagulation (Mansoorian et al. 2014). This technique is known to reduce treatment time but demands a lot of energy which boosts up capital as well as maintenance cost and requires trained personnel. Adsorption turns out to be a viable option due to its simplicity, utility, safety, feasibility and more importantly its economical nature (Chen et al. 2012). In addition to this, adsorption is the individual process that can be employed to extract a soluble material from solution as it provides an option for desorbing the adsorbed materials for the further use of adsorbent and adsorbate. It is important to study this possible method, which is discussed in the following sections.

15.4 Adsorption

Adsorption means the mass transfer of a molecule from the aqueous or gaseous matrix to a solid interface but a complex process involving many factors (Fig. 15.1). It is a phenomenal surface- interplay between two components namely adsorbent and adsorbate. A solid surface acts as the adsorbent on which an adsorbate occupies its position depending on the favorable conditions. Adsorption takes place in two different means in which the adsorbate associates with adsorbent either through a physical attraction (physisorption) or forms a chemical bond with the latter (chemisorption). The former process is a rapid process involving multilayer adsorption and is of reversible nature whereas the latter is an irreversible reaction with a single layer adsorbed specifically on the adsorbent. This purely depends on the nature of affinity between adsorbate and adsorbent specific sites as few metal ions occupy specific sites whereas the rest compete for non-specific sites (Jiang et al. 2010). Adsorption continues to occur till equilibrium is reached between metal ions (adsorbate) and the adsorbent. The equilibrium condition relies on whether the adsorption process is in static mode or dynamic mode (Goel et al. 2005).

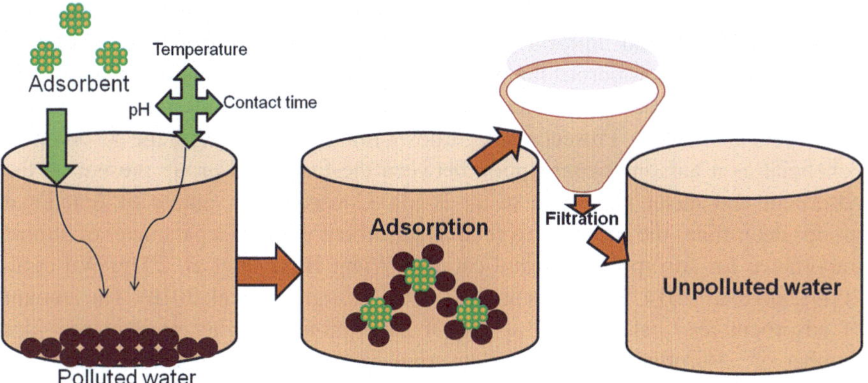

Fig. 15.1 Overview of adsorption

15.4.1 Factors Affecting Adsorption

The complexity of the adsorption process relies typically on the internal and external factors surrounding the adsorbent and adsorbate. The governance of the internal factors such as nature of adsorbent and adsorbate necessitates the proper selection of both whereas the pH, temperature and contact time being the external factors assist in driving the adsorption towards the equilibrium condition. These factors are discussed in the subsequent sections.

15.4.1.1 Nature of Adsorbate

The size and charge of the metal ions have great influence on its adsorption on a substrate. Metal with higher ionic radius subside on the surface of adsorbent and introduce steric hindrances resulting in fast saturation of adsorbent sites (Kadirvelu et al. 2000). Heavy metal ions have a positive charge on their surface and are likely to get attracted to an adsorbent surface possessing negative charge and vice versa (Pillay et al. 2009). Adsorption generally increases with increase in initial adsorbate concentration since the driving force for diffusion is larger in the initial stage and continues until the adsorbent sites are saturated with the adsorbate (Najafi et al. 2012).

15.4.1.2 Nature of Adsorbent

An adsorbent is characterized by several physicochemical properties that determine its adsorption capacity for a metal ion. Primarily, the surface area and pores size play a major role in deciding the adsorbent capacity. The high surface area represents number of adsorption sites available for the attachment of metal ions and enables faster kinetics. Mesoporous materials have gained popularity in adsorption due to its large surface area, and uniform pores distribution (Aguado et al. 2009; Li et al. 2011). However, microporous materials present much more efficiency in adsorption when compared to mesoporous materials as it exhibits faster film transfer diffusion (Kadirvelu et al. 2000). Presence of significant functional groups on the adsorbent is beneficial as it aids in chemisorption between the functional group present on the adsorbent and metal ions (Jain et al. 2013). Consequently, nature of functional groups determines the surface charge of an adsorbent which is a parameter of utmost importance for adsorption of metal cations (Wang Hongjie et al. 2009; Wu et al. 2009; Jain et al. 2014; Karthik et al. 2016, 2017; Zarghami et al. 2016). The amount of adsorbent also determines the rate of adsorption as higher amount provides number of adsorption sites thereby ensuring higher removal of cations (Motsi et al. 2009).

15.4.1.3 pH

pH is one of the crucial factors that govern the rate of adsorption of metal ions in solution. Generally, at lower pH the concentration of H^+ ions is very high; hence, the competition between metal ions for the adsorbent would be high. This greatly reduces adsorption rate at low pH whereas at higher pH, the reduction in competition ultimately favors adsorption of metal ions efficiently (Kadirvelu et al. 2000; Wingenfelder et al. 2005; Wu et al. 2009). Adsorption trend is severely changed if the pH goes beyond neutral, as the metals tend to precipitate due to the presence of more hydroxyl ions. This reduces the solubility of metals in water and hence a higher amount of metal uptake in hydrated forms (Malandrino et al. 2006). Moreover, functional groups present on the adsorbent alter the surface charge at various pH as ionization occurs and aid in the binding of metal ions process via ion exchange (Argun et al. 2007; WANG Hongjie et al. 2009; Karthik et al. 2016, 2017). The changes in adsorption pattern under various pH conditions are well depicted in Fig. 15.2.

15.4.1.4 Contact Time

Time is a factor that helps determine the strength of interaction between adsorbent and adsorbate (Jain et al. 2009). Generally, the initial process of adsorption is assumed to be very rapid but the time required to reach equilibrium varies with adsorbent and adsorbate (Meena et al. 2008; Abdel Salam et al. 2011). A good adsorbent with the large surface area, porosity, and ideal pH and temperature conditions drive the reaction in the forward reaction to attain equilibrium at a faster pace.

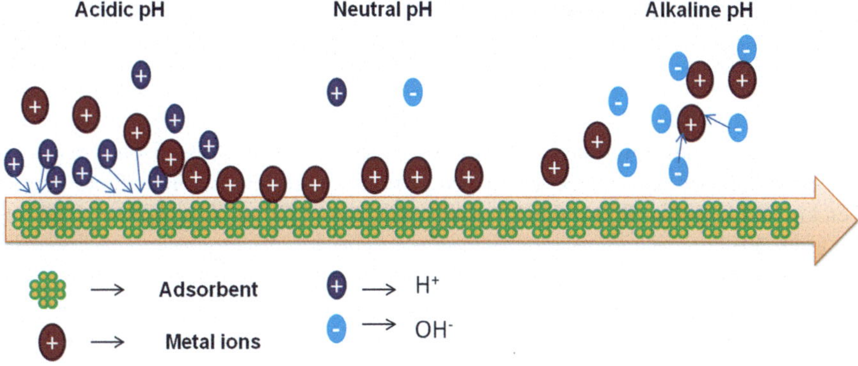

Fig. 15.2 Adsorption of metal ions at various pH

15.4.1.5 Temperature

The dependence of temperature on the adsorption of heavy metals on a substrate is being studied using thermodynamic parameters. It gives insight knowledge on the energy changes occurring during adsorption. The equations relate the equations relate the Gibbs free energy (ΔG), enthalpy change (ΔH), entropy change (ΔS) and temperature (T) in KelvinGibbs free energy (ΔG), enthalpy change (ΔH), entropy change (ΔS) and temperature (T) in Kevin.

$$\Delta G = -RT\ln K_l \left(\text{R is the molar gas constant } 8.314 \text{ J mol}^{-1} K^{-1} \text{ and } K_l \text{ is the Langmuir constant}\right)$$

$$\text{Also} \ln K_l = -\Delta HRT + C \text{and}$$

$$\Delta G = \Delta H - T\Delta S$$

It is imperative to determine the spontaneity of a reaction as it very well predicts the reaction rate which indirectly gives adsorption capacity for heavy metals with change in a temperature. A negative value of ΔG indicates that the reaction is spontaneous and it increases with increase in temperature (Eloussaief et al. 2009; Karthik et al. 2016, 2017). This suggests that the rate of heavy metal removal increases with temperature and vice versa. For a positive value of ΔS, an increase in adsorption was reported as the circulation of rotational and transitional energy increases (Argun et al. 2007; An et al. 2017). Positive value enthalpy is a strong indication of endothermic the nature of metal adsorption process. This can be explained in terms of regarding the reaction between water molecules and metal cations (Shaker 2014). However, the amount of water being displaced varies with different metals.

15.4.2 Adsorption Isotherms

Isotherms are linear graphs that give accurate information about the adsorbent and adsorbate. It usually represents the adsorption capacity of a material in terms of regarding the equilibrium concentration of metal ions. The classic isotherm equations that are used to describe adsorption process are Langmuir, Freundlich, Sipps, Temkin and Dubinin, Radushkevich (DR) (Xiang et al. 2016; Zarghami et al. 2016; An et al. 2017).

Langmuir model is a theoretical model and describes monolayer adsorption since it assumes adsorption takes place on a homogenous surface with all the sites equally available for metal ions (Jain et al. 2013). It is expressed according to the following equation: $q_e = \frac{q_m K_l C_e}{1+K_l C_e}$, where q_e is the equilibrium adsorption capacity in mmol g^{-1}, q_m is the maximum monolayer adsorption capacity in mmol g^{-1}, C_e is the equilibrium concentration in mmol L^{-1}, K_l is the Langmuir constant. Furthermore, a dimensionless parameter R_l could be related with Langmuir constant by the equation: $R_l = (1 + K_l C_0)$. If the value of R_l lies between 0 and 1, it represents favorable adsorption whereas $R_l > 1$ represents unfavorable adsorption.

Freundlich model is an empirical model, which describes multilayer adsorption on heterogeneous surfaces with different adsorption sites on the adsorbent possessing different adsorption energy (Jain et al. 2009, 2013). It is represented by the equation: $\ln q_e = \frac{1}{n} \ln C_e + \ln K_f$ where q_e is the amount of solute on adsorbent in mmol g^{-1}, C_e is the equilibrium concentration in mmol L^{-1}, K_f is the Freundlich constant which indicates adsorption capacity whereas n is constant indicative of adsorption intensity. Adsorption process is favorable if $n > 1$, whereas $n = 1$shows linear adsorption and $n < 1$depicts unfavorable conditions for adsorption.

The Sips model is a combination of Langmuir and Freundlich and is described by the equation as follows: $q_e = \frac{q_m K_s\, C_e^m}{1 + K_s C_e}$, Ks is the sips equilibrium constant and m is the dissociation parameter. Sips isotherm becomes Langmuir isotherm when the value of m is close to 1 whereas when the value is closer to 0, it becomes Freundlich isotherm.

Temkin isotherm describes adsorption as a uniform distribution of bonding energies up to some maximum binding energy and heat of adsorption decreases linearly with an increase in adsorbent-adsorbate interactions. The equation is generally represented as $q_e = BlnK_T + BlnC_e$ where $B = \frac{RT}{b_T}$ and it is related to heat of adsorption the n, K_T is the Temkin equilibrium binding constant, R is the molar gas constant, T is the temperature in Kelvin, b_T is the Temkin isotherm constant.

Dubinin- Radushkevich (DR) model is used to express the adsorption in terms of free energy of adsorption. This model helps to differentiate the physisorption and chemisorptions of metal ions onto a substrate. The prediction can determine from the following equation: $lnq_e = ln\ q_m - K_{DR}\varepsilon^2$, where K_{DR} is the Dubinin-Radushkevich constant, $\varepsilon = RTln\left(1 + \frac{1}{C_e}\right)$.

The apparent free energy is given by $E = (2K_{DR})^{-0.5}$, if the value of E is in between 1 and 8 $kJmol^{-1}$ then adsorption is governed by the physical attraction between adsorbent and adsorbate but chemisorptions happens if the value is exceeds above 8 $kJ\ mol^{-1}$.

15.5 Adsorption on Conventional Sources

Conventional adsorbents have been successfully used for the last few decades by the industries in the large-scale removal of contaminants. These adsorbents are known for their excellent surface adsorption properties due to its high surface area and porous nature they possess. Some of the conventional sources as well as its modified forms are discussed in the following sections.

15.5.1 Activated Carbon

Among the carbon family, activated carbon (AC) is the most prominently used porous adsorbent for the removal of contaminants from wastewater due to its large adsorption capacity. It is available in different forms including powder, granular, fiber and felt. A greater part of the investigations on this resourceful adsorbent has been done using granular in the batch model due to its ease of availability. However, column studies have also been found to be successful in removing heavy metals from water (Goel et al. 2005; Park et al. 2007; Natale et al. 2015). Modifications on AC has out lifted the adsorption capacity for the removal of Cd^{2+}, Pb^{2+}, Zn^{2+}, Mn^{2+}, and Cu^{2+} in batch as well as column mode (Park et al. 2007). Hachao Liu et al. (2014) has investigated the effects of humic acid modified AC and reported that increase in humic acid on AC increased its adsorption capacity for copper. Modification on AC by either sulfur or oxidation by nitric acid has proven to be advantageous for the removal of Cd^{2+} as the introduced functional aid in proper bonding of AC with the metal ion (Liang et al. 2016; Tajar et al. 2009). Advancement in the carbon industry gave rise to the addition of activated carbon in fiber form possessing high flexibility and low mass transfer resistance (Kadirvelu et al. 2000).

15.5.2 Silica Gel

Silica gel is the amorphous version of silicon dioxide and present in the form of granules or beads. The distinct feature of silica gel is its highly porous nature providing large surface area necessary for adsorption. Moreover, it is non-toxic, odorless, thermally resistant and does not undergo any side reactions with the adsorbate. Among the porous form, mesoporous are known to possess excellent features including large surface area, distinct pore size, and pore volume. This prompted the researchers to exploit them as an adsorbent for the removal of heavy metals. Functionalized silica is more prominently used to improve the affinity and specificity towards a particular metal ion (WANG Hongjie et al. 2009; Mureseanu et al. 2008). Silica gel modified with amine groups has shown significant adsorption of Cu, Cd, Pb, Ni and Zn from aqueous solution (Aguado et al. 2009). Chen et al. (2012) introduced amidoxime groups on silica gel by homogenous and heterogeneous methods have proved it to be an excellent material for the removal of specific removal Hg^{2+} ions from an aqueous solution containing Hg-Pb, Hg-Ag, Hg-Ni, Hg-Cu binary systems. Thiol-functionalized reusable silica adsorbent has been proved well effective for the removal of Hg and Pb from both acid and alkaline conditions (Li et al. 2011). Later, Ren et al. (2013) attempted to develop magnetic EDTA modified silica for the adsorption of Cu, Pb and Cd ions and it could be reused up to 12 cycles, and thereafter its efficiency decreased. Silica gel acts as an excellent support material for chelating agents for solid phase extraction as well as recovery of noble metals (Kondo et al. 2015).

15.5.3 Zeolites

Zeolites are porous aluminosilicates exhibiting crystalline nature with honeycomb structure formed by the assemblage of SiO_4 and AlO_4 via sharing of oxygen atoms. Due to the introduction of Si by Al, it possesses a net negative charge, which must be balanced by some cations. Low density, hydrothermal stability, resistance to the acidic solution and its abundance in nature makes it a widely used material for adsorption of heavy metals. A typical composition of the zeolite may contain SiO_2 (45.09), Al_2O_3 (14.43), Fe_2O_3 (10.59), CaO (5.79) and MgO (4.49) (Abdel Salam et al. 2011). The most commonly use zeolite clinoptilolite as it has gained immense research interest due to its higher thermal stability than other zeolites. The affinity of different zeolites towards metal ions varies depending on the composition of the former. The dependence of pH on Pb^{2+} removal varies with the type of zeolites used for adsorption (Kabwadza-Corner et al. 2015). Ibrahami and Sayyadi (2015) have well explained the use of natural and synthetic zeolites for the removal of heavy metal ions. Murthy et al. (2013) investigated the significance of different adsorption isotherms for the removal of Hg2+ using different zeolites.

15.5.4 Clay

Clay is composed of finely grained particles containing phyllosilicates, organic matter and few other matters that give plasticity to the clay. They are known to be hydrous aluminosilicates containing silicon-oxygen tetrahedron and aluminum octahedron arranged in hexagonal patterns. The four major types of clay materials used for the adsorption of metal ions from aqueous media are bentonite, montmorillonite, kaolinite, and sepiolite (Adeyemo et al. 2017). Abollino et al. (2008) investigated the adsorption behavior of Cd^{2+}, Pb^{2+}, Mn^{2+}, Cu^{2+} and Zn^{2+} on montmorillonite and vermiculite and reported the latter exhibited more adsorption capacity than the former. Kaolinite presents various adsorption sites responsible for the competition that exists in multielement system (Srivastava et al. 2005). For an equilibration time of 30 min, kaolinite presented strong ion exchange capacity towards Pb preceded by Ni, Cd, and Cu (Jiang et al. 2010). De-Pablo et al. (2011) reported that the adsorption capacity of montmorillonite and Ca-montmorillonite varies for different metal ions. Ca montmorillonite was found to be a superior adsorbent for Pb, Zn, Cr, Ba, Hg, Mn, and Ag, whereas montmorillonite exhibited good adsorption capacity for Cd, Cu, and Ni. Eloussaief et al. 2009 investigated the significance of pH and temperature for the adsorption of copper on pristine and acid treated clays.

15.6 Non – conventional Technologies for Heavy Metal Removal

Various conventional adsorbents have been found as potential adsorbents for metal removal from the aqueous environment. On the other hand, usage of these adsorbents is restricted because of its high operating cost. In this circumstance, where functional cost factor plays a crucial role, researchers are in need of low-cost eco-friendly adsorbent for wastewater treatment. Previously, few attempts have been made towards the development of cost-effective non-conventional adsorbents for heavy metals removal. These adsorbents can be classified based on their availability (agricultural waste, biological material, and industrial waste/by-products) and nature of the adsorbents (organic and inorganic materials). Figure 15.3 provides information about various non-conventional adsorbents for heavy metal removal.

15.6.1 Agricultural Waste/Byproducts

Recently, utilization of low-cost agriculture waste and byproducts as adsorbents for heavy metals is a growing interest for researchers. Agro-materials are considered as a promising material for heavy metal adsorption, due to its ease of availability and low cost (Jain et al. 2009, 2013, 2014; Meena et al. 2008). Various agro-materials such as rice husk, sugarcane bagasse, coconut waste, orange peel, sawdust, maize cob, jackfruit, sunflower head waste, sawdust, hazelnut and almond shell can be utilized as a low cost adsorbent for heavy metal removal after chemical treatment or conversion by heating into activated carbon (Ibrahim et al. 2006; Hameed et al. 2008; Hameed 2009). These agro-materials are mainly composed of cellulose,

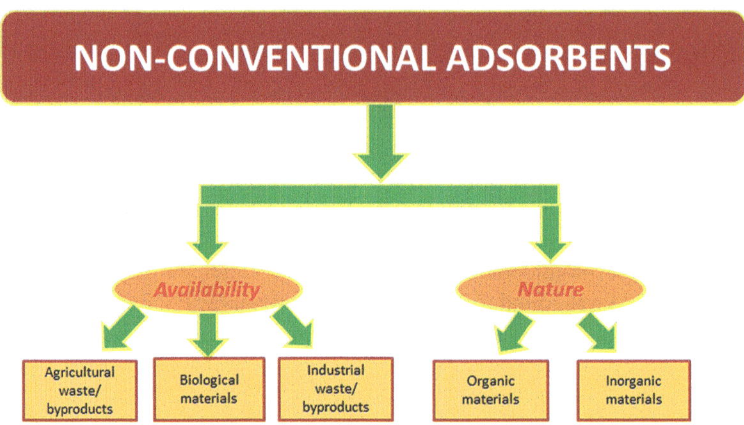

Fig. 15.3 Classification of non-conventional adsorbents

hemicellulose, lipids, protein, lignin, sugars components, water, and hydrocarbons, which consists of various functional groups such as carboxy, hydroxyl, sulphydryl, amide, and amine (Bhatnagar and Sillanpää 2010). In general, the adsorption process occurs through complexation, hydrogen bonding, and ion exchange. Easy availability, the presence of large surface area and functional groups make different agricultural waste and byproducts good alternative to the expensive commercial adsorbents for heavy metal removal.

These agro-materials can be used directly (natural form) or after some physical/chemical modifications. Different types of agents such as organic compounds (ethylenediamine, formaldehyde, epichlorohydrin and methanol), organic acids (hydrochloric acid, nitric acid, sulfuric acid, tartaric acid and citric acid) oxidizing agents (hydrogen peroxide), base solutions (sodium hydroxide, calcium hydroxide and sodium carbonate) and dyes have been used for pretreatment process (Grassi et al. 2012). Previously, various researchers demonstrated the heavy metal adsorption ability of agro-materials. Jain et al. (2009, 2013) successfully removed Cr from industrial wastewater by utilizing sunflower head waste as an adsorbent. Similarly, chemically modified sunflower waste has been used for Ni removal from aqueous solution (Jain et al. 2014). Zuorro et al. (2013) investigated the heavy metal adsorption ability of spent tea leaves for heavy metal adsorption. Spent tea leaves have adsorbed almost 2 g/l of Pb at 40 °C. Hegazi (2013) results highlighted that rice husk could be a better candidate for simultaneous removal of Fe, Pb, and Ni. Earlier, Meena et al. (2008) the suggested that mustard husk would be a better component for Pb and Cd removal from aqueous solution.

15.6.2 Biological Materials

Adsorption of heavy metals by biological materials is an emerging area in the field of water treatmentt as it is a passive uptake process and is mostly reversible and metabolism-independent processes. The adsorption of heavy metals onto the cell surface of biological materials (bacteria, fungi and algal biomass) is known as biosorption (Arivalagan et al. 2014). The significant advantages of microbial biosorption are their low operational cost, simplicity, highly effective and environmentally friendly methodology (Fomina and Gadd 2014).

Cell wall composition of the microorganisms plays a crucial role in metal adsorption. Functional groups such as hydroxyl, carboxyl, sulfonate, amide and phosphate groups are mainly the metal uptake process from aqueous solutions (Karthik et al. 2016, 2017). The anionic nature of microbial surface enables them to bind metal cations through electrostatic forces. Gram-positive bacteria contain a thicker cell wall composed of peptidoglycan, teichoic and teichuronic acid, whereas in Gram-negative bacteria, teichoic, teichuronic acid is absent and the peptidoglycan layer is thinner. Gram-positive bacteria are more efficient in trapping metal ions compared to Gram-negative bacteria. Moreover, funguses are well known to tolerate and detoxify heavy metal contaminated effluents. Fungal cell wall consists of chitin and other polysaccharides along with proteins, lipids, polyphosphates as well as inorganic ions cementing the cell wall. Owing to greater cell-to-surface ratio, fungi have a greater tendency to come in physical and enzymatic contact with the surroundings. Fungal biomass is a good sorption material, as it can be easily cultured on a large scale using simple fermentation technique (Congeevaram et al. 2007). Similarly, the living and non-living biomass of algae is a predominant candidate for metal adsorption. Metal adsorption in living algae biomass is more complex than nob-living biomass since adsorption and intracellular localization occur during the growth phase. On the other hand, uptake by non-living biomass takes place on the surface of the cell membrane, and it is considered as an extracellular process. Non-living algal biomass consists of sugar, cellulose, pectins, glycoproteins which are capable of binding to heavy metals as a cost-effective wastewater treatment. Several researchers have studied the usage of living and non-living microorganisms as adsorbent extensively (Arief et al. 2008).

15.6.3 Industrial Waste/by-Products

Industries are producing a large number of solid waste materials and byproducts, due to the numerous industrial activities. Various waste materials such as fly ash, red mud, magnetite, activated slags, iron waste and slags, hydrous titanium oxide and bagasse are being released into the environment, which can be utilized for heavy metal removal from wastewater (Ahmaruzzaman 2011). Industrial wastes are available easily, with free of cost and cause significant disposal problem. Utilization of these industrial waste/byproducts will provide twofold benefits in environmental pollution management. Previously, many researchers reported that the use of various industrial waste and byproducts in heavy metal adsorption from aqueous solution. Pengthamkeerati et al. (2008) utilized biomass fly ash as an adsorbent for wastewater treatment. Mohammed et al. (2017) reported the heavy metal adsorption ability of various industrial wastes such as slag, fly ash and biomass ash in acid mine drainage in continuous stirred tank experiments. Among the industrial waste materials, fly ash showed maximum adsorption efficiency of Mn, Fe, Ni, Cu, Pb, Cd, Zn and As meanwhile slag and biomass ash showed significant adsorption efficiency of heavy metals. Similarly, Hegazi (2013) also documented the heavy metal (Cd and Cu) adsorption efficiency of fly ash.

15.6.3.1 Conclusion and Future Prospects

Heavy metal toxicity induces serious health risk on all kind of living forms. Several, remediation strategies have been implemented in the treatment of heavy metal ions. Technologies such as physicochemical have their advantages over metal remediation from environmental matrices. In recent years, development of eco-friendly, low-cost adsorbents for heavy metal removal from aqueous solutions have gained potential interest among the public and research community. Several conventional and non-conventional adsorbents have been investigated for their heavy metal adsorption ability from industrial wastewater. These adsorbents found to be more suitable for heavy metal adsorption from the aqueous environment.

On the other hand, before concluding them as a potential adsorbents need to carryout extensive research in following points: (i) enhance the removal efficiency of these adsorbents after appropriate treatment or modification, (ii) before going field application, cost factor should be analyzed and (iii) very few research literatures available about safe disposal of spent adsorbents. It should dispose of in an eco-friendly way. Thus, more research work should be made in this direction. If we could manage the above-stated characteristics, then these conventional and non-conventional adsorbents may offer significant advantages over the currently existing expensive physicochemical approaches.

References

Abdelatey LM, Khalil WKB, Ali TH, Mahrous KF (2011) Heavy metal resistance and gene expression analysis of metal resistance genes in gram positive and gram negative bacteria present in Egyptian soil. J Appl Sci Environ Sanit 6:201–211

Abollino O, Giacomino A, Malandrino M, Mentasti E (2008) Interaction of metal ions with montmorillonite and vermiculite. Appl Clay Sci 38(3):227–236

Adeyemo AA, Adeoye IO, Bello OS (2017) Adsorption of dyes using different types of clay: a review. Appl Water Sci 7(2):543–568

Aelion CM, Davis HT, McDermott S, Lawson AB (2008) Metal concentrations in rural topsoil in South Carolina: potential for human health impact. Sci Total Environ 402:149–156

Aguado J, Arsuaga JM, Arencibia A, Lindo M, Gascón V (2009) Aqueous heavy metals removal by adsorption on amine-functionalized mesoporous silica. J Hazard Mater 163(1):213–221

Ahmaruzzaman M (2011) Industrial wastes as low-cost potential adsorbents for the treatment of wastewater laden with heavy metals. Adv Colloid Interf Sci 166(1–2):36–59

An FQ, Wu RY, Li M, Hu TP, Gao JF, Yuan ZG (2017) Adsorption of heavy metal ions by iminodiacetic acid functionalized D301 resin: kinetics, isotherms, and thermodynamics. React Funct Polym 118:42–50

Argun ME, Dursun S, Ozdemir C, Karatas M (2007) Heavy metal adsorption by modified oak sawdust: thermodynamics and kinetics. J Hazard Mater 141(1):77–85

Arief VO, Trilestari K, Sunarso J, Indraswati N, Ismadji S (2008) Recent progress on biosorption of heavy metals from liquids using low cost biosorbents: characterization, biosorption parameters, and mechanism studies. Clean Soil Air Water 36(12):937–962

Arivalagan P, Singaraj D, Haridass V, Kaliannan T (2014) Removal of cadmium from aqueous solution by batch studies using *Bacillus cereus*. Ecol Eng 71:728–735

Bhatnagar A, Sillanpää M (2010) Utilization of agro-industrial and municipal waste materials as potential adsorbents for water treatment – a review. Chem Eng J 157:277–296

Bissen M, Frimmel FH (2003) Arsenic – a review. Part I: occurrence, toxicity, speciation, mobility. Acta Hydrochim Hydrobiol 31:9–18

Bruins MR, Kapil S, Oehme FW (2000) Microbial resistance to metals in the environment. Ecotoxicol Environ Safe 45:198–207

Cervantes C, Campos-García J, Devars S, Gutiérrez-Corona F, Loza-Tavera H, Torres-Guzmán JC, Moreno-Sánchez R (2001) Interactions of chromium with microorganisms and plants. FEMS Microbiol Rev 25:335–347

Chen J, Qu R, Zhang Y, Sun C, Wang C, Ji C, Yin P, Chen H, Niu Y (2012) Preparation of silica gel supported amidoxime adsorbents for selective adsorption of Hg (II) from aqueous solution. Chem Eng J 209:235–244

Chibuike G, Obiora S (2014) Heavy metal polluted soils: effect on plants and bioremediation methods. Appl Environ Soil Sci 2014:1–12

Congeevaram S, Dhanarani S, Park J, Dexilin M, Thamaraiselvi K (2007) Biosorption of chromium and nickel by heavy metal resistant fungal and bacterial isolates. J Hazard Mater 146(1–2):270–277

De-Pablo L, Chávez ML, Abatal M (2011) Adsorption of heavy metals in acid to alkaline environments by montmorillonite and Ca-montmorillonite. Chem Eng J 171(3):1276–1286

Di Natale F, Erto A, Lancia A, Musmarra D (2015) Equilibrium and dynamic study on hexavalent chromium adsorption onto activated carbon. J Hazard Mater 281:47–55

Duruibe JO, Ogwuegbu MO, Egwurugwu JN (2007) Heavy metal pollution and human biotoxic effects. Intern J Phy Sci 2(5):112–118

Eloussaief M, Jarraya I, Benzina M (2009) Adsorption of copper ions on two clays from Tunisia: pH and temperature effects. Appl Clay Sci 46(4):409–413

Fashola M, Ngole-Jeme V, Babalola O (2016) Heavy metal pollution from gold mines: environmental effects and bacterial strategies for resistance. Int J Environ Res Public Health 13:1047

Ferhat M, Kadouche S, Lounici H (2016) Immobilization of heavy metals by modified bentonite coupled coagulation/flocculation process in the presence of a biological flocculant. Desalination Water Treat 57(13):6072–6080

Fomina M, Gadd GM (2014) Biosorption: current perspectives on concept, definition, and application. Bioresour Technol 160:3–14

Gherasim CV, Křivčík J, Mikulášek P (2014) Investigation of batch electrodialysis process for removal of lead ions from aqueous solutions. Chem Eng J 256:324–334

Goel J, Kadirvelu K, Rajagopal C, Garg VK (2005) Removal of lead (II) by adsorption using treated granular activated carbon: batch and column studies. J Hazard Mater 125(1):211–220

Gogoasa I, Gergen I, Rada MA, Pârvu D, Ciobanu CA, Bordean D, Marunoiu C, Moigradea D (2006) AAS detection of heavy metals in sheep cheese (the Banat area,Romania). Buletinul USAMV-CN 62:240–245

Gopal R, Dube BK, Sinha P, Chatterjee C (2003) Cobalt toxicity effects on growth and metabolism of tomato. Commun Soil Sci Plant Anal 34(5–6)
Grassi M, Kaykioglu G, Belgiorno V, Lofrano G (2012) Removal of emerging contaminants from water and wastewater by adsorption process. In: Lofrano G (ed) Emerging compounds removal from wastewater, SpringerBriefs in green chemistry for sustainability, pp 15–37
Gumpu MB, Sethuraman S, Krishnan UM, Rayappan JBB (2015) A review on detection of heavy metal ions in water – an electrochemical approach. Sens Actuators B Chem 213:515–533
Hameed BH (2009) Removal of cationic dye from aqueous solution using jackfruit peel as non-conventional low-cost adsorbent. J Hazard Mater 162:344–350
Hameed BH, Mahmoud DK, Ahmad AL (2008) Equilibrium modeling and kinetic studies on the adsorption of basic dye by a low-cost adsorbent: coconut (*Cocos nucifera*) bunch waste. J Hazard Mater 158:65–72
Hegazi HA (2013) Removal of heavy metals from wastewater using agricultural and industrial wastes as adsorbents. HBRC J 9(3):276–282
Hongjie WA, Jin KA, Huijuan LI, Jiuhui QU (2009) Preparation of organically functionalized silica gel as an adsorbent for copper ion adsorption. J Environ Sci 21(11):1473–1479
Ibrahim SC, Hanafiah MAKM, Yahya MZA (2006) Removal of cadmium from aqueous solution by adsorption on sugarcane bagasse. Am-Euras. J Agric Environ Sci 1:179–184
Ibrahimi MM, Sayyadi AS (2015) Application of natural and modified zeolites in removing heavy metal cations from aqueous media: an overview of including parameters affecting the process. Int J Geo Earth Env Sci 3:1–7
Jain M, Garg VK, Kadirvelu K (2009) Equilibrium and kinetic studies for sequestration of Cr (VI) from simulated wastewater using sunflower waste biomass. J Hazard Mater 171(1–3):328–334
Jain M, Garg VK, Kadirvelu K (2013) Chromium removal from aqueous system and industrial wastewater by agricultural wastes. Biorem J 17(1):30–39
Jain M, Garg VK, Kadirvelu K (2014) Removal of Ni (II) from aqueous system by chemically modified sunflower biomass. Desalination Water Treat 52(28–30):5681–5695
Jiang MQ, Jin XY, Lu XQ, Chen ZL (2010) Adsorption of Pb (II), Cd (II), Ni (II) and Cu (II) onto natural kaolinite clay. Desalination 252(1):33–39
Kabwadza-Corner P, Johan E, Matsue N (2015) Ph dependence of lead adsorption on zeolites. J Environ Protect 6(01):45
Kadirvelu K, Faur-Brasquet C, Cloirec PL (2000) Removal of Cu (II), Pb (II), and Ni (II) by adsorption onto activated carbon cloths. Langmuir 16(22):8404–8409
Karthik C, Oves M, Thangabalu R, Sharma R, Santhosh SB, Arulselvi PI (2016) *Cellulosimicrobium funkei*-like enhances the growth of Phaseolus vulgaris by modulating oxidative damage under Chromium (VI) toxicity. J Adv Res 7(6):839–850
Karthik C, Ramkumar VS, Pugazhendhi A, Gopalakrishnan K, Arulselvi PI (2017) Biosorption and biotransformation of Cr (VI) by novel *Cellulosimicrobium funkei* strain AR6. J Taiwan Inst Chem Eng 70:282–290
Kondo K, Kanazawa Y, Matsumoto M (2015) Adsorption of Noble metals using silica gel modified with surfactant molecular assembly containing an Extractant. Sep Sci Technol 50(10):1453–1460
Li G, Zhao Z, Liu J, Jiang G (2011) Effective heavy metal removal from aqueous systems by thiol functionalized magnetic mesoporous silica. J Hazard Mater 192(1):277–283
Liang J, Liu M, Zhang Y (2016) Cd (II) removal on surface-modified activated carbon: equilibrium, kinetics and mechanism. Water Sci Technol 74(8):1800–1808
Lim SR, Schoenung JM (2010) Human health and ecological toxicity potentials due to heavy metal content in waste electronic devices with flat panel displays. J Hazard Mater 177:251–259
Liu H, Feng S, Zhang N, Du X, Liu Y (2014) Removal of Cu (II) ions from aqueous solution by activated carbon impregnated with humic acid. Front Environ Sci Eng 8(3):329–336
Malandrino M, Abollino O, Giacomino A, Aceto M, Mentasti E (2006) Adsorption of heavy metals on vermiculite: influence of pH and organic ligands. J Colloid Interface Sci 299(2):537–546
Malik A (2004) Metal bioremediation through growing cells. Environ Int 30:261–278

Mansoorian HJ, Mahvi AH, Jafari AJ (2014) Removal of lead and zinc from battery industry wastewater using electrocoagulation process: influence of direct and alternating current by using iron and stainless steel rod electrodes. Sep Purif Technol 35:165–175

Marchioretto MM, Bruning H, Rulkens W (2005) Heavy metals precipitation in sewage sludge. Sep Sci Technol 40(16):3393–3405

Meena AK, Kadirvelu K, Mishraa GK, Rajagopal C, Nagar PN (2008) Adsorption of Pb (II) and Cd (II) metal ions from aqueous solutions by mustard husk. J Hazard Mater 150(3):619–625

Mehdipour S, Vatanpour V, Kariminia HR (2015) Influence of ion interaction on lead removal by a polyamide nanofiltration membrane. Desalination 362:84–92

Mohammed NH, Atta M, Yaacub WZW (2017) Remediation of heavy metals by using industrial waste by products in acid mine drainage. Am J Eng Appl Sci 10(4):1001.1012

Mohsen-Nia M, Montazeri P, Modarress H (2007) Removal of Cu^{2+} and Ni^{2+} from wastewater with a chelating agent and reverse osmosis processes. Desalination 217(1–3):276–281

Moroni J, Scott B, Wratten N (2003) Differential tolerance of high manganese among rapseed genotypes. Plant Soil 253:507–519

Motsi T, Rowson NA, Simmons MJ (2009) Adsorption of heavy metals from acid mine drainage by natural zeolite. Intern J Miner Process 92(1):42–48

Mukesh KR, Kumar P, Singh M, Singh A (2008) Toxic effect of heavy metals in livestock health. Veterin World 1:28–30

Mureseanu M, Reiss A, Stefanescu I, David E, Parvulescu V, Renard G, Hulea V (2008) Modified SBA-15 mesoporous silica for heavy metal ions remediation. Chemosphere 73(9):1499–1504

Murthy ZV, Parikh PA, Patel NB (2013) Application of β-Zeolite, Zeolite Y, and Mordenite as adsorbents to remove mercury from aqueous solutions. J Dispers Sci Technol 34(6):747–755

Nagajyoti P, Lee K, Sreekanth T (2010) Heavy metals, occurrence and toxicity for plants: a review. Environ Chem Lett 8:199–216

Najafi M, Yousefi Y, Rafati AA (2012) Synthesis, characterization and adsorption studies of several heavy metal ions on amino-functionalized silica nano hollow sphere and silica gel. Sep Purif Technol 85:193–205

Park HG, Kim TW, Chae MY, Yoo IK (2007) Activated carbon-containing alginate adsorbent for the simultaneous removal of heavy metals and toxic organics. Process Biochem 42(10):1371–1377

Pengthamkeerati P, Satapanajaru T, Chularuengoaksorn P (2008) Chemical modification of coal fly ash for the removal of phosphate from aqueous solution. Fuel 87(12):2469–2476

Pillay K, Cukrowska EM, Coville NJ (2009) Multi-walled carbon nanotubes as adsorbents for the removal of parts per billion levels of hexavalent chromium from aqueous solution. J Hazard Mater 166(2):1067–1075

Ren Y, Abbood HA, He F, Peng H, Huang K (2013) Magnetic EDTA-modified chitosan/SiO_2/Fe_3O_4 adsorbent: preparation, characterization, and application in heavy metal adsorption. Chem Eng J 226:300–311

Salam OE, Reiad NA, Elshafei MM (2011) A study of the removal characteristics of heavy metals from wastewater by low-cost adsorbents. J Adv Res 2(4):297–303

Salmani MH, Davoodi M, Ehrampoush MH, Ghaneian MT, Fallahzadah MH (2013) Removal of cadmium (II) from simulated wastewater by ion flotation technique. Iran J Environ Health Sci Eng 10(1):16

Shaker MA (2014) Dynamics and thermodynamics of toxic metals adsorption onto soil-extracted humic acid. Chemosphere 111:587–595

Sharma RK, Agrawal M, Marshall F (2007) Heavy metal contamination of soil and vegetables in suburban areas of Varanasi, India. Ecotoxic Environ Safe 66(2):258–266

Sharma RK, Agrawal M, Marshall FM (2009) Heavy metals in vegetables collected from production and market sites of a tropical urban area of India. Food Chem Toxic 47(3):583–591

Sofinska-Chmiel W, Kolodynska D (2017) Application of ion exchangers for purification of galvanic wastewater from heavy metals. Sep Sci Technol. Accepted

Srivastava P, Singh B, Angove M (2005) Competitive adsorption behavior of heavy metals on kaolinite. J Colloid Interface Sci 290(1):28–38

Tajar AF, Kaghazchi T, Soleimani M (2009) Adsorption of cadmium from aqueous solutions on sulfurized activated carbon prepared from nut shells. J Hazard Mater 165(1):1159–1164

USEPA (United States Environmental Protection Agency) (1997) Mercury study report to congress, vol 3. USEPA, Washington, DC

Vinodhini PA, Sudha PN (2017) Removal of heavy metal chromium from tannery effluent using ultrafiltration membrane. Text Cloth Sustain 2(1):5

Wingenfelder U, Hansen C, Furrer G, Schulin R (2005) Removal of heavy metals from mine waters by natural zeolites. Environ Sci Technol 39(12):4606–4613

Wu P, Wu W, Li S, Xing N, Zhu N, Li P, Wu J, Yang C, Dang Z (2009) Removal of Cd^{2+} from aqueous solution by adsorption using Fe-montmorillonite. J Hazard Mater 169(1):824–830

Wuana RA, Okieimen FE (2011) Heavy metals in contaminated soils: a review of sources, chemistry, risks and best available strategies for remediation. ISRN Ecol:1–20

Xiang B, Fan W, Yi X, Wang Z, Gao F, Li Y, Gu H (2016) Dithiocarbamate-modified starch derivatives with high heavy metal adsorption performance. Carbohydr Polym 136:30–37

Yasri NG, Gunasekaran S (2017) Electrochemical Technologies for Environmental Remediation. In: Enhancing cleanup of environmental pollutants. Springer, Cham, pp 5–73

Zarghami Z, Akbari A, Latifi AM, Amani MA (2016) Design of a new integrated chitosan-PAMAM dendrimer biosorbent for heavy metals removing and study of its adsorption kinetics and thermodynamics. Bioresour Technol 205:230–238

Zhao H, Wu L, Chai T, Zhang Y, Tan J, Ma S (2012) The effects of copper, manganese and zinc on plant growth and elemental accumulation in the manganese-hyperaccumulator Phytolacca americana. J Plant Physiol 169(13):1243–1252

Zuorro A, Lavecchia R, Medici F, Piga L (2013) Spent tea leaves as a potential low-cost adsorbent for the removal of Azo dyes from wastewater. Chem Eng Trans 32:19

Chapter 16
Graphene Based Composites of Metals/Metal Oxides as Photocatalysts

Asim Jilani, Mohammad Omaish Ansari, Mohammad Oves, Syed Zajif Hussain, and Mohd Hafiz Dzarfan Othman

Abstract Graphene based metal and metal oxide composites have attracted great attention towards curing and solving various environmental issues. Further, the Honeycomb structured of graphene is an ideal nominee for various advanced application such as photovoltaics and optoelectronics etc. However, the performance of graphene based material strongly depends on the synthesis method. So, the selection of appropriate synthesis technique is important for targeted application. In the chapter we reviewed the synthesis and properties of graphene and further its role as a photocatalyst.

Keywords Graphene · Metal/metal oxides composites · Photocatalyst

---

A. Jilani
Advanced Membrane Technology Research Centre, Universiti Teknologi Malaysia, Johor Bahru, Johor, Malaysia

School of Chemical and Energy Engineering, Faculty of Engineering, Universiti Teknologi Malaysia, Johor Bahru, Johor, Malaysia

Center of Nanotechnology, King Abdul-Aziz University, Jeddah, Saudi Arabia

M. O. Ansari
Center of Nanotechnology, King Abdul-Aziz University, Jeddah, Saudi Arabia
e-mail: moansari@kau.edu.sa

M. Oves
Centre of Excellence in Environmental Studies, King Abdul-Aziz University, Jeddah, Saudi Arabia

S. Z. Hussain
Department of Chemistry and Chemical Engineering, SBA School of Science & Engineering (SBASSE), Lahore University of Management Sciences (LUMS), Lahore, Pakistan

M. H. D. Othman (✉)
Advanced Membrane Technology Research Centre, Universiti Teknologi Malaysia, Johor Bahru, Johor, Malaysia

School of Chemical and Energy Engineering, Faculty of Engineering, Universiti Teknologi Malaysia, Johor Bahru, Johor, Malaysia
e-mail: hafiz@petroleum.utm.my

M. Oves et al. (eds.), *Modern Age Waste Water Problems*,
https://doi.org/10.1007/978-3-030-08283-3_16

16.1 Introduction

Metals/metal oxides based composites of graphene (GN) have attracted a tremendous attention worldwide due to their potential applications in various fields such as electrical, electrochemical, supercapacitor, photochemical, and biomedical domains. The prominence to keep the pristine characteristics of GN in its composites is rationalized here for their maximum efficiency in photocatalysis. This chapter also highlights some important key points to be considered while fabricating these materials.

16.2 Design Rule for the Synthesis of Photocatalytic Material

The photoactive materials uses nature's sunlight or other light sources to degrade the toxic pollutants in the environment. However, some key factors like tuning of bang gap in range 0–4.5 eV and minimization of charge recombinations can control the efficiency of these materials (Niu et al. 2014).

Giacomo Ciamician conducted the first ever experimental process "light and light alone" to investigate the ability of light to enable the chemical reactions under blue and red light which was named as "photocatalysis" in early 1911. (Albini and Fagnoni 2008; Bruner and Kozak 1911) Afterwards in 1924, ZnO was used as photocatalyst for the reduction of Ag^+ to Ag (Baur and Perret 1924).

The basic principle of photocatalysis is; when the light having wavelength $\leq \frac{1240}{E_g}$ (Eg = Band gap of investigated material) falls onto the under-investigated material then charge acceleration starts from their valence to conduction band (Fig. 16.1). This transfer of charge induces some surface reactions that starts adsorption/degradation of pollutants through oxidation/reduction processes occurring at interface of subjected catalyst. Moreover, the photocatalytic process is directly proportional to the charge recombination ratio and specific surface area of the subjected material (Huang et al. 2011). The high specific surface area of the material provides more active sites for the absorption of material which is to be degraded such as dyes, industrial wastes, etc. Moreover, the surface oxygen vacancies can also provide active absorption site to accelerate the photocatalytic process (Djurišić et al. 2014). Further, inducing the oxygen defects can also change the visible spectral range of the photocatalyst. The morphology of the photocatalyst also alters its absorption activity of dyes as reported by Joo et al. (2013).

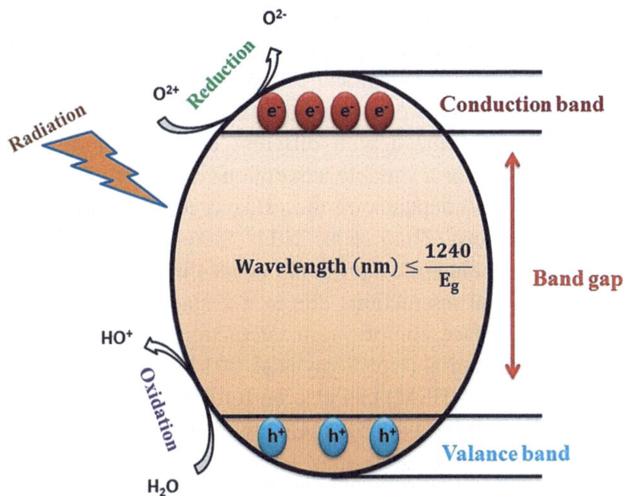

Fig. 16.1 Schematic representation of the photocatalytic process showing charge transfer from valance to conduction band and oxidation/reduction process (Ref: Jilani et al. 2018)

16.3 Synthesis and Properties of Graphene

Graphene (GN) was discovered by Novoselov et al. in 2004, and has attracted universal attention due to its prospective applications in different fields such as electrical, electrochemical, supercapacitor, photochemical, and optics. GN comprises of sp2-hybridized, single layered, two dimensional carbon atoms arranged in a honeycomb lattice. The planer orbitals are energetically stable and localized sigma bonded with the three nearest neighbor atoms are arranged in a honeycomb lattice. This structure of GN is responsible for its high surface area and electrical conductivity as well as its other exceptional properties. The breaking strength of the GN is ~42 Nm^{-1} and has the tensile strength (Young modules) of 1.0 TPa. (Huang et al. 2011) Furthermore, the large surface area (2630 m^2 g^{-1}), electronic transportation i.e. high intrinsic mobility of (200,000 cm^2v^{-1} s^{-1}), thermal conductivity (~5000 $Wm^{-1}K^{-1}$) and extra ordinary mechanical properties makes it a prominent material for applications in the above mentioned fields (Lee et al. 2008; Li et al. 2009). GN can be synthesized by various techniques such as chemical vapor deposition (CVD) (Li et al. 2009; Reina et al. 2009), exfoliation processes (Lotya et al. 2009; Liu et al. 2008), and direct growth through carbon (Sun et al. 2010; Ruan et al. 2011). The choice of methodology for the synthesis of GN is directly related to its quality and thus the synthesis methodology can be customized as per its application.

Organic pollutants such as dyes and other chemical wastes of industry, that are discharged into the water streams, are the major cause of water borne diseases and have a colossal negative impact on the aquatic life and human beings. To counter this, heterogeneous photocatalysts and adsorption have proven to be one of the easiest and efficient way to combat these problems. The absorbent engages the environmental pollutants through physico-chemical interactions and thus the high surface area and further specific functionalization of adsorbent are very important for an efficient and selective adsorption of pollutant (Dąbrowski 2001). Thus, GN composites can be functionalized with different functional moieties in order to enhance the mesoporosity for a suitable adsorption. However, the efficiency of GN to absorb the pollutant also depends on the pH, oxygen functional groups and the nature of functional moiety (Zhao et al. 2011; Sitko et al. 2013). In GN based composite materials the adsorption process occurs through two different kind of reactions: namely physical interactions due to π-π stacking interaction of the GN composites materials, surface complication interactions between molecular ion of the GN and the foreign material (Upadhyay et al. 2014). Furthermore, the GN based semiconducting materials (GBSM) can also be further exploited to fabricate nano-structured photocatalysts (Liu et al. 2014; Chang et al. 2015). Such photocatalysts, due to their highly conducting nature may act as an outstanding electron acceptors and, thus may find potential applications in photocatalysis. The band gap of these GBSM can be further engineered by changing the amount of metal or metal oxide dopant.

16.3.1 Synthesis of Graphene Based Composites

The GN based material can be fabricated through chemical and physical routes. In general, these routes are classified as "bottom up" and "top down approach". (Wang et al. 2013) The bottom-up methods involves the fabrication of GN based material through chemical reactions whereas top down approach represents the physical routes such as chemical vapor deposition system (Jilani et al. 2018). The optical and electrical properties of GN based materials (Omiciuolo et al. 2014) greatly depends on number and folding of GN layers (Somani et al. 2006) that are reliant on the fabrication methods (Jilani et al. 2017). The Fig. 16.2 shows the synthesis of GN based materials from bulk solids (top down approach) and from molecular levels (Bottom up approach) (Habiba et al. 2014).

16.3.2 Role of Graphene for Photocatalytic

The photoactive materials should have four basic characteristics i.e. light harvesting, charge excitation, charge separation and should be able to participate in surface reactions (Zhu and Wang 2017). GN is one of the supportive materials to enhance the photocatalytic activity in its photoactive composites owing to the π-π* electronic transitions due to presence of sp2 network of carbon atoms arranged in two dimensional structure. In combination with different materials like TiO_2, GN is a promising material for charge excitation and absorption processes (Yeh et al. 2013).

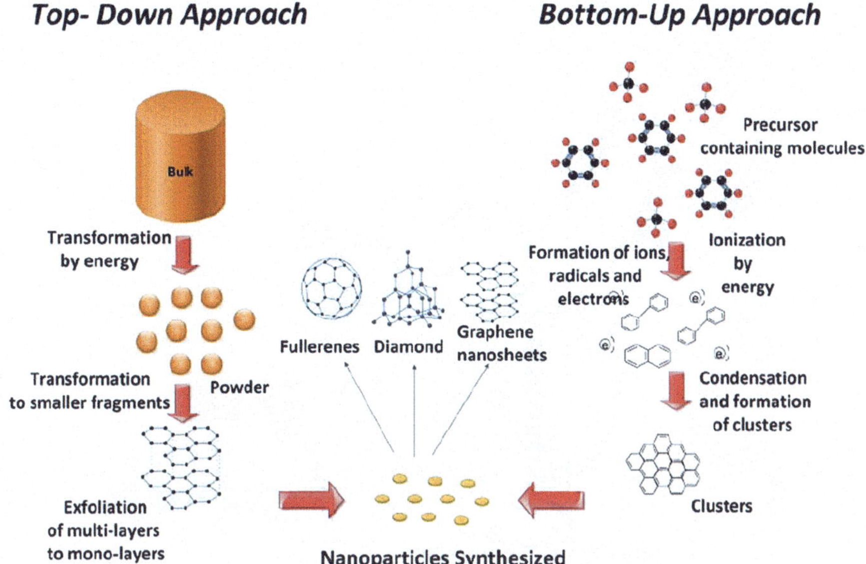

Fig. 16.2 The top down and bottom approaches for the synthesis of graphene and its composites (Habiba et al. 2014)

Moreover, tuning the functional groups (like Oxygen) of GN attached to its basal plane can change its charge excitation properties which are highly desirable for enhanced photocatalytic process. The tuning/reduction of graphene oxide can be done by various physical, chemical and thermal techniques such as laser ablation, electromagnetic radiation. However, the low temperature reduction method is commonly used to obtain less defective GN sheets which is highly useful for the synthesis of GN based photocatalyst as reported by Jilani et al. (2017).

Keeping in view the extra ordinary properties of GN, various research groups (Zhang et al. 2013; Yahia et al. 2016; Guo et al. 2017; Tang et al. 2015; Xian et al. 2014) have fabricated the composites of GN with ZnO (Zhang et al. 2013), Ag_3PO_4 (Yahia et al. 2016), Fe_2O_3 (Guo et al. 2017), SnO_2 (Tang et al. 2015) and $SrTiO_3$ (Xian et al. 2014). The known approaches to fabricate the composites of GN are microwave assisted reduction (Hassan et al. 2009), solution mixing process, (Du et al. 2011) electro-less metallization process (Zhou et al. 2009) and vapor deposition technique (Kim et al. 2009). The Fig. 16.3 illustrates the photocatalytic activity of ZnO based composites of reduced graphene oxide (rGO/ZnO composites) in comparison with pure ZnO for the degradation of 2-Chlorophenol (2-CP). The presence of π bond in reduced graphene oxides (rGO) causes the reduction of band gap of rGO/ZnO composites as compared to pure ZnO. Further, the presence of sp2 bonds in rGO causes the absorption of more light photons which accelerates the transfer of charge carrier from valance to conduction band in its composites while

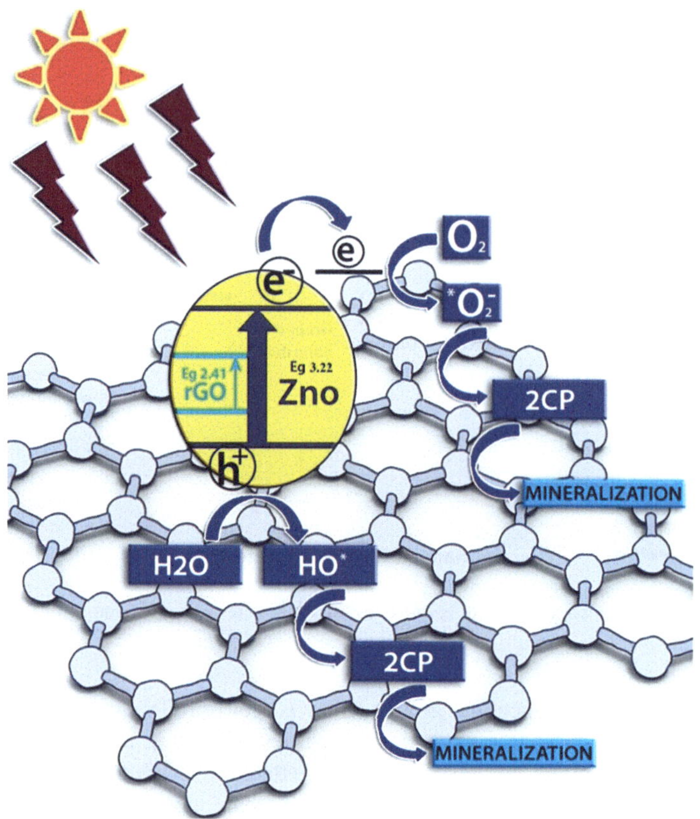

Fig. 16.3 Schematic illustration for the role of GN based composites to enhance the photocatalytic activity. Reprinted with permission from (Jilani et al. 2017)

reducing the charge recombination ratios which eventually resulted in the enhancement of photocatalytic activity as compared to pure ZnO (Jilani et al. 2017).

Graphene based photocatalyst material is also being used for the reduction of heavy metals such as Cr (VI). Recent study shows that GN doped ZnO nanoparticles prepared by Liu et al. efficiently reduced the Cr (VI) up to 98% (Liu et al. 2011). The reduction of Cr (VI) was associated through the alteration of band gap while incorporating the GN into ZnO nanoparticles.

TiO_2 is also commonly used material for the degradation of organic and inorganic dyes because of its low cast, ease of synthesis and its strong oxidizing capability. Further, GN and rGO doped TiO_2 have been used for photocatalytic degradation process. The presence of sp2 network in GN interacts with the aromatic molecules of the dye through π-π stacking which enhances the photocatalytic activity of GN based TiO_2 as compared to pure TiO_2. Moreover, Lee et al. observed the reduction in the band gap of GN doped TiO_2 while lowering the charge recombination ratio (Lee et al. 2012). The literature (Zhang et al. 2011; Chen et al. 2010) shows that the concentration of GN, its morphology and the number of GN sheets present in GN/ TiO_2 nanoparticles affects the photocatalytic efficiency.

16.4 Future Recommendation

Graphene based composites have extraordinary physical and chemical properties, which makes them highly efficient materials in the field of photocatalysis. However, the properties of these composites are accredited to the electronic interactions between the GN and individual components of metals/metal oxides along with the nature of its sp2- bonding. Further, the number of GN sheets in its composites and their degree of dispersion is responsible to alter the absorption and amount of reduction of organic and inorganic pollutants. Further, the control over the van der Waals and π- π interaction among the GN sheets are important for its highly efficient performance in the field of photocatalysis.

Acknowledgment The authors gratefully acknowledge the financial support from the Ministry of Higher Education Malaysia under the Higher Institution Centre of Excellence Scheme (Project Number: R.J090301.7846.4 J201) and Universiti Teknologi Malaysia under Tier 1 Research University Grant (Project Number: Q.J130000.2546.16H40). The authors would also like to thank Research Management Centre, Universiti Teknologi Malaysia for the technical support.

References

Albini A, Fagnoni M (2008) 1908: Giacomo Ciamician and the concept of green chemistry. ChemSusChem 1(1–2):63–66

Baur E, Perret A (1924) Über die Einwirkung von Licht auf gelöste Silbersalze in Gegenwart von Zinkoxyd. Helv Chim Acta 7(1):910–915

Bruner L, Kozak J (1911) Zur Kenntnis der Photokatalyse. I. Die Lichtreaktion in Gemischen: Uransalz + Oxalsäure. Z Elektrochem Angew Phys Chem 17(9):354–360

Chang Y-H, Wang C-M, Hsu Y-K, Pai Y-H, Lin J-Y, Lin C-H (2015) Graphene oxide as the passivation layer for Cu x O photocatalyst on a plasmonic Au film and the corresponding photoluminescence study. Opt Express 23(19):A1245–A1252

Chen C, Cai W, Long M, Zhou B, Wu Y, Wu D, Feng Y (2010) Synthesis of visible-light responsive graphene oxide/TiO2 composites with p/n heterojunction. ACS Nano 4(11):6425–6432

Dąbrowski A (2001) Adsorption—from theory to practice. Adv Colloid Interf Sci 93(1):135–224

Djurišić AB, Leung YH, Ching Ng AM (2014) Strategies for improving the efficiency of semiconductor metal oxide photocatalysis. Mater Horiz 1(4):400–410

Du J, Lai X, Yang N, Zhai J, Kisailus D, Su F, Wang D, Jiang L (2011) Hierarchically ordered macro−mesoporous TiO2−graphene composite films: improved mass transfer, reduced charge recombination, and their enhanced photocatalytic activities. ACS Nano 5(1):590–596

Guo Y, Guo Y, Wang X, Li P, Kong L, Wang G, Li X, Liu Y (2017) Enhanced photocatalytic reduction activity of uranium(vi) from aqueous solution using the Fe2O3–graphene oxide nanocomposite. Dalton Trans 46(43):14762–14770

Habiba K, Makarov VI, Weiner BR, Morell G (2014) Fabrication of nanomaterials by pulsed laser synthesis. In: Manufacturing nanostructures. One Central Press, Manchester

Hassan HMA, Abdelsayed V, Khder AERS, Abou Zeid KM, Terner J, El-Shall MS, Al-Resayes SI, El-Azhary AA (2009) Microwave synthesis of graphene sheets supporting metal nanocrystals in aqueous and organic media. J Mater Chem 19(23):3832–3837

Huang X, Yin Z, Wu S, Qi X, He Q, Zhang Q, Yan Q, Boey F, Zhang H (2011) Graphene-based materials: synthesis, characterization, properties, and applications. Small 7(14):1876–1902

Jilani A, Othman MHD, Ansari MO, Kumar R, Alshahrie A, Ismail AF, Khan IU, Sajith VK, Barakat MA (2017) Facile spectroscopic approach to obtain the optoelectronic properties of few-layered graphene oxide thin films and their role in photocatalysis. New J Chem 41 (23):14217–14227

Jilani A, Othman MHD, Ansari MO, Hussain SZ, Ismail AF, Khan IU (2018) Inamuddin, Graphene and its derivatives: synthesis, modifications, and applications in wastewater treatment. Environ Chem Lett 16:1301–1323

Joo JB, Lee I, Dahl M, Moon GD, Zaera F, Yin Y (2013) Controllable synthesis of mesoporous TiO2 hollow shells: toward an efficient photocatalyst. Adv Funct Mater 23(34):4246–4254

Kim KS, Zhao Y, Jang H, Lee SY, Kim JM, Kim KS, Ahn J-H, Kim P, Choi J-Y, Hong BH (2009) Large-scale pattern growth of graphene films for stretchable transparent electrodes. Nature 457:706–710

Lee C, Wei X, Kysar JW, Hone J (2008) Measurement of the elastic properties and intrinsic strength of monolayer graphene. Science 321(5887):385–388

Lee E, Hong J-Y, Kang H, Jang J (2012) Synthesis of TiO2 nanorod-decorated graphene sheets and their highly efficient photocatalytic activities under visible-light irradiation. J Hazard Mater 219-220:13–18

Li X, Zhu Y, Cai W, Borysiak M, Han B, Chen D, Piner RD, Colombo L, Ruoff RS (2009) Transfer of large-area graphene films for high-performance transparent conductive electrodes. Nano Lett 9(12):4359–4363

Liu N, Luo F, Wu H, Liu Y, Zhang C, Chen J (2008) One-step ionic-liquid-assisted electrochemical synthesis of ionic-liquid-functionalized graphene sheets directly from graphite. Adv Funct Mater 18(10):1518–1525

Liu X, Pan L, Lv T, Lu T, Zhu G, Sun Z, Sun C (2011) Microwave-assisted synthesis of ZnO–graphene composite for photocatalytic reduction of Cr(vi). Cat Sci Technol 1(7):1189–1193

Liu J, Durstock M, Dai L (2014) Graphene oxide derivatives as hole-and electron-extraction layers for high-performance polymer solar cells. Energy Environ Sci 7(4):1297–1306

Lotya M, Hernandez Y, King PJ, Smith RJ, Nicolosi V, Karlsson LS, Blighe FM, De S, Wang Z, McGovern I (2009) Liquid phase production of graphene by exfoliation of graphite in surfactant/water solutions. J Am Chem Soc 131(10):3611–3620

Niu M, Cheng D, Cao D (2014) Understanding the mechanism of photocatalysis enhancements in the graphene-like semiconductor sheet/TiO2 composites. J Phys Chem C 118(11):5954–5960

Omiciuolo L, Hernández ER, Miniussi E, Orlando F, Lacovig P, Lizzit S, Menteş TO, Locatelli A, Larciprete R, Bianchi M, Ulstrup S, Hofmann P, Alfè D, Baraldi A (2014) Bottom-up approach for the low-cost synthesis of graphene-alumina nanosheet interfaces using bimetallic alloys. Nat Commun 5:5062

Reina A, Jia X, Ho J, Nezich D, Son H, Bulovic V, Dresselhaus MS, Kong* J (2009) Layer area, few-layer graphene films on arbitrary substrates by chemical vapor deposition. Nano Lett 9 (8):3087–3087

Ruan G, Sun Z, Peng Z, Tour JM (2011) Growth of graphene from food, insects, and waste. ACS Nano 5(9):7601–7607

Sitko R, Turek E, Zawisza B, Malicka E, Talik E, Heimann J, Gagor A, Feist B, Wrzalik R (2013) Adsorption of divalent metal ions from aqueous solutions using graphene oxide. Dalton Trans 42(16):5682–5689

Somani PR, Somani SP, Umeno M (2006) Planer nano-graphenes from camphor by CVD. Chem Phys Lett 430(1):56–59

Sun Z, Yan Z, Yao J, Beitler E, Zhu Y, Tour JM (2010) Growth of graphene from solid carbon sources. Nature 468(7323):549–552

Tang L, Nguyen VH, Lee YR, Kim J, Shim J-J (2015) Photocatalytic activity of reduced graphene oxide/SnO2 nanocomposites prepared in ionic liquid. Synth Met 201:54–60

Upadhyay RK, Soin N, Roy SS (2014) Role of graphene/metal oxide composites as photocatalysts, adsorbents and disinfectants in water treatment: a review. RSC Adv 4(8):3823–3851

Wang H, Yuan X, Wu Y, Huang H, Peng X, Zeng G, Zhong H, Liang J, Ren M (2013) Graphene-based materials: fabrication, characterization and application for the decontamination of wastewater and wastegas and hydrogen storage/generation. Adv Colloid Interf Sci 195–196:19–40

Xian T, Yang H, Di L, Ma J, Zhang H, Dai J (2014) Photocatalytic reduction synthesis of SrTiO3-graphene nanocomposites and their enhanced photocatalytic activity. Nanoscale Res Lett 9 (1):327

Yahia IS, Jilani A, Abdel-wahab MS, Zahran HY, Ansari MS, Al-Ghamdi AA, Hamdy MS (2016) The photocatalytic activity of graphene oxide/Ag3PO4 nano-composite: loading effect. Optik 127(22):10746–10757

Yeh T-F, Cihlář J, Chang C-Y, Cheng C, Teng H (2013) Roles of graphene oxide in photocatalytic water splitting. Mater Today 16(3):78–84

Zhang H, Xu P, Du G, Chen Z, Oh K, Pan D, Jiao Z (2011) A facile one-step synthesis of TiO2/graphene composites for photodegradation of methyl orange. Nano Res 4(3):274–283

Zhang L, Du L, Cai X, Yu X, Zhang D, Liang L, Yang P, Xing X, Mai W, Tan S, Gu Y, Song J (2013) Role of graphene in great enhancement of photocatalytic activity of ZnO nanoparticle–graphene hybrids. Physica E 47:279–284

Zhao G, Li J, Ren X, Chen C, Wang X (2011) Few-layered graphene oxide nanosheets as superior sorbents for heavy metal ion pollution management. Environ Sci Technol 45(24):10454–10462

Zhou X, Huang X, Qi X, Wu S, Xue C, Boey FYC, Yan Q, Chen P, Zhang H (2009) In situ synthesis of metal nanoparticles on single-layer graphene oxide and reduced graphene oxide surfaces. J Phys Chem C 113(25):10842–10846

Zhu S, Wang D (2017) Photocatalysis: basic principles, diverse forms of implementations and emerging scientific opportunities. Adv Energy Mater 7(23):1700841

Chapter 17
Microbial Electrochemical Cell: An Emerging Technology for Waste Water Treatment and Carbon Sequestration

Abdul Hakeem Anwer, Mohammad Danish Khan, Mohammad Zain Khan, and Rajkumar Joshi

Abstract Recently, treatment of waste water using biofuel technology has gained more attention because of its bio-sustainable resource by generating powering microbes (electrical energy) which exponentially reducing dependence of fossil fuels. In the last one decade, one of the bioelectro-chemical approach; microbial electrolysis cell (MEC) has been developed to treat waste water and energy production. It is considered as a potential green technology to tackle the issues of energy shortage and global warming. This technique employs conversion of waste water (which contain organic matter) into hydrogen or a variety of value-added products (acetate, hydrogen peroxide, methane, ethanol) via electrochemically active bacteria (electrogenes). Significant outcomes of MECs offers a new solution to emerging environmental issues related to waste water treatment, energy and resource recovery as well. In future, it is expected that treatment of industrial waste water using MECs has become a promising renewable green technology to manage waste water and biofuels production. The present chapter mainly reviews utilization of various polymer-based electrode materials in MECs for treatment of waste water along with their future potential substrates.

Keywords Bioelectrochemical system · Biofuel · Greenhouse gas · Hydrogen production rates · Microbial electrolysis cell · Waste water treatment

---

A. H. Anwer (✉) · M. D. Khan · M. Z. Khan
Department of Chemistry, Aligarh Muslim University, Aligarh, Uttar Pradesh, India

R. Joshi
Department of Civil Engineering, Jamia Millia Islamia, New Delhi, India
e-mail: rkjoshi@nic.in

M. Oves et al. (eds.), *Modern Age Waste Water Problems*,
https://doi.org/10.1007/978-3-030-08283-3_17

17.1 Introduction

Nowadays, two most serious issues; energy crisis and water pollution are facing globally. Energy is a basic segment to improve the economy any country. Worldwide energy utilization was approx. 520 quadrillion BTUs (British thermal Unit) in 2010 and is relied upon to increment by 56% (820 quadrillion BTU) by 2040. Now, 78% of worldwide energy utilization originates from non-renewable sources like, petroleum, natural gas, coal and that exclusive 19% of the current energy originates from renewable power sources (Khanal et al. 2010; Khan et al. 2018). Since the non-renewable energy sources are exhausting day by day and petroleum products generate pollution. Reducing renewable resources and quickly expanding energy utilization obviously recommend that we should act sincerely and conclusively to grow spotless, maintainable, moderate and sustainable power sources. Resources which are biodegradable and inexhaustible can be utilized and can make energy and minimize reliance on non-renewable energy sources and do not create pollution. A large body of researchers around the world are searching out a renewable alternatives source for satisfy world growing energy demand such as wind energy, solar energy, geothermal energy, & waste to energy etc. (International energy outlook 2013). The exhausting petroleum products may not be main concern. Global increasing pollution is also concern which deviate the atmospheric climate that might harmful to the people and other living beings. As the expanding population and progression in the way of life has prompt an expanded advancement produce a various kinds of waste materials like municipal waste, dairy waste, industrial waste, and agricultural. Different countries are producing extensive amount of waste water and their treating capability is also different for example in high income countries they treat 70% of waste water, at the same time waste water treating capability for lower income countries drop to 38% and only 28% for lower middle-income countries (Sato et al. 2013). The world economic forum has regularly analyze that crises of water is one of the biggest global crises since last 5 year. In 2016, the crisis of water was set as the global highest risk for people as well as economies for the next ten years (WEF World Economic Forum 2016). Current studies have exhibited that 66% of the total population right now lives in regions that experience water shortage for no less than one month a year. It is remarkable that almost half of the general population going through this level of water shortage live in China and India. Around 4.3 billion of the worldwide population lives under states of moderate to serious water shortage by a factor of two. A large portion of a people around half of billion meet serious water shortage throughout the year. Of those half-billion individuals are live in India (180 million), Egypt (27 million), Mexico (20 million), Saudi Arabia (20 million), Pakistan (73 million), and in Yemen (18 million) (Mekonnen and Hoekstra 2011; Hoekstra et al. 2011). A huge value of water are utilized in different sectors such as agricultural, industries and domestic utilization in which agricultural is only responsible for 70% water utilization. However, In developing countries, industries consume more than half of water available for human use and produce a large amount of waste water. Waste water generation is one of the greatest challenges related to the development of informal slums in developing countries. In India (urban areas) approx. 20 billion cubic meter (bm^3) wastewater is generated every year and sewage treatment plants have capacity to treat only 8 bm^3 of wastewater (Khan et al. 2017b). By the use of suitable technology, the potential chemical energy contained in the organic compounds found in waste water might help to improve the energy and financial balance of waste water treatment plants. During the previous decade another age of bio-based advances, known as bio electrochemical system (BESs), with awesome potential for wastewater treatment and resources recovering have developed. BESs can be comprehensively named microbial fuel cell (MFCs) or microbial electrolysis cells (MECs) depending upon whether they work in electrolytic mode (MECs) or galvanic (MFCs). After deliberation as to which technique is the most helpful method for recovering energy from Waste Water (Cusick et al. 2010), it appears that from economic, natural, and technical points of view MECs offer significant favorable circumstances over MFCs (Sleutels et al. 2012; Zhang and Angelidaki (2014)). Clearly, the validity of this announcement depends on a valuable product got from the MEC (hydrogen, ethanol, hydrogen peroxide, methane and so forth) Scholz (1993).

A **microbial fuel cell (MFC)** is a bio electrochemical device which has a ability to generate current by respiring microbes to convert organic substrates into electrical energy (Khan et al. 2015a, b, 2017a, b; Sultana et al. 2015). Microbes producing current by transforming chemical energy into electricity by using oxidation and reduction reactions shown in Fig. 17.1. A **microbial electrolysis cell (MEC)** is an innovation identified with MFC. MEC is a reverse process of MFC to produce a valuable product such as hydrogen, methane and ethanol by applying a small amount

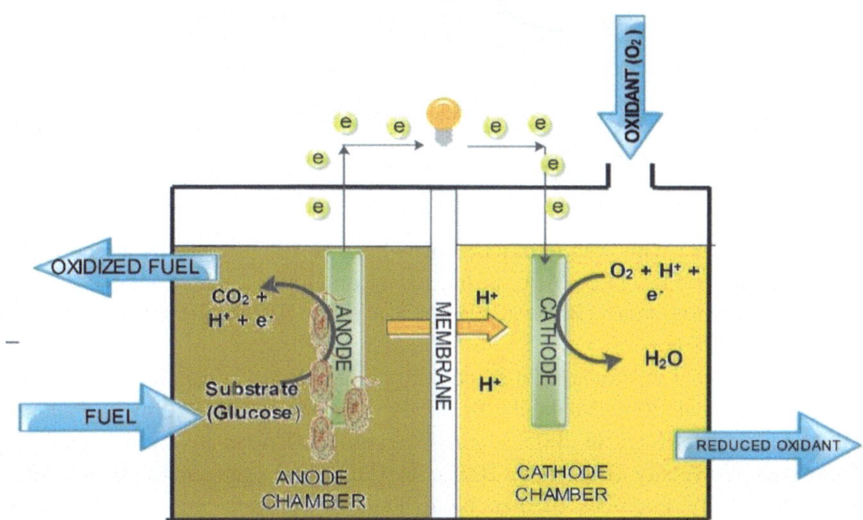

Fig. 17.1 Schematic representation of dual-chamber of MFCs

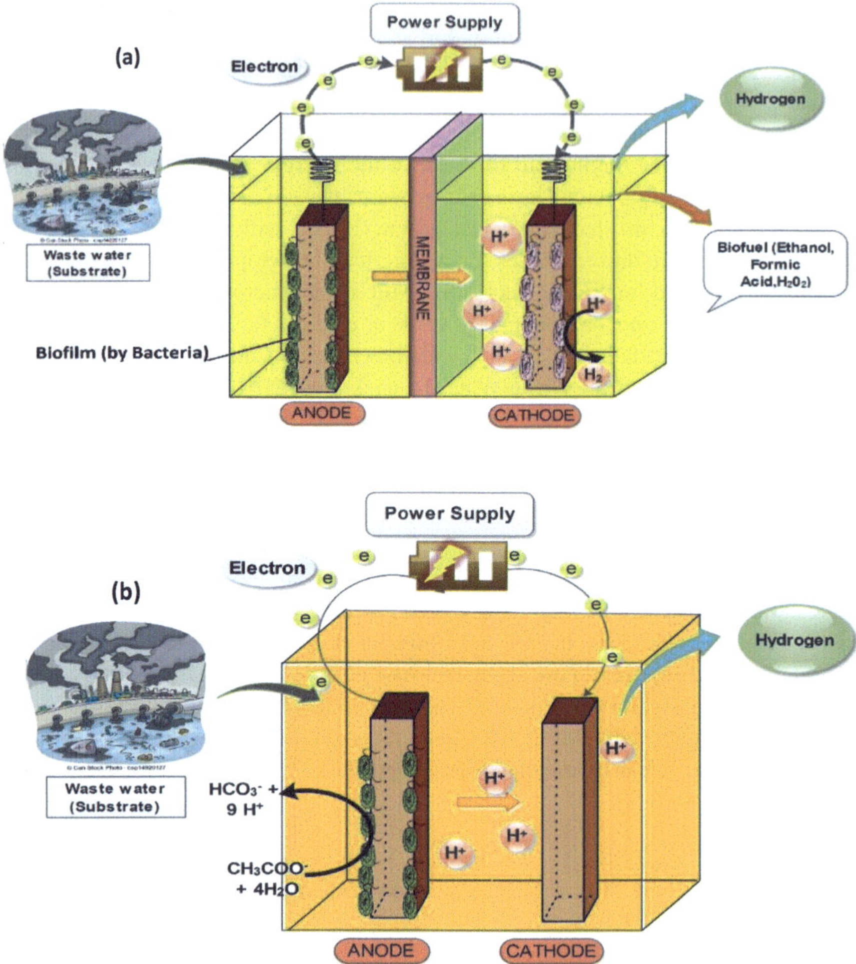

Fig. 17.2 Generalized Schematic diagram of dual (**a**) & single (**b**) chamber of MEC

of voltage (> 1 V). This energy in terms of voltage ideally produce by a renewable source of power Fig. 17.2. MEC is the most supportive strategy for recovering energy from waste water (Badwal 2014; Khan et al. 2017a; Nizami et al. 2017; Miandad et al. 2017).

MEC is another encouraging methodology for hydrogen generation from waste water (Logan et al. 2008; Meda 2015). MECs were first constructed by two autonomous researchers body in 2005, one at Wageningen University and the other at Penn State University (Liu et al. 2005a, b; Call and Logan 2008). In a MEC, electrochemically microbes oxidize biodegradable substrate and create protons electrons and CO_2.Electrons are transferred to the anode and flow through an external circuit to the cathode and proton are diffuse to cathodic chamber by proton exchange membrane (PEM).

For hydrogen production in the cathode chamber, the mixing of protons and these electrons, MEC needs an additionally (≥0.2 V) voltage under an microbial optimized environment such as T = 30 °C, pH = 7, P = 1 atm (10^5 Pa) (Liu et al. 2005a, b). This is completed by the contribution of a voltage by means of a D.C (direct current) source. As compare to water electrolysis 1.23–1.8 V voltage needed for hydrogen production but in MEC generally require low power input 0.2–0.8 Volt. Diagram of dual-chamber MEC is appeared in Fig. 17.2 and all the reactions are occurred in both the chambers if we taken acetate as a substrate in MEC. The mechanism of MEC is given below:

Anode:

$$C_2H_4O_2 + 2H_2O \longrightarrow 2CO_2 + 8e^- + 8H^+ \quad (1)$$

Cathode:

$$8H^+ + 8e^- \longrightarrow 4H_2 \quad (2)$$

Hydrogen gas production rate is directly related to a current as the more electron is travel to the cathode chamber and more hydrogen will be produced (Call and Logan 2008; Rozendal et al. 2007). Hydrogen gas production rate is directly related to a current as the more electron is travel to the cathode chamber more will be the hydrogen are produced. The utilization of close spacing electrode, large surface area of electrode and enhanced reactor designs has rapidly multiplied both current densities and recoveries of hydrogen in MECs.

17.2 Configurations & Design of MEC

The MEC configuration straightforwardly influences the current density and hydrogen yield through internal resistance (Rin). In the previous a couple of years, different arrangements of MECs have been created utilizing a large number of different types of materials. Numbers of MEC reactor design had been proposed for laboratory scale research, however they all offer the same working standards. They are worked under various conditions to enhance the efficiency, hydrogen production rate (HPR) and decrease the general cost of MECs.

17.2.1 Dual Chamber MEC

Most generally utilized design of MEC is two-chamber which is separate by ion exchange membrane Fig. 17.2a one is an anodic and other is cathodic both chambers are under anaerobic condition. In anodic chamber electron and proton are produce by the microbial oxidation of substrate & proton are diffuses to cathodic chamber via separating ion exchange membrane. (Khan et al. 2018). By applying small amount of voltage to pull out the electron from anodic chamber to transfer cathodic chamber for reducing the substrate and proton present in cathodic chamber for producing hydrogen and biofuel. Different types of dual chamber were developed with two liquid chambers. It can be a rectangular-shaped (Rozendal et al. 2008a, b; Jeremiasse et al. 2009), disc-shaped (Rozendal et al. 2007), cylindrical or cube type (Ditzig et al. 2007), dual chambered H-type or bottle-type (Liu et al. 2005a, b).

17.2.2 Single-Chamber MEC

MECs are totally anaerobic instead of MFCs, Single chamber MEC have both cathodic and anodic electrolyte in shared electrolyte and no separating ion exchange membrane between them is found as shown in Fig. 17.2. To beat the inconveniences and challenges being used of membrane in MECs, various single-chamber MECs were created by eliminating from dual chamber.

17.3 Materials Used for Construction of MECs

17.3.1 Anode

Similar materials can also be utilized for anodes in MECs as utilized in MFCs (Logan et al. 2006). Materials used as anode in MECs are graphite brushes (Call and Logan 2008), Carbon paper (Ditzig et al. 2007), Graphite Felt (Rozendal et al. 2006; Rozendal et al. 2007). Carbon Cloth (Liu et al. 2005a, b) and Graphite Granules (Cheng and Logan 2007a; Ditzig et al. 2007). Different international companies provide the carbon materials such as Alfa Aesra (Germany), National Electrical Carbon BV (The Netherlands) &FMI Composites Ltd. (UK) (Logan et al. 2007) To enhance the overall performance of anode electrode especially carbon materials can be pretreated with ammonia gas at high temperature (Cheng and Logan 2007b). This treatment can increase current densities in MECs and brings about speedier start-up, which is believed to be caused by the greater attachment of microbes to electrode and enhanced transfer of electron to surface modified by chemical means.

17.3.2 Cathode

In MEC, hydrogen generation occurs at cathode. If plain carbon electrodes are utilized, a high over potential is required to generate hydrogen gas because hydrogen evaluation reaction (HER) is very slow to overcome this problem platinum catalyst is utilized it. Difference manufacturers supply the Platinum catalyzed electrodes (e.g. Alfa Aesa, Germany Magneto Special Anodes) and can also be prepared in laboratory (Cheng et al. 2006a, b) by mixing 10 wt% pt./c with 2% PTFE solution or 5% nafion solution and paste is applied to a single side of cathode on (carbon paper) and after that dried at normal temp for 24 h before utilize. But there are many negative aspects came to the usage of platinum, along with the high cost and the terrible environmental influences (Freguia et al. 2007). Also, platinum can be harmed by Chemicals contained in wastewater like sulfide. Recently, it was discovered that bacteria can catalyzed the HER without using metal catalyst (Rozendal et al. 2008a, b). Ideal strategies to build up this biocathode have not been completely researched (Rozendal et al. 2008a, b) for the first-time build-up biocathode by developing electrochemically active culture by enriching a biofilm on anode of hydrogen-oxidizing bacteria. Utilization of electrode bacteria and by switching the polarity of electrode, a dynamic biocathode is obtained for hydrogen generation. Before utilization of biocathode in MEC should be needed more research. Better performance and low cost of biocathode may make it a suitable substitution of platinum.

17.3.3 Separator or Membrane

The membrane is used to enhance the quality of hydrogen produced and to avoid hydrogen consumption by microbes at the anode. It minimizes the chance of any short circuit. Therefore, membrane is proton exchange membrane (PEM) called nafion (Logan et al. 2007; Call and Logan 2008; Chae et al. 2009). Different membranes have additionally been tried in MEC like anion exchange membrane (AMI-7001), (Cheng and Logan 2007a; Rozendal et al. 2007), forward osmosis membrane (Rozendal et al. 2008a, b), nanofiber- reinforced composite proton exchange membrane (NFR-PEM) (181), charge mosaic and bipolar membrane (Rozenda et al. 2008). As an informational note, there are few downsides of utilizing a membrane in MEC a potential loss, a pH gradient is developed because of membrane which lead to high pH at cathode an low pH at anode which reduce overall performance of MEC. Almost 0.06 V of potential loss occurs by change in one unit of pH. Similar result was observed by Rozendal and coworkers, (Rozendal et al. 2007). They found that on increase pH from 6.4, corresponded to 0.38 V of potential loss of the 1 V applied potential was found (Rozendal et al. 2007). Also, the cost of membrane is high which incease the cost to the MEC framework.

17.4 Substrates Used in MEC

Substrate is one of the most important factors for generating H_2 and biofuel production. Using different substrate such as glucose, lactic acid, butyric acid, cellulose, acetic acid and different types of wastewater can be used in MEC for Hydrogen and Biofuel production. In a renewable way this innovation can be a alternative technique to supply an increasingly number of need fuel (hydrogen) at the same time eliminating waste compounds from waste effluents in an efficient manner. A wide range of substrates which have been utilized in MEC are listed in Table 17.1. It is very difficult to compare the performance of MEC in literature because of various optimized condition, types of electrode, concentration, different types of substrate and applied voltage. To make it simpler, this table consist of a few key data about the MEC are mentioned.

17.5 MEC Working Factors

17.5.1 Applied Voltage

In MEC, the energy in the form of hydrogen that affect the formation and evolvement of the microbial system can be obtained by the applied voltage. The applied voltage must be in range of 0.3–1.0 V as the number of MECs was used to operate this voltage (Cheng and Logan 2007a; Hu et al. 2008; Call and Logan 2008). Above this voltage electrical energy input is quite large enough that made the water electrolysis process in MEC. However applied voltage has significant effect on the growth of the micro-organism which impacts on microbial anode potential or methane generation. The applied voltage at lower value i.e. less than 0.3 V is somewhat less and that results in low HPR and erratic system performance (Rozendal, et al. 2006), so a moderate voltage of 0.7 V is quite appropriate applied voltage for the fast cycle times as compared to obtained results on the lower applied voltage (Call and Logan 2011).

17.5.2 Electrolyte or Ion Strength

Electrolytic condition generally effect on the performance of MECs are conductivity and pH which influence the electrochemically active bacterial growth and used to control the redox potential of the electrode and internal resistance (Merrill and Logan 2009). Merrill MD revealed that particular electrolytes at specific pH of solution will improve the performance of MECs by decreasing the solution resistance and cathode over potential. For example phosphate and acetate electrolytes at pH 5 improve the performance by reducing the overpotential but at higher pH (pH 9), increased the

Table 17.1 Different types of substrates used in MECs for hydrogen production

Types of substrate	Types of electrode used in MEC		Production rate hydrogen (m^3H^2 m^{-3} day^{-1})	Applied voltage (V)	Reference
	Anode	Cathode			
Industrial (IN) wastewater food processing (FP) wastewater	Heat-treated graphite fiber brushe	Stainless steel (flat (MoS2)carbon cloth/Pt (10 wt% molybdenum disulfide on Vulcan XC-72)catalyst sheets of type 304)	0.12 ± 0.02	0.7	Tenca et al. (2013)
			0.12 ± 0.02		
			0.5 ± 0.06		
			0.17 ± 0.03		
			0.41 ± 0.02		
			0.08 ± 0.01		
Domestic wastewater	Carbon paper (non wet proofed)	Carbon paper/Pt 0.5 mg/cm^2 catalyst	0.154 L-H_2/g (hydrogen production based on COD removal)	0.5	Ditzig et al. (2007)
Swine wastewater	Graphite fiber brush	Carbon cloth/Pt 0.5 mg-^{1} cm^2 catalyst	0.9–1.0	0.5	Wagner et al. (2009)
Synthetic effluent	Graphite fiber brush	Carbon cloth/Pt (10% Pt/C) catalyst	1.11 ± 0.13 1.00 ± 0.19	0.5	Lalaurette et al. (2009)
Lignocelluloses Cellobiose					
Fermentation effluent	Graphite plates	Stainless steel mesh	79% (hydrogen production based on CODremoval)	0.7	Lijiao et al. (2013)
Potato wastewater (COD)	Ammonia-treated graphite fiber brushe	Carbon cloth/Pt 0.5 mg/cm2 catalyst	0.74	0.9	
Acetate (sodium acetate)	Disc-shaped piece of graphite felt	Ti mesh/Pt 0.5 mg/cm2 catalyst	0.02	0.5	
Acetate (sodium acetate)	Graphite felt	Ti plate/Pt 0.5 mg/cm2	0.052	0.8	Chae et al. (2008)

(continued)

Table 17.1 (continued)

Types of substrate	Types of electrode used in MEC		Production rate hydrogen ($m^3H^2\ m^{-3}\ day^{-1}$)	Applied voltage (V)	Reference
	Anode	Cathode			
Acetate (sodium acetate)	Ammonia-treated graphite fiber brushe	Stainless steel brush	1.7	0.5	Call et al. (2009)
	Graphite fiber brush	Stainless steel þ NiOx	0.76	0.6	
Acetate (sodium acetate)	Carbon cloth type A (without wetproofing)	Carbon cloth/NiMo	2.0	0.6	Hu et al. (2009)
		Carbon cloth/NiW	1.5		
Acetate (sodium acetate)	Carbon cloth type A (without wet proofing)	Carbon cloth/Pt 0.5 mg/cm2 catalyst	0.53	0.6	Hu et al. (2008)
Acetate (sodium acetate)	Ammonia-treated graphite fiber brushe	Stainless steel brush	1.7	0.5	Call et al. (2009)
	Graphite fiber brush	Stainless steel þ NiOx	0.76	0.6	
Valeric acid	Graphite granules	Carbon cloth/Pt 0.5 mg/cm2 catalyst	0.14	0.6	Cheng and Logan (2007a)
Acetic acid			1.23		
Glucose			0.45		
Butyric acid			1.04		
Lactic acid			0.72		
Propionic acid			1.10		
Glucose	Ammonia- treated graphite brush	Carbon cloth/Pt (10 wt% on Vulcan XC-72)	0.83 ± 0.18	0.5	Selembo et al. (2009)
Glucose			1.87 ± 0.30	0.9	
P-glycerol			0.80 ± 0.08	0.5	
P-glycerol			2.01 ± 0.41	0.9	
B-glycerol			0.14 ± 0.06	0.5	
B-glycerol			0.41 ± 0.13	0.9	

Acetate (sodium acetate)	Carbon brush	Carbon cloth (CC) CNT Pt/C (10% Pt in carbon black) Fe/Fe3C@C	0.00877 ± 0.0007	0.8	
			0.00757 ± 0.0006		
			0.02307 ± 0.0031		
			0.01817 ± 0.0011		
Acetate (sodium acetate)	Graphite granules	Mipor titanium tube/Pt	1.58	1.0	Guoa et al. (2010)
Acetate (sodium acetate)	Heat-treated graphite brushe	Bio-cathode		0.7	Chae et al. (2008)
Acetate (sodium acetate)	Ammonia-treated graphite brushe	SS 304	0.59 ± 0.01	0.9	Selembo et al. (2009)
		SS 316	0.35 ± 0.08		
		SS 420	0.58 ± 0.07		
		SS A286	1.50 ± 0.04		
Winery wastewater	Heat-treatedgraphite brushe	SS 304 (mesh#60)	0.19 ± 0.04	0.9	Cusick et al. (2011)
Proteins	Graphite fiber brushe	Carbon cloth/Pt 0.5 mg/cm2 catalyst	0.42 ± 0.07	0.6	Lu et al. (2010)
Acetate (sodium acetate)	Graphite felt	Ni foam	50	1.0	Jeremiasse et al. (2010)
Acetate (sodium acetate)	Ammonia- treated graphite brush	Carbon cloth/Pt 0.5 mg/cm2 catalyst	1.99 ± 0.02	0.6	Call and Logan (2008)
			3.12 ± 0.02	0.8	
Glucose	Ammonia- treated graphite brush	Carbon cloth/Pt 0.5 mg/cm2 catalyst	0.25 ± 0.03	0.6	Lu et al. (2012)
			0.37 ± 0.04	0.8	

performance of carbonate electrolytes by decreasing the internal resistance. Munoz et al. 2010 showed that the presence of phosphate species in an electrolyte enhanced the current density for production of hydrogen due to cathodic deprotonation reaction. By using phosphate buffer, deionized water, NaCl solution, tap and acidified water have been already tested by Yossan and coworkers in which 100 mM phosphate catholytes exhibited higher production rate of hydrogen due to high buffer capacity, followed by NaCl solution due to its higher conductance. Acidified water is also found effective for hydrogen production which investigated the optimal anolytes pH in MECs. (Yossan et al. 2013) Liu and coworkers found that the pH 9 was for a most extreme HPR of 0.55 m^3 H_2/m^3d (Liu et al. 2014).

17.5.3 Electrode Physico-Chemical Properties

Many factors which affect the internal resistance of MEC can be reduced by the electrode physico-chmeical process such as surface morphology, conductance, surface area, activation resistance, distance, electrode position, etc. Already many improvements have been done such as Call and Logan (Call and Logan 2008) improved the HPR by increasing the anodic surface area using graphitic granules. Wang et al. (Wang et al. 2010) reported that internal resistance of the MEC can be decreased by shortening the distance between the anode and cathode electrode from 14 to 4 cm. Hydrogen production rate can also be increased by reducing the electrode spacing. Cheng and Kadier coworkers have achieved the maximum hydrogen production rate of MEC using 2 cm electrode spacing (Cheng and Logan 2011). Liang and coworkers similarly showed that optimizing the electrode arrangement successfully for decreasing the inner resistance and increase the higher current density (Liang et al. 2011).

17.5.4 Temperature

Temperature is the most prominent thermodynamic parameter in MECs which affects the activity and selection of microorganisms. Omidi and Sathasivan (Omidi and Sathasivan 2013) investigated that is based on the COD removal rate and amount of biomass in MECs, at 31 °C (optimum condition for MEC operation). MEC performance has been conducted under controlled temperature at 30 °C. Hydrogen can also be successfully generated at lower temperature in a single chambered MEC as 4 °C and 9 °C (Lu et al. 2011). At this cryophilic circumstance, methane manufacturing by means of methanogen is effectively inhibited.

17.5.5 Hydraulic Retention Time (HRT) and Organic Loading Rate

As the performance of MECs depend on many factors described earlier in which the effects of hydrodynamic force and dissolve oxygen and anode film have been evaluated as the operational factors (Ajayi et al. 2010). Production of hydrogen in an MEC markedly affected by the hydrodynamic force, but less affected by the anode biofilm exposure to dissolve oxygen. So, it can be concluded that organic loading rate and hydraulic retention time can be the operational factors under the continuous mode operation of MECs (Gil-Carrera et al. 2013; Lee and Rittmann 2009). Moreover, hydrogen can be obtained by optimizing the aforementioned operational factors.

17.6 Value-Added Bioproduct Are Produce from MEC

With the advent of latest practical scale opportunities of MECs, this multipurpose technology promises plethora of applications. Beside generating hydrogen, MEC cathodes led the way for the expansion of several other important applications. To be precise, the current advancement of microbial electrosynthesis (MES) offers a novel alternative for sustainable and effectual chemical production. MESs consists of two alleyways in the cathodic chamber; in which one pathway is for H_2O reduction and other for proton reduction. Generation of green hydrogen is another imperative application of MECs. Apart from hydrogen gas, diverse valuable compounds are produced through the cathode reactions of MECs. Table 17.1 depicts various reported value-added compounds obtained via MECs, including acetate, methane, ethanol, H_2O_2 and formic acid.

17.6.1 Methane

Currently, methane is being employed as an exemplary fuel all over the world. It is generally noticed in the MECs amidst hydrogen generation owing to the growth of methanogens. It has been reported that the generation of methane in the cathode of MECs often coincides with the production of hydrogen (Cusick et al. 2010). This will significantly diminish the viable value of H_2 and upsurge the energy and cost-effectiveness for its refinement. An innovative substitute for this problem is to employ MECs in producing methane as an alternative energy source. A group led by Cheng and coworkers for the very first time demonstrated the synthesis of methane from carbon dioxide by means of two-chamber MECs comprising a methanogens-attached biocathode. A methane production rate of 0.06 mmol/L/h was achieved at voltage of 1.2 V when the cathode potential was set at less than

0.7 V (vs. Ag/AgCl) (Cheng et al. 2009). Similarly, Villano and coworkers found a methane production ratio of 0.05570.002 mmol/D-mg VSS from carbon dioxide in a two-chamber MECs when a biocathode was stored with a hydrogenophilic menthanogenic culture at potentials more negative than 0.65 V (vs. SHE) (corresponding to 0.8 vs. Ag/AgCl) (Villano et al. 2010). Furthermore, Cheng and coworker reported that the methane production from MECs is varied with substrate, inoculum and reactor configuration. The presence of methanogens is unsuspected in hydrogen-producing MECs, as it lessens the hydrogen generation. Numerous methods have been used to prevent the growth of methanogens in MECs (Call and Logan 2008; Wang et al. 2009; Clauwaert and Verstraete 2009). Still, most of the approaches are futile or energy exhaustive. Contrary to inhibiting the methanogens, direct generation of methane in MECs holds various benefits in comparison to classical anaerobic digestion procedures. Firstly, methane production and organic matter oxidation are two distinct procedures in MECs which permits the high content of methane in biogas. Secondly, the process happens at ambient temperature, i.e. thermal heating is not essential which saves energy. Thirdly, methanogens can receive electrons straight from cathode, which might possibly make the procedure more stable to toxic materials like ammonia. Fourthly, MECs can utilize waste streams having less organic matter content, where anaerobic digestion doesn't happen (Villano et al. 2011). Earlier, production of methane in the cathode of MECs was conciliated by hydrogen with abiotic cathode. Clauwaert observed generation of hydrogen from cathode of MECs could be further transformed to methane in an external anaerobic digester, where the procedure was not inhibited even at ammonium concentration of 5 g-N/L (Clauwaert et al. 2008). The utilization of biocathode has significantly lessened the costs of electrode catalyst in MECs thereby preserving the cost-effectiveness. Furthermore, for a better understanding of the fundamental mechanisms for methane bioelectrosynthesis, a two-chamber MECs encompassing a carbon biocathode was fabricated and analysed (Zhen et al. 2015). It was found that a substantial methane yield achieved at a potential of 0.9 V (vs. Ag/AgCl), reaching 2.30 ± 0.34 mL after 5 h of operation with a faradaic efficacy of $24.2 \pm 4.7\%$. One of the chief principle that directs electron exchange and methane production yields is the electrode substance. To increase methane generation, a plain carbon stick modified with a layer of graphite felt (GF) (hereafter referred as "hybrid GF-biocathode") was formulated as a bioelectrode and assessed in a two-chamber MEC (Zhen et al. 2016). Production of methane with this hybrid GF biocathode touched 80.9 mL/L at the potential of 1.4 V when it was incubated for 24 h having C_E of 194.4%.

17.6.1.1 Acetate

In the last one decade, wind and solar energy have attained remarkable attention as they provide renewable sources of energy. Nevertheless, the irregular nature of these energy sources requires resourceful storage tactics to preserve the ceased electrical energy. One of the ploy is to capture the electric energy in covalent chemical bonds, as products can be easily stored and supplied whenever required through prevailing infrastructures. In one of the breakthrough experiment, Nevin and Coworkers illustrated the probability of decrement of carbon dioxide to acetate by acetogenic microorganism *Sporomusa ovata* when electrons were supplied straight from a graphite electrode (Nevin et al. 2010). It was found that that *S. ovata* biofilms on the graphite cathode surfaces drained electrons from electrode and transformed carbon dioxide to acetate along with small amounts of 2-oxobutyrate. As many as 85% of provided electrons were apprehended into these products. Thus, first time that the idea of microbial electrosynthesis was presented, thereby providing a fascinating and innovative way proposing to transform solar energy for valuable organic products efficiently compared to classical methods. A study performed by Rabaey and coworkers unraveled the challenges, principles and prospects of microbial electrosynthesis, illustrating vital point of view on this thrilling and novel discipline at the nexus of electrochemistry and microbiology (Rabaey and Rozendal 2010).

17.6.1.2 Hydrogen Peroxide (H_2O_2)

One of the very important industrial chemical produced by MECs is hydrogen peroxide. In a work performed by Rozendal and coworkers report the viability of H_2O_2 generation in context to the microbial oxidation of organic matter in the anode with synergism to oxygen reduction in the cathode of MECs. This arrangement having an E ap = 0.5 V was proficient of generating H_2O_2 with a rate of 1.17 mmol/L/h in an aerated cathode, outcoming to an overall efficacy of 83% based on acetate oxidation (Rozendal et al. 2009). In comparison to traditional electrochemical process, the H_2O_2 generation in MECs entails much lesser energy, which was 0.93 kWh/kg H_2O_2 in the above-mentioned work. Typically, H_2O_2 can be generated in MFCs coupled to electricity production, which has been confirmed by various studies (Foley et al. 2010). Though, the rate at which H_2O_2 is produced in MFCs has been found much low compared to that of MECs. H_2O_2 generation has significantly prolonged the application opportunities of MECs. The utmost remarkable application is the amalgamation of Fenton reaction with MECs, as the MECs could assist as moderately cost-effective H_2O_2 source for the Fenton-reaction (Nidheesh et al. 2013; Wu and Englehardt 2012). In order to transform into an established technology, more exertions should be put on upgrading of H_2O_2 concentration. H_2O_2concentration which had been attained in MECs at current is only 0.13 wt% (Rozendal et al. 2009), Thus level is still comparatively lesser than the anticipated level for applied industrial applications. A life-cycle investigation had suggested that production of H_2O_2 in MECs was more feasible than classical manufacturing ways (Foley et al. 2010). Nevertheless, the technique is still restricted courtesy to numerous challenges, amongst which H_2O_2 source and occurrence of residual H_2O_2 after the Fenton reaction are two main problems (Nidheesh et al. 2013; Wu and Englehardt 2012). A novel Bioelectro Fenton system having proficiency of alternate switching amid MEC and MFC approach of operation was established to overcome there problems (Zhang et al. 2015).

17.6.2 Ethanol (C_2H_5OH)

The sustainability of producing ethanol by employing an electrode in place of hydrogen as electron donor in a biocathode MEC has been reported by Steinbusch and coworker. It was achieved via two-chamber MEC in which acetate was reduced to ethanol with the aid of electron mediator like methyl viologen (MV). The cathode potential was adapted to 0.55 V having a optimal current density of 1.33 A/m^2 after MV addition which led to 1.82 mM ethanol production. The ethanol generation was chiefly reliant on the MV concentrations, and the synthesis stop following 5 days when MV was exhausted (Steinbusch et al. 2010). MECs technology suggests an innovative means for overcoming classical biological ethanol generation limitations. Though, there are fundamental problems that are essential to be addressed. The mechanism of acetate reduction in the cathode is still unconfirmed. As hydrogen (0.0035 N m3/m2/d) was detected in the cathode, it might also be associated to acetate reduction. Furthermore, necessity of irreversibly electron acceptors would cumulate up to the operation economies, which is an uphill task for the practical implications. Selecting electroactive microorganisms could receive electrons straight from cathode rather than through mediators for ethanol generation might possibly be an interesting future work. Moreover, the ethanol synthesis rate and the final concentration attained in the reported studies are still less, which will necessitate extensive energy for distillation. Additional reduction in electrode overpotential, energy losses and system internal resistance could upsurge the ethanol synthesis thereby making the technique industrially applicable (Steinbusch et al. 2010).

17.6.3 Formic Acid

The formic acid production is accomplished through CO_2 reduction in the cathode and organic matter oxidation in the anode, which holds an utmost importance in paper and pulp production as well as in pharmaceutical syntheses (Zhao et al. 2012). Besides assisting the greenhouse (CO_2) gas reduction., this technology will aid in recovery and recycling of the released carbon dioxide during wastewater or waste treatment without energy input. Though, in the present scenario the concentration of formic acid finally attained including the production rate is still low. Thus, in the near future approaches like gas diffusion through hollow fiber membrane can possibly be adapted to enhance the dissolution of CO_2 and thereby improving the mass transfer. Moreover, the upgrades in the electrode materials of MFCs like nanofabrication of electrode surface may also fetch advantage to decrease the cathode overpotential in MECs.

17.7 Needs and Challenges in the Development of Practical MEC Technology

Wide variety of of MECs application is shown in Fig. 17.3. It is important to note are many hurdles that need to be sorted out before economically affordable application of this technology, to evaluate the industrial affordability, system upscaling of MECs is very important. For the purpose of hydrogen production, the performance of MECs looks good and satisfying. It has been reported recently that, for the purpose of hydrogen production scale up of MECs has been conducted although there are many new challenges respect to the cost and efficiency of these scaling up processes. The maximum value that was obtained in the bench is still not possible to scaling up processes. Only 70% of the hydrogen production and Coulombic efficiency is being reproduced. There are two other major issues associated with scaling up of MECs such as electrode material and reactor design. There are many other factors that are still have to move out of the lab bench to the scale up process such as pollutants removal, chemical synthesis and metal recovery. One theory is that the carbon fiber brush, carbon mesh or activated carbon could be an ideal option for the electron transfer from bacteria to electrode in MFCs as MECs and MFCs share the same anode reaction. The cathode electrode of MECs is very important in reducing overpotential and improving conversion rate.

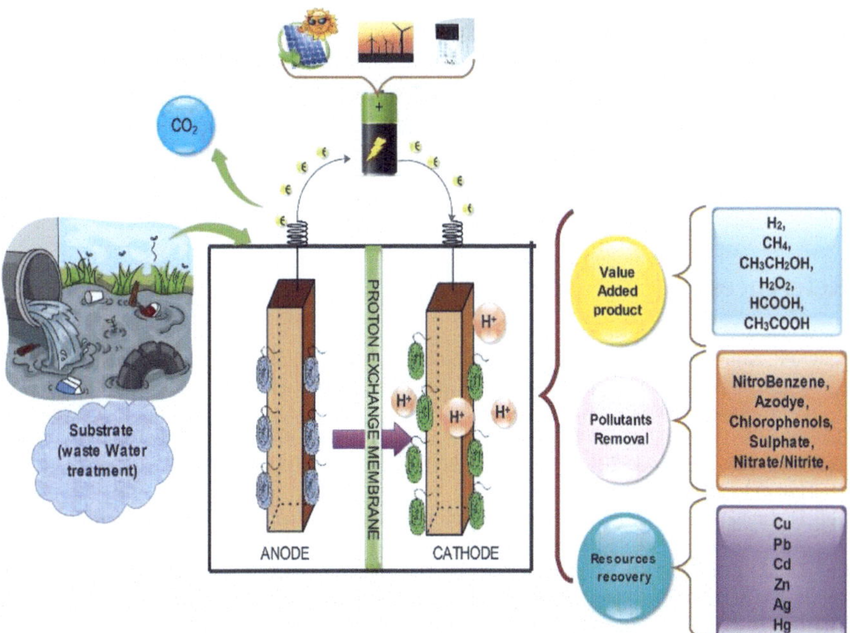

Fig. 17.3 Different Application of MEC as versatile platform

17.8 Conclusion and Future Perspectives

The present chapter summarizes various substances that have been used in MECs to get high yield energy hydrogen gas production and also the waste water treatment.

At the beginning, many simple substrates like glucose and acetate (sodium acetate) but nowadays researchers are mostly using unconventional substrates with two main aims.

Firstly, to utilize waste biomass or curing wastewater and secondly improving MEC performance. Energy obtained from waste biomass that is obtained as a hydrogen gas has also shown great potential. It is believed that in coming years, with more improvement in the technology and reduced prices, more options will be available as the substrates to be used and thus leading to a better and economical bioenergy. So, the final conclusions that can be drawn are:

Nowadays, substrates that are being used have greater complexity and strength. A complicated substrate also supports in establishing a diverse and electrochemically active microbial community in the system while the less complex substrates are easier to degrade and also are better in improving the hydrogen production of the MECs.

One last point which is important to mention is wastewater treatment. Different varieties of wastewater can be brought as substrates under the MEC setups. This includes the wastewater obtained from molasses-based distilleries rich in natural issue and delivered in large volumes such as a food processing (FP) wastewater, winery wastewater, methanol-rich industrial (IN) wastewater, dairy manure wastewater, potato processing wastewater, refinery wastewaters, domestic wastewater and swine wastewater. For large-scale applications this system is still a very far away due to the performance of this system. More innovative improvement regarding material, microbial optimal condition and substrates being utilized are important to bring these system at a level in which they may be commercially exploited.

References

Ajayi FF, Kim KY, Chae KJ, Choi MJ, Kim IS (2010) Effect of hydrodynamic force and prolonged oxygen exposure on the performance of anodic biofilm in microbial electrolysis cells. Int J Hydrog Energy 35(8):3206–3213

Badwal SPS (2014) Emerging electrochemical energy conversion and storage technologies. Front Chem 2:79. Bibcode:2014FrCh....2...79B. https://doi.org/10.3389/fchem.2014.00079. PMC 4174133 Freely accessible

Call D, Logan BE (2008) Hydrogen production in a single chamber microbial electrolysis cell (MEC) lacking a membrane. Environ Sci Technol 42:3401–3406

Call DF, Logan BE (2011) A method for high throughput bioelectrochemical research based on small scale microbial electrolysis cells. Biosens Bioelectron 26:4526–4531

Call DF, Merrill MD, Logan BE (2009) High surface area stainless steel brushes ascathodes in microbial electrolysis cells. Environ Sci Technol 43:2179e83

Chae KJ, Choi MJ, Lee J, Arayi FF, Kim IS (2008) Biohydrogen production viabiocatalyzed electrolysis in acetate-fed bioelectrochemical cells and microbial community analysis. Int J Hydrog Energy 33:5184e92

Chae KJ, Choi MJ, Kim KY, Ajayi FF, Chang IS, Kim IS (2009) A solar-powered microbial electrolysis cell with a platinum catalyst-free cathode to produce hydrogen. Environ Sci Technol 43:9525–9530

Cheng S, Logan BE (2007a) Sustainable and efficient biohydrogen production via electrohydrogenesis. Proc Natl Acad Sci U S A 104:18871–18873

Cheng S, Logan BE (2007b) Ammonia treatment of carbon cloth anodes to enhance power generation of microbial fuel cells. Electrochem Commun 9:492–496

Cheng S, Logan BE (2011) High hydrogen production rate of microbial electrolysis cell (MEC) with reduced electrode spacing. Bioresour Technol 102:3571–3574

Cheng S, Liu H, Logan BE (2006a) Increased performance of single chamber microbial fuel cells using an improved cathode structure. Electrochem Commun 8:489–494

Cheng S, Liu H, Logan BE (2006b) Power densities using different cathode catalysts (Pt and CoTMPP) and polymer binders (Nafion and PTFE) in single chamber microbial fuelcells. Environ Sci Technol 40:364–369

Cheng S, Xing D, Call DF, Logan BE (2009) Direct biological conversion of electrical current into methane by electro methanogenesis. Environ Sci Technol 43(10):3953–3958

Clauwaert P, Verstraete W (2009) Methanogenesis in membraneless microbial electrolysis cells. Appl Microbiol Biotechnol 82(5):829–836

Clauwaert P, Toledo R, van der Ha D, Crab R et al (2008) Combining biocatalyzed electrolysis with anaerobic digestion. Water Sci Technol 57(4):575–579

Cusick RD, Kiely PD, Logan BE (2010) A monetary comparison of energy recovered from microbial fuel cells and microbial electrolysis cells fed winery or domestic waste waters. Int J Hydrog Energy 35:8855–8861

Cusick RD, Bryan B, Parker DS, Merrill MD, Mehanna M et al (2011) Performance of a pilot-scale continuous flow microbial electrolysis cell fed winery wastewater. Appl Microbiol Biotechnol 89:2053e63

Ditzig J, Liu H, Logan BE (2007) Production of hydrogen from domestic waste water using a bioelectrochemically assisted microbial reactor (BEAMR). Int J Hydrog Energy 32 (13):2296–2304

Energy Information Administration. International energy outlook (2013). http://www.eia.doe.gov/oiaf/ieo/index.html

Foley JM, Rozendal RA, Hertle CK, Lant PA, Rabaey K (2010) Life cycle assessment of high rate anaerobic treatment, microbial fuel cells, and microbial electrolysis cells. Environ Sci Technol 44(9):3629–3637

Freguia S, Rabaey K, Yuan Z, Keller J (2007) Non-catalyzed cathodic oxygen reduction at graphite granules in microbial fuel cells. Electrochim Acta 53:598–603

Gil-Carrera L, Escapa A, Carracedo B, Mora'n A, Gomez X (2013) Performance of a semi-pilot tubular microbial electrolysis cell (MEC) under several hydraulic retention times and applied voltages. Bioresour Technol 146:63–69

Guoa K, Tang X, Du Z, Li H (2010) Hydrogen production from acetate in a cathode-on-top single chamber microbial electrolysis cell with a mipor cathode. Biochem Eng J 51:48e52

Hoekstra AY, Chapagain AK, Aldaya MM, Mekonnen MM (2011) The water footprint assessment manual: setting the global standard. Earthscan, London/Washington, DC. waterfootprint.org/media/downloads/TheWaterFootprintAssessmentManual_2pdf

Hu H, Fan Y, Liu H (2008) Hydrogen production using single-chamber membrane-free microbial electrolysis cells. Water Res 42:4172e8

Hu H, Fan Y, Liu H (2009) Hydrogen production in single-chamber tubular microbial electrolysis cells using non-precious metal catalysts. Int J Hydrogen Energy:8535e42

Jeremiasse AW, Hamelers HV, Kleijn JM (2009) Buisman CJN. Use of biocompatible buffers to reduce the concentration overpotential for hydrogen evolution. Environ Sci Technol 43 (17):6882–6887

Jeremiasse AW, Hamelers HVM, Saakes M, Buisman CJN (2010) Ni foam cathode enables high volumetric H_2 production in a microbial electrolysis cell. Int J Hydrog Energy 35:12716e23

Khan MZ, Singh S, Sultana S et al (2015a) Studies on the biodegradation of two different azo dyes in bioelectrochemical systems. New J Chem 39:5597–5604

Khan MD, Abdulateif H, Ismail IM, Sabir S, Khan MZ (2015b) Bioelectricity generation and bioremediation of an azo-dye in a microbial fuel cell coupled activated sludge process. PLoS One 10:e0138448

Khan MD, Khan N, Sultana S, Joshi R, Ahmed S, Yu E, Scott K, Ahmad A, Khan MZ (2017a) Bioelectrochemical conversion of waste to energy using microbial fuel cell technology. Proc Biochem 57:141–158

Khan N, Khan MD, Sultana S, Khan MZ, Ahmad A (2017b) Bioelectrochemical systems for transforming waste to energy. Modern age environmental problems and their remediation. https://doi.org/10.1007/978-3-319-64501-8

Khan MD, Khan N, Sultana S, Khan MZ, Sabir S, Azam A (2018) Microbial fuel cell: waste minimization and energy generation. Modern age environmental problems and their remediation. https://doi.org/10.1007/978-3-319-64501-8_8

Khanal KS, Surampali YR, Zhang PB et al (2010) Bioenergy and biofuels from biowastes and biomass. EWRI of ASCE,USA

Lalaurette E, Thammannagowda S, Mohagheghi A, Maness P-C, Logan BE (2009) Hydrogenproduction from cellulose in a two-stage process combining fermentation and electrohydrogenesis. Int J Hydrog Energy 34:6201–6210

Lee HS, Rittmann BE (2009) Significance of biological hydrogen oxidation in a continuous single chamber microbial electrolysis cell. Environ Sci Technol 44(3):948–954

Liang DW, Peng SK, Lu SF, Liu YY, Lan F, Xiang Y (2011) Enhancement of hydrogen production in a single chamber microbial electrolysis cell through anode arrangement optimization. Bioresour Technol 102(23):10881–10885

Lijiao R, Michael S, Ivan I et al (2013) Treatability studies on different refinery wastewater samples using highthroughput microbial electrolysis cells (MECs). Bioresour Technol 136:322e8

Liu H, Grot S, Logan BE (2005a) Electrochemically assisted microbial production of hydrogen from acetate. Environ Sci Technol 39:4317–4320

Liu H, Grot S, Logan BE (2005b) Electrochemically assisted production of hydrogen from acetate. Environ Sci Technol 39:4317–4320

Liu YP, Wang YH, Wang BS, Chen QY (2014) Effect of anolyte pH and cathode Pt loading on electricity and hydrogen co-production performance of the bioelectrochemical system. Int J Hydrog Energy 39(26):14191–14195

Logan BE, Hamelers B, Rozendal R, Schröder U, Keller J, Freguia S, Aelterman P, Verstraete W, Rabaey K (2006) Microbial fuel cells: methodology and technology. Environ Sci Technol 40:5181–5192

Logan BE, Cheng S, Watson V, Estadt (2007) Graphite fiber brush anodes for increased power production in air cathode microbial fuel cells. Environ Sci Technol 41:3341–3346

Logan BE, Call D, Cheng S, Hamelers HV et al (2008) Microbial electrolysis cells for high yield hydrogen gas production from organic matter. Environ Sci Technol 42:8630–8640

Lu L, Xing D, Xie T, Ren N, Logan BE (2010) Hydrogen production from proteins via electrohydrogenesis in microbial electrolysis cells. Biosens Bioelectron 25:2690e5

Lu L, Ren N, Zhao X, Wang H, Wu D, Xing D (2011) Hydrogen production, methanogen inhibition and microbial community structures in psychrophilic single-chamber microbial electrolysis cells. Energy Environ Sci 4(4):1329–1336

Lu L, Xing D, Ren N, Logan BE (2012) Syntrophic interactions drive the hydrogen production from glucose at low temperature in microbial electrolysis cells. Bioresour Technol 124:68e76

Meda US (2015) Bio-hydrogen production in microbial electrolysiscell using waste water from sugar industry. Int J Eng Sci Res Technol 4:452–458

Mekonnen MM, Hoekstra AY (2011) National Water Footprint Accounts: the green, blue and Grey water footprint of production and consumption. UNESCO-IHE Institute for Water Education, Delft, The Netherlands. Waterfootprint.org/media/downloads/Report50- National Water Footprints Vol1.pdf

Merrill MD, Logan BE (2009) Electrolyte effects on hydrogen evolution and solution resistance in microbial electrolysis cells. J Power Sources 191:203–208

Miandad R, Rehan M, Ouda OKM, Khan MZ, Shahzad K, Ismail IMI, Nizami AS (2017) Waste-to-hydrogen energy in Saudi Arabia: challenges and perspectives: in biohydrogen production: sustainability of current technology and future perspective, vol 23. Springer, New Delhi, pp 7–252

Munoz LD, Erable B, Etcheverry L, Riess J, Basséguy R, Berge A (2010) Combining phosphate species and stainless steel cathode to enhance hydrogen evolution in microbialelectrolysis cell (MEC). Electrochem Commun 12:183–186

Nevin KP, Woodard TL, Franks AE, Summers ZM, Lovley DR (2010) Microbial electrosynthesis feeding microbes electricity to convert carbon dioxide and water to multicarbon extracellular organic compounds. MBio 1(2):e00103–e00110

Nidheesh PV, Gandhimathi R, Ramesh ST (2013) Degradation of dyes from aqueous solution byFenton processes: a review. Environ Sci Pollut Res 20:2099–2132

Nizami AS, Shahzad K, Rehan M, Ouda OKM, Khan MZ, Ismail IMI (2017) Developing waste biorefineryA in Makkah: a way forward to convert urban waste into renewable energy. Appl Energy 186:189–196

Omidi H, Sathasivan A (2013) Optimal temperature for microbes in an acetate fed microbial electrolysis cell (MEC). Int Biodeterior Biodegradation 85:688–692

Rabaey K, Rozendal RA (2010) Microbial electrosynthesis revisiting the electrical route for microbial production. Nat Rev Microbiol 8(10):706–716

Rozendal RA, Hamelers HVM, Euverink GJW, Metz SJ, Buisman CJN (2006) Principle and perspectives of hydrogen production through biocatalyzed electrolysis. Int. J. Hydrogen Energy 31:1632–1640

Rozendal RA, Hamelers HVM, Molenkamp RJ, Buisman CJN (2007) Performance of single chamber biocatalyzed electrolysis with different types of ion exchange membranes. Water Res 41:1984–1994

Rozendal RA, Jeremiasse AW, Hamelers HVM, Buisman CJN (2008a) Hydrogen production with a microbial biocathode. Environ Sci Technol A 42:629–634

Rozendal RA, Jeremiasse AW, Hamelers HVM (2008b) Effect of the type of ion exchange membrane on performance ion transport and pH in biocatalyzed electrolysis of wastewater. Water Sci Technol 57:1757–1762

Rozendal RA, Leone E, Keller J, Rabaey K (2009) Efficient hydrogen peroxide generationfrom organic matter in a bioelectrochemical system. Electrochem Commun 11(9):1752–1755

Sato T, Qadir M, Yamamoto S, Endo T, Zahoor A (2013) Global regional, and country level need for data on wastewater generation, treatment, and use. Agric Water Manag 130:1–13. https://doi.org/10.1016/j.agwat.08.007

Scholz WH (1993) Processes for industrial production of hydrogen and associated environmental effects. Gas Sep Purif 7:131–139

Selembo PA, Perez JM, Lloyd WA, Logan BE (2009a) High hydrogen production from glycerolor glucose by electrohydrogenesis using microbial electrolysis cells. Int J Hydrog Energy 34:5373–5381

Selembo PA, Merrill MD, Logan BE (2009b) The use of stainless steel and nickel alloys as low-cost cathodes in microbial electrolysis cells. J Power Sources 190:271–278

Sleutels TH, Ter HA, Buisman CJN, Hamelers HVM (2012) Bioelectrochemical systems: anoutlook for practical applications. ChemSusChem 5:1012–1019

Steinbusch KJJ, Hamelers HVM, Schaap JD, Kampman C, Buisman CJN (2010) Bioelectrochemical ethanol production through mediated acetate reduction by mixed cultures. Environ Sci Technol 44(1):513–517

Sultana S, Khan MD, Sabir S et al (2015) Bio-electro degradation of azo-dye in a combined anaerobic–aerobic process along with energy recovery. New J Chem 39:9461–9470

Tenca A, Cusick RD, Schievano A, Oberti R, Logan BE (2013) Evaluation of low cost cathode materials for treatment of industrial and food processing wastewater using microbial electrolysis cells. Int J Hydrog Energy 38:1859e65

Villano M, Aulenta F, Ciucci C, Ferri T, Giuliano A, Majone M (2010) Bioelectrochemical reduction of CO2 to CH4 via direct and indirect extracellular electron transfer by a hydrogenophilic methanogenic culture. Bioresour Technol 10:3085–3090

Villano M, Monaco G, Aulenta F, Majone M (2011) Electrochemically assisted methane production in a biofilm reactor. J Power Sources 196(22):9467–9472

Wagner RC, Regan JM, Oh SE, Zuo Y, Logan BE (2009) Hydrogen and methane production from swine wastewater using microbial electrolysis cells. Water Res 43:1480e8

Wang X, Cheng S, Feng Y, Merrill MD, Saito T, Logan BE (2009) Use of carbon mesh anodes and the effect of different pretreatment methods on power production in microbial fuel cells. Environ Sci Technol 43(17):6870–6874

Wang A, Liu W, Ren N, Cheng H, Lee DJ (2010) Reduced internal resistance of microbial electrolysis cell (MEC) as factors of configuration and stuffing with granular activate carbon. Int J Hydrog Energy 35(24):13488–13492

WEF (World Economic Forum) (2016) The global risks report 2016. WEF, Geneva, Switzerland. wef.ch/risks2016

Wu T, Englehardt JD (2012) A new method for removal of hydrogen peroxide interference in the analysis of chemical oxygen demand. Environ Sci Technol 46:2291–2298

Yossan S, Xiao L, Prasertsan P, He Z (2013) Hydrogen production in microbial electrolysis cells: choice of catholyte. Int J Hydrog Energy 38:9619–9624

Zhang Y, Angelidaki I (2014) Microbial electrolysis cells turning to be versatile technology: recent advances and future challenges. Water Res 56:11–25

Zhang Y, Wang Y, Angelidaki I (2015) Alternate switching between microbial fuel cell andmicrobial electrolysis cell operation as a new method to control H_2O_2 level in bioelectro-Fenton system. J Power Sources 291:108–116

Zhao HZ, Zhang Y, Chang YY, Li ZS (2012) Conversion of a substrate carbon source to formic acid for carbon dioxide emission reduction utilizing series-stacked microbial fuel cells. J Power Sources 217:59–64

Zhen G, Kobayashi T, Lu X, Xu K (2015) Understanding methane biocathode. Bioresour Technol 186:141–148

Zhen G, Lu X, Kobayashi T, Kumara G, Xu K (2016) Promoted electromethanosynthesis in a two chamber microbial electrolysis cells (MECs) containing a hybrid biocathode covered with graphite felt (GF). Chem Eng J 284:1146–1155

The manufacturer's authorised representative in the EU is Springer Nature Customer Service Centre GmbH, Europaplatz 3, 69115 Heidelberg, Germany. If you have any concerns regarding our products, please contact ProductSafety@springernature.com

Printed and bound by CPI Group (UK) Ltd, Croydon, CR0 4YY
15/07/2026
02167623-0004